Pharmaceutical Discovery, Development and Manufacturing Forum 2019

Held at the 2019 AIChE Annual Meeting

Orlando, Florida, USA
10 - 15 November 2019

ISBN: 978-1-7138-0547-2

Printed from e-media with permission by:

Curran Associates, Inc.
57 Morehouse Lane
Red Hook, NY 12571

Some format issues inherent in the e-media version may also appear in this print version.

Printed with permission by Curran Associates, Inc. (2020)

For permission requests, please contact AIChE
at the address below.

AIChE
120 Wall Street, FL 23
New York, NY 10005-4020

Phone: (800) 242-4363
Fax: (203) 775-5177

www.aiche.org

Additional copies of this publication are available from:

Curran Associates, Inc.
57 Morehouse Lane
Red Hook, NY 12571 USA
Phone: 845-758-0400
Fax: 845-758-2633
Email: curran@proceedings.com
Web: www.proceedings.com

Pharmaceutical Discovery, Development and Manufacturing Forum 2019

Held at the 2019 AIChE Annual Meeting

Orlando, Florida, USA
10 - 15 November 2019

AIChE
The Global Home of Chemical Engineers

TABLE OF CONTENTS

565296 Collaborative Simulation Development Accelerated By Cloud Based Computing and Software As a Service Model

Simulations are increasingly used in pharmaceutical development to deliver medicines to patients more quickly; more efficiently; and with better designs, safety, and effect. Many of the risks in product design, including scale up can be mitigated by modeling and simulation. Improved product understanding and design space delineation, as well as the possibility for control of complex continuous manufacturing equipment are possible.

These pharmaceutical development oriented simulations need high performance computing resources as well as a variety of software to model everything from molecular dynamics to entire processes trains for the effects on the pharmaceutical product at various scales of scrutiny. The demand curve for these resources has many peaks and can shift in a time scale much faster than a typical procurement process.

Both on demand cloud based computing capability and software as a service models have been growing in use. This presentation describes the efforts of the Enabling Technology Consortium to apply these information technology models to pharmaceutical simulations which have special needs for documentation and security. It is expected that the environment will have more benefits as the cloud can be configured for collaborative work among companies in the non-competitive space and work can be made available for collaboration and use by contract service vendors or health authorities.

The expected benefits of this computing environment include economies of scale for both the providers and the consumer, increased speed, and a higher level of value generation for providers by the use of increased resources and for consumers by the information available to accelerate and improve delivery of pharmaceutical products.

563390 Rational Solvent Selection Guided By Machine Learning and Molecular Descriptors in Asymmetric Catalytic Reactions

Rational solvent selection guided by machine learning and molecular descriptors in asymmetric catalytic reactions

Yehia Amar,[a] Artur M. Schweidtmann,[b] Liwei Cao,[a,d] Paul Deutsch[c] and Alexei Lapkin [a,d]

[a] Department of Chemical Engineering and Biotechnology, University of Cambridge, Philippa Fawcett Drive, Cambridge, CB3 0AS, United Kingdom

[b] Aachener Verfahrenstechnik – Process Systems Engineering, RWTH Aachen University, Aachen, Germany

[c] UCB Pharma S.A. Allée de la Recherche, 60 1070 Brussels, Belgium

[d] Cambridge Centre for Advanced Research and Education in Singapore Ltd, 1 Create Way, CREATE Tower #05-05, 138602, Singapore

Email: aal35@cam.ac.uk

Abstract

Rational solvent selection remains a significant challenge in process development, especially within pharmaceutical applications. Therefore, a hybrid mechanistic - machine learning approach, geared towards automated process development work-flow was developed, and successfully applied on a $Rh(CO)_2(acac)$/Josiphos (R_1=cyclohexyl, R_2=4-methoxy,3,5-dimethylphenyl) catalyzed asymmetric hydrogenation of a chiral a-ß unsaturated γ-lactam reaction. The mechanistic part of the model is based on molecular descriptors of physico-chemical properties of solvents, including the reaction-specific descriptors, substrate and hydrogen solubility. A library of 400 solvents was used with 17 molecular descriptors.

The algorithm, which is based on a Gaussian process surrogate model, was trained to learn and optimize for both conversion and diastereomeric excess simultaneously, ultimately identifying better solvents.

In addition to being a powerful design of experiments methodology, the resulting statistical surrogate model is predictive, with a cross-validation correlation coefficient of 0.83. Furthermore, a solvent-mixing strategy based on the black box approach was also investigated. These methods open the door for process chemists to use enhanced process development workflows for optimization and discovery.

571487 Application of Multivariate Statistical Process Control to a Drying Process for Continuous Pharmaceutical Manufacturing

1. Introduction

In the pharmaceutical industry, the method for the improvement of the production efficiency has been discussed and the documents on Quality by Design (QbD) and Process Analytical Technology (PAT) [1-4] were published by Food and Drug Administration (FDA) and International Conference on Harmonisation of Technical Requirements for Registration of Pharmaceuticals for Human Use (ICH). Since then, online process monitoring and control technologies have attracted much attention. These technologies are important to realize the continuous manufacturing, and a new draft guideline that addresses quality aspects in the continuous manufacturing of medicinal products are published by the FDA in 2019 [5].

Data-driven multivariate analysis is an important part of PAT and many research papers were published [6, 7]. Latent variable method such as principal component analysis (PCA) and partial least squares (PLS) are popular, and they are applied to various kinds of processes such as crystallization, mixing, granulation, drying, and coating [6]. In this research, multivariate statistical process control (MSPC) is applied to the drying process. The research on MSPC in the drying process is rarely reported. As far as the authors know, only one paper [8] reported the application results of MSPC to a drying process and A.F. Silva et al [8] concluded that the abnormal conditions were successfully detected by MSPC. However, the rate of false positive is quite high: Q statistics are always over the threshold in some cases with normal operation conditions. This research aims to realize more accurate abnormal detection.

2. Experiments

Eight experiments were conducted under the normal operation conditions, and six experiments were conducted under the abnormal operation conditions. The abnormal operation conditions were caused by changing the flowrate, the temperature, and the humidity of the inlet air. Five variables, i.e. the inlet air flowrate, the inlet air temperature, the outlet air temperature, the product temperature, and the NIR spectrum of the product, were recorded every ten seconds.

3. PCA based MSPC

The data from seven experiments with normal operation conditions were used for model construction. Multiway PCA with one principal component was used since the drying process is batch process, and the model was constructed for every batch time. The maximum values of Hotteling's T^2 and Q statistics for the model construction data are selected as the threshold for the abnormal detection. The model validation was conducted by using the data from one experiment with normal operation condition and six experiments with abnormal operation conditions. The MSPC method could classify the operation condition with the accuracy of more than 95%.

References

[1] ICH, ICH harmonised tripartite guideline - pharmaceutical development Q8 (R2) (2005)

[2] ICH, ICH harmonised tripartite guideline - quality risk management Q9 (2005)

[3] ICH, ICH harmonised tripartite guideline - pharmaceutical quality system Q10 (2008)

[4] FDA, Pharmaceutical cGMPs for the 21st century—a risk-based approach final report (2004)

[5] FDA, Quality considerations for continuous manufacturing guidance for Industry (2019)

[6] T. Rajalahti and O. M. Kvalheim "Multivariate data analysis in pharmaceutics: a tutorial review", International journal of pharmaceutics, 417, (2011) 280-290

[7] G. M. Troup, C. Georgakis, "Process systems engineering tools in the pharmaceutical industry", Computers and chemical engineering, 51 (2013) 157-171

[8] A.F. Silva et al. "Multivariate statistical process control of a continuous pharmaceutical twin-screw granulation and fluid bed drying process", International journal of pharmaceutics, 528, (2017) 242-252

571490 Multivariate Data Analysis to Drive Process Understanding and Optimization

Chemical and bio processes can be sensitive to a wide range of parameters. Significant amounts of data are generated to find certain Critical Quality Attributes (CQAs) not meeting specifications. In such cases, a retrospective analysis helps identify trends and patterns in the data. Multivariate data analysis (MVDA) is a very powerful tool which uses various algorithms to elucidate inter-parameter interactions; as well as correlations with output variables. This presentation involves four case-studies wherein MVDA has been implemented to enhance process understanding.

- Spectroscopic methods are widely used for tracking reactions and crystallizations in pharmaceutical processes. In many cases, entire spectra (rather than a single peak specific to a molecule) need to be tracked to accurately represent the process, thus creating a complex multivariate data-set. In this example, a Partial Least Squares (PLS) regression was performed on Raman spectra to understand thermodynamic dependence of crystal forms over a range of temperatures and solvent compositions.
- A lipid nanoparticle formulation was manufactured in a recent manufacturing campaign, with several process parameters and output variables. Principal Component Regression (PCR) provided correlations between parameters and each variable, while PLS regression provided overall scores and loadings plots to eliminate outliers amongst batches.
- Process parameters for a cell-culture process were investigated to quantify influence on cell growth and product quality. Trajectories were generated using the scores plot to track evolution of the batches over time and projection models showed how failed batches deviated from ideal runs. The correlation loadings plot highlighted the continuous variables which had the strongest influence at each stage of the process.
- Data was generated from a series of product stability experiments for two monoclonal antibodies. Multivariate Linear Regression (MLVR) was used to correlate independent variables to CQAs. Results were verified using a PLS regression and a predictive mechanistic model was created.

In each of the above examples, MVDA was able to extract useful insights and successful recommendations were made for process improvement.

577656 Dissolution Prediction By Process Analytical Technology, Machine Learning and Mathematical Modeling: Toward the Real-Time Release Testing of Pharmaceuticals

Continuous pharmaceutical manufacturing, both at upstream (API production) and downstream (formulation technology) processing, is rapidly gaining interest as a promising way to improve efficiency, reduce operating costs and ensure easier scale-up and shorter time-to-market [1]. However, the risk mitigation of process failure is higher when the unit operations are integrated into one continuous process line as the failure can cumulatively disturb the whole line [2]. To fully exploit the system-wide performance of a continuous manufacturing line, while providing the required product quality, the entire process must be flexibly controlled to compensate the process variations. Consequently, the conversion from batch to continuous manufacturing also entails the necessity of changing the current quality assurance approaches. Addressing these challenges, the Process Analytical Technology (PAT) and Quality-by-Design (QbD) initiatives emerged over the last few years to support the design, analysis and control of critical quality attributes (CQA) and critical process parameters (CPP) of the pharmaceutical processes. These can also lead to the feasibility of real-time release. The real-time release testing (RTRT) methods [3] utilize a combination of measured material attributes and in-process analysis to demonstrate that the product conforms to the quality attributes defined by the regulations, without the need for end-product quality tests.

In this work, the prediction of *in vitro* dissolution properties of pharmaceutical tablets is studied using PAT data, machine learning and mathematical modeling of the continuous process. *In vitro* dissolution testing is one of the most substantial analytical methods in the pharmaceutical industry. It is widely applied during the research and development stage, such as for formulation optimization, prediction of in vivo performance or can be used as a surrogate measurement for bioequivalence. Dissolution measurement also serves as a routine quality control (QC) test during the manufacturing process to assess product performance, stability or batch-to-batch variability [4]. In this study, the dissolution of two model pharmaceutical tablet formulations was studied, an immediate release acetylsalicylic acid formulation as well as caffeine tablets with extended-release dissolution properties. The dissolution curves of the formulations were simultaneously influenced by several critical factors: the concentration of the active pharmaceutical ingredient (API), the concentration of the polyethylene oxide, used to create the extended-release matrix, the tableting compression force and the particle sizes of raw materials. Reflection and transmission NIR and Raman spectra were measured on tablets prepared by following experimental designs as well process data were acquired during the manufacturing.

Several modeling approaches were tested and compared to predict the dissolution of the tablets based on the PAT and process data. Partial Least Square (PLS) regression and feedforward backpropagation Artificial Neural Network (ANN) models were developed for the different spectroscopic measurements individually as well as by combining them together. The decrease of the root mean square errors (RMSE) of the models and the improved prediction of the dissolution curves belonging to independent validation tablets indicated a consistently enhanced performance of the ANN models, which could be explained by their capability of modeling non-linearity between the process parameters and dissolution curves. To move forward the real-time release testing approach, the developed dissolution prediction models could also be fitted into a system-wide process model of a continuous manufacturing line, which describes the variation of the identified critical parameters (*e.g.* API concentration, particle size) throughout the manufacturing process steps. In this way, the effect of mitigation of process parameter variability on the dissolution properties could be studied, which can also serve as a basis of process control.

The results of this work indicate that data-driven models could provide a non-destructive and fast alternative for the current quality control standard, in line with the real-time release initiative. Machine learning also served as a straightforward data fusion method of the PAT data without the need for additional preprocessing steps. The method could significantly increase the efficacy of dissolution testing, advance data processing in the PAT environment and contribute to an enhanced real-time release testing procedure, especially when the dissolution model is incorporated into a system-wide modeling approach.

[1] S.L. Lee, T.F. O'Connor, X. Yang, C.N. Cruz, S. Chatterjee, R.D. Madurawe, C.M.V. Moore, L.X. Yu, J. Woodcock, Modernizing pharmaceutical manufacturing: From batch to continuous production, Journal of Pharmaceutical Innovation, 10 (2015) 191-199

[2] A. Mesbah, J.A. Paulson, R. Lakerveld, R.D. Braatz, Model predictive control of an integrated continuous pharmaceutical manufacturing pilot plant, Organic Process Research & Development, 21 (2017) 844-854.

[3] European Medicines Agency, 2012. Guideline on Real Time Release Testing (formerly Guideline on Parametric Release)

[4] Dressman, Jennifer B., and Johannes Krämer, eds. Pharmaceutical dissolution testing. Boca Raton, FL:: Taylor & Francis, 2005.

556721 An Integrated Approach to Modelling, Data, and Control for Pharmaceutical Manufacturing

An integrated Quality-by-Design approach for pharmaceutical manufacturing is described. Continuous manufacturing is an approach that brings process research and development into closer alignment and execution with manufacturing, by significantly reducing the amount of scale up needed to go from laboratory scale to production scale. Dynamic process and plant models are used for integration across different value streams, from a seamless connection from raw materials through drug substance and drug product. The streamlined process development workflow incorporates best practices from the relevant ICH guidelines [1-4], including the selection of control strategies for each critical quality attribute, and the effective use of modular plant-wide simulation. The main concepts are illustrated for small-molecule pharmaceutical pilot plants and bench-scale biologic drug manufacturing platforms. Additional challenges associated with biologic drugs are discussed. Aided with in-line process analytical technology, automated plant-wide control systems are designed and implemented to satisfy specifications on the critical quality attributes. The presentation draws from Quality by Design strategies that were developed and applied in the pharmaceutical industry from before the ICH guidelines were published [e.g., 5-7], up to recent advances [8-22].

References

1. ICH Q8(R2) Pharmaceutical Development, ICH Harmonised Tripartite Guideline, International Conference on Harmonisation of Technical Requirements for Registration of Pharmaceuticals for Human Use, 2009
2. ICH Q9 Quality Risk Management, ICH Harmonised Tripartite Guideline, International Conference on Harmonisation of Technical Requirements for Registration of Pharmaceuticals for Human Use, 2005
3. ICH Q10(R2) Pharmaceutical Quality System, ICH Harmonised Tripartite Guideline, International Conference on Harmonisation of Technical Requirements for Registration of Pharmaceuticals for Human Use, 2009
4. ICH Q11(R2) Development and Manufacture of Drug Substances (Chemical Entities and Biotechnological/Biological Entities), ICH Harmonised Tripartite Guideline, International Conference on Harmonisation of Technical Requirements for Registration of Pharmaceuticals for Human Use, 2012
5. Togkalidou, R. D. Braatz, B. K. Johnson, O. Davidson, and A. Andrews. Experimental design and inferential modeling in pharmaceutical crystallization. AIChE J., 47:160-168, 2001
6. Togkalidou, H.-H. Tung, Y. Sun, A. Andrews, and R. D. Braatz. Solution concentration prediction for pharmaceutical crystallization processes using robust chemometrics and ATR FTIR spectroscopy. Org. Process Res. Dev., 6:317-322, 2002
7. Togkalidou, H.-H. Tung, Y. Sun, A. Andrews, and R. D. Braatz. Parameter estimation and optimization of a loosely-bound aggregating pharmaceutical crystallization using in-situ infrared and laser backscattering measurements. Ind. Eng. Chem. Res., 43:6168- 6181, 2004
8. S. Hong, K. A. Severson, M. Jiang, A. E. Lu, J. C. Love, and R. D. Braatz. Challenges and opportunities of biopharmaceutical manufacturing control. Computers & Chemical Engineering, 110:106-114, 2018
9. Lakerveld, P. L. Heider, K. D. Jensen, R. D. Braatz, K. F. Jensen, A. S. Myerson, and B. L. Trout. End-to-end continuous manufacturing: Integration of unit operations. In Continuous Manufacturing of Pharmaceuticals, edited by P. Kleinebudde, J. Khinnast, and J. Rantanen, Wiley, New York, Chapter 13, pages 447-483, 2017
10. Mo Jiang and Richard D. Braatz. Integrated control of continuous (bio)pharmaceutical manufacturing. Am. Pharm. Rev., 19(6):110-115, 2016
11. Lakerveld, B. Benyahia, P. L. Heider, H. Zhang, A. Wolfe, C. J. Testa, S. Ogden, D. R. Hersey, S. Mascia, J. M. B. Evans, R. D. Braatz, and P. I. Barton. The application of an automated control strategy for an integrated continuous pharmaceutical pilot plant. Organic Process Research & Development, 19(9):1088-1100, 2015

12. Jiang, Y. E. D. Li, H. H. Tung, and R. D. Braatz. Effect of jet velocity on crystal size distribution from antisolvent and cooling crystallizations in a dual impinging jet mixer. Chemical Engineering and Processing: Process Intensification, 97:242-247, 2015
13. Jiang, Y. E. D. Li, H. H. Tung, and R. D. Braatz. Effect of jet velocity on crystal size distribution from antisolvent and cooling crystallizations in a dual impinging jet mixer. Chemical Engineering and Processing: Process Intensification, 97:242-247, 2015
14. Amos E. Lu, Joel A. Paulson, Nicholas J. Mozdzierz, Alan Stockdale, Ashlee N. Ford Versypt, Kerry R. Love, J. Christopher Love, and Richard D. Braatz. Control systems technology in the advanced manufacturing of biologic drugs. Proceedings of the IEEE Conference on Control Applications, Sydney, Australia, 1505-1515, 2015.
15. T. Zhang, R. Lakerveld, P. L. Heider, M. Y. Tao, M. Su, C. J. Testa, A. N. D'Antonio, P. I. Barton, R. D Braatz, B. L. Trout, A. S. Myerson, K. F. Jensen, and J. M. B. Evans. Application of continuous crystallization in an integrated continuous pharmaceutical pilot plant. Crystal Growth & Design, 14(5):2148-2157, 2014
16. Mascia, P. L. Heider, H. Zhang, R. Lakerveld, B. Benyahia, P. I. Barton, R. D. Braatz, C. L. Cooney, J. M. B. Evans, T. F. Jamison, K. F. Jensen, A. S. Myerson, and B. L. Trout. End-to-end continuous manufacturing of pharmaceuticals: Integrated synthesis, purification, and final dosage formation. Angewandte Chemie Int. Ed., 52(47):12359-12363, 2013
17. Lakerveld, B. Benyahia, R. D. Braatz, and P. I. Barton. Model-based design of a plant-wide control strategy for a continuous pharmaceutical plant, AIChE Journal, 59:3671-3685, 2013
18. C. S. Kee, R. B. H. Tan, and R. D. Braatz. Semiautomated identification of the phase diagram for enantiotropic crystallizations using ATR-FTIR spectroscopy and laser backscattering. Ind. Eng. Chem. Res., 50:1488-1495, 2011
19. C. S. Kee, P. D. Arendt, L. M. Goh, R. B. H. Tan, and R. D. Braatz. Nucleation and growth kinetics estimation for L-phenylalanine hydrate and anhydrate crystallization. CrystEngComm, 13:1197-1209, 2011
20. C. S. Kee, P. D. Arendt, R. B. H. Tan, and R. D. Braatz. Selective crystallization of the metastable anhydrate form in the enantiotropic pseudo-dimorph system of L-phenylalanine using concentration feedback control. Crystal Growth & Design, 9:3052-3061, 2009
21. C. S. Kee, R. B. H. Tan, and R. D. Braatz. Selective crystallization of the metastable alpha-form of L-glutamic acid using concentration feedback control. Crystal Growth & Design, 9:3044-3051, 2009
22. X. Zhou, M. Fujiwara, X. Y. Woo, E. Rusli, H.-H. Tung, C. Starbuck, O. Davidson, Z. Ge, and R. D. Braatz. Direct design of pharmaceutical antisolvent crystallization through concentration control. Crystal Growth & Design, 6:892-898, 2006

555508 Combining Big Data Analytics and Computational Chemistry in Target-Based Drug Discovery

A new target-based drug screening method is proposed exploiting the synergy effect of big data analytics and computational chemistry. The method was illustrated in the development of a target-based drug using 45 sulphonamide derivatives (33 training compounds and 12 testing compounds) with carbonic anhydrase IX (CA IX) as a drug target. For each sulphonamide compound, about 5,000 molecular descriptors were calculated using Gaussian (Gaussian Inc.), and lipophilicity ($\log k_w$) and inhibitory activity ($\log K_i$) were measured using RP-LC as drug properties. Genetic algorithm-partial least squares (GA-PLS) provided a quantitative structure-property-relationship (QSPR) model with high prediction capability employing only 7 descriptors. The ideal drug structure was obtained by inverting the QSPR model numerically in respect to reference properties of a reference drug (one of testing compounds). 12 testing compounds were ranked using the ideal drug structure and the rank was further validated through molecular docking and molecular dynamic (MD) simulation of drug candidate-CA IX complexes. Parameterization of the simulation was verified experimentally using structural analyses of the complexes including MALDI-TOF/TOF-MS. The results proved that the drug screening method proposed has great flexibility and can provide drug candidates with lower risk.

567740 Utilizing Data Systems and Modeling to Evaluate Equipment Fit and Technical Capabilities in External Drug Substance Manufacturing

Both clinical and commercial small molecule pharmaceutical intermediates and API are increasingly manufactured at Contract Manufacturing Organizations (CMO's) relative to internal manufacturing plants. This paradigm shift implies some loss of ability and flexibility to ensure optimal process fit within the plant equipment. Internal manufacturing plants allow for self-selection of equipment for process fit. At an external plant, the CMO selects the equipment for a process. Particularly in the clinical space where processes are new and CMO experience with the processes is low, this leads to a higher risk of poor equipment selection, potentially causing timeline delays, quality insufficiencies, and yield fluctuation. To mitigate this risk, our team developed data systems and a suite of tools to assess process fit and technical capabilities throughout our CMO network. For the data systems, we used a combination of vendor survey and historical operating experience to assemble a consolidated CMO equipment database and a plant capability log. Our suite of tools, including various Dynochem utilities, interactive data visualization, and statistical modeling, leverages our data systems and allows us to cross-reference CMO equipment and capabilities against individual API or intermediate processes. Typical products of this evaluation include proactive equipment fit, decision analysis around vendor selection, and evaluations of equipment impact on process performance. As a result, we can proactively address issues with vendor capabilities and equipment fit, which saves us from delays and cost increases. In this talk, we will describe the systems that we implemented, the workflows that were developed, and several case studies describing how we used these tools to address individual problems associated with technology transfer and scale up in an external manufacturing network for clinical API.

581317 Artificial Intelligence for Pharma 4.0: Challenges and Opportunities

In the pharmaceutical industry, the urgency to increase productivity, shorten development cycle time, reduce costs, and get it right the first time has never been more serious than it is now. The old empirical paradigm of guess-and-test, learning by doing, does not seem to be working as well as it used to. As the industry embraces its version of Industry 4.0 - Pharma 4.0 – to address these challenges, exciting recent advances in artificial intelligence (AI) offer important opportunities.

Decision-making in pharmaceutical product development and manufacturing involves the integration of process modelling tools, effective use of laboratory-generated information, use of knowledge from the scientific literature, as well as development of technical specifications and a knowledge base to satisfy regulatory requirements. The amount and complexity of information of different types, ranging from raw experimental data to lab reports to sophisticated mathematical models, that need to be stored, accessed, validated, manipulated, managed, and used for decision-making is staggering. Pharmaceutical discovery and manufacturing has long been a "Big Data" discipline. In fact, a typical New Drug Application (NDA) contains more than 100,000 pages of data, information, and knowledge.

To address the challenges, we need a systematic, integrated, informatics framework, pharmaceutical informatics, based on formal and explicit models of information. In addition, we also need tools that would support rapid extraction of mechanistic, first principles, knowledge from raw data gathered from PAT-like techniques. The information models need to be accessed easily by humans and software tools, and should provide a common understanding for information sharing. A critical challenge is the scarcity of semantically rich, properly populated, ontologies for the pharma domain.

Recent progress in machine learning, natural language processing, ontology engineering, etc., offers potential solutions to these challenges. In this talk, I will discuss a few case studies in structure-property prediction, hybrid-AI models, and a "Watson-like" system for pharmaceutical manufacturing.

581599 Positioning for a Sustainable Future – Continuous Processing, Automation and Predictive Modeling for Efficient Biopharmaceutical Process Development and Manufacturing

The focus on development of new modalities for advanced therapies and challenges with complex drug candidates has driven adoption of novel concepts and technologies in pharmaceutical development and manufacturing. Industry is focusing more on flexible design concepts involving smaller footprints, shorter cycle times, high degree of automation, digitalization, and improved process control. Enabling technologies such as continuous processing, high throughput experimentation and predictive modeling through first-principles, as well as machine learning can expedite process development, scale-up and technology transfer while ensuring enhanced process efficiency and delivering quality products. This presentation will share AbbVie's experiences in integrating these technologies into process development and manufacturing facilities and discuss the challenges and opportunities associated with our approaches.

*Shailendra Bordawekar is an employee of AbbVie and may own AbbVie stock. AbbVie sponsored and funded the study; contributed to the design; participated in the collection, analysis, and interpretation of data, and in writing, reviewing, and approval of the final publication.

581471 Systems-Based Pharmaceutics – an End-of-Decade Report

Systems-based Pharmaceutics (SbP) is a systems engineering methodology for the pharmaceutical industry that encompasses drug substance and drug product manufacturing, as well as *in vivo* drug performance. The approach aims to capture available relevant knowledge in the form of validated mathematical models. By integrating knowledge across the product lifecycle, SbP allows the quantification of the impact of critical process parameters and other decisions and environmental factors on the product's critical quality attributes and the process key performance indicators.

This paper assesses the extent to which the above vision, originally formulated at the start of this decade, has been translated into reality. Aspects considered include the extent to which advances in scientific understanding have been incorporated into mathematical models that can be deployed in support of product and process design and process operations; the formal validation of these models; and the use of models in quantifying and managing the risk in model-based decisions, and in identifying practically important gaps in current knowledge. The paper also considers how current trends towards digitalization, and the associated advances in IT technologies, impact the SbP vision and its ongoing realization.

581427 What Challenges Need to be Addressed, and What New Technological Approaches Are Needed to Put Life Sciences on Path to Pharma 4.0?

No matter where one looks today, one reads about Industry 4.0, Internet of Things, and Digital Transformation. In the Life Sciences industry this is being referred to as Pharma 4.0. The speaker, based on more than 30 years of experience in Pharmaceutical Manufacturing will put forward perspectives on why Pharma is lagging behind significantly, and what it needs to do ready itself for Pharma 4.0. Over the past 40 years, the way of doing things in Pharma has been dominated by a culture of compliance that has fed an ultra-conservative approach to adopting new, innovative approaches to enabling the manufacturing supply chain. How can this culture be changed, and what needs to happen with foundational technologies in order to enable Digital Transformation, from Process R&D through manufacturing? What constitutes the core 'Digital Thread' that is needed to enable Life Sciences companies to transform to a digitally enabled enterprise, across the entire product lifecycle? Addressing these challenges is fundamental to Pharma catching up to Industry 4.0

581719 Quantum-on-Demand Biofabrication of 3D Tissue Test Systems

Scientific modeling can be used to conceptualize, test, understand, and direct complex natural systems. The ability to create accurate models of three-dimensional (3D), functioning tissues is a growing reality. Benchtop tissue models, i.e. tissue test systems, have many potential advantages over current modeling technologies, including two-dimensional (2D) cell culture and animal models, for biomedical research. Three-dimensional fabricated tissues are more representative of natural tissues than 2D cell cultures, are easier to control, replicate, and discern than animal models, and are potentially a cost saving for biomedical research. Quantum-on-demand biofabrication, i.e. the deposition of biomaterials such as cells and granules in a specific quantity at a specific location, offers control over cell distribution, biomolecule deposition, and biomaterial geometries and surfaces. Tissue models can be used to investigate tissue function and disease progression, discover new pharmaceuticals, teach, and personalize treatments for patients.

580862 Robustness of Cellular Signaling with Respect to Dosage

Introduction: How tissues are patterned in a natural context can serve as the foundation for our future methods of tissue engineering and regeneration. In a developing tissue, the timing and placement of gene expression is crucial to the proper assignment of cell fate, and subsequent differentiation of the cells into a reproducible pattern of tissues and organs. Proper gene expression patterns are often achieved through the action of spatial gradients of chemicals, typically a class of proteins called morphogens, which signal to cells in a concentration-dependent fashion. However, morphogen-mediated patterning models face a paradox: spatial patterns of gene expression are surprisingly robust to variations in morphogen dosage despite the fact that, by definition, cells must be sensitive to morphogen concentration. In this work we investigate the robustness of the Dorsal/NF-κB signaling module with respect to perturbations to the dosage of maternally-supplied *dorsal* (*dl*). Through the action of Toll receptor signaling, which is high on the ventral side of the embryo, Dorsal (Dl) becomes free of its cytoplasmic inhibitor Cactus (Cact)/IκB and is able to enter the nucleus to regulate gene expression[1]. In this manner, a spatial gradient (ventral high, dorsal low) in Dl nuclear concentration is established, which acts to pattern the dorsal/ventral (DV) axis of the early embryo. Although these biophysical interactions have been characterized in detail, how their function is integrated to produce robust gene expression patterns to variations in the *dl* maternal dosage is still unclear.

Results: We measured gene expression of two Dl target genes: *sna*, which is on the ventral-most 20% of the embryo, and *sog*, which resides in a broad lateral region, in embryos carrying 1, 2, or 4 copies of *dl*. The gene expression boundaries have small, but detectable differences depending on *dl* dosage. However, naïve models of Dl distribution, which assume that Dl concentration globally scales with *dl* gene dosage, predict high sensitivity gene expression boundaries. This implies there must be mechanisms that alter the shape of the Dl gradient when *dl* dosage is perturbed. Previous work[2,3] has suggested three such mechanisms: facilitated diffusion of Dl by Cact, saturation of active Toll receptors, and the presence of Dl/Cact complex in the nucleus. We perform a parameter screen using a mechanistic model of the Dl/Cact system to show that these mechanisms are together necessary for robustness of gene expression boundary with respect to *dl* dosage.

Conclusions: In the Dl gradient system, gene expression is robust due to the action of three mechanisms: facilitated diffusion, Toll receptor saturation, and the action of Dl/Cact complex in the nucleus. Morphogen gradients must be able to determine cell fates in a concentration-dependent fashion, while at the same time maintain robustness with respect to morphogen dose. The Dl/Cact system shows an example of this engineering principle. Ultimately, we may be able to use our engineering knowledge gained from the basic study of tissue patterning to benefit the field of regenerative medicine.

This research was supported by NSF CBET-1254344 and NIH R21HD092830.

References:

1. Reeves, GT & Stathopoulos, A. *Cold Spring Harb Perspect Biol* (2009) **1**: a000836.
2. Carrell, SN et al. *Development* (2017) **144**: 4450-4461.
3. O'Connell, MD & Reeves GT. *PLoS Comput. Biol.* (2015) **11**: e1004159.

564816 Synergistically Enhance Bone Cell Proliferation and Osteogenesis on 3D-Printed Scaffolds Using Two-Dimensional Black Phosphorus and Graphene Oxide Nanosheets

Synergistically Enhance Bone Cell Proliferation and Osteogenesis on 3D-Printed Scaffolds Using Two-Dimensional Black Phosphorus and Graphene Oxide Nanosheets

Xifeng Liu[ab], A. Lee Miller II[b], Sungjo Park[c], Matthew N. George[a], Brian E. Waletzki[b], Haocheng Xu[a], Andre Terzic[c], and Lichun Lu*[ab]

a Department of Physiology and Biomedical Engineering, Mayo Clinic, Rochester, MN 55905, USA

b Department of Orthopedic Surgery, Mayo Clinic, Rochester, MN 55905, USA.

c Department of Cardiovascular Diseases and Center for Regenerative Medicine, Mayo Clinic, Rochester, Minnesota 55905, USA

* Corresponding author. Email address: Lu.Lichun@mayo.edu; Phone: 507-284-2267; Fax: 507-284-5075.

Keywords: 2D Materials; 3D-Printed Scaffolds; Bone Cell Osteogenesis

1. Introduction: Incorporation of two©\dimensional (2D) materials to tissue scaffolds for improved biocompatibility and cell affinity has emerged as a promising new strategy in tissue engineering. For example, 2D graphene oxide (GO) has been widely reported to enhance cell adhesion and proliferation. Recently, a newly emerged black phosphorus (BP) 2D material has attracted attention in biomedical applications due to its unique mechanical and electrochemical characteristics. In this study, we investigated the synergistic effect of these two types of 2D materials on cell osteogenesis on three-dimensional (3D) printed poly(propylene fumarate) (PPF) scaffolds for bone tissue engineering (Fig. 1a).

2. Materials and Methods: Graphene oxide sheets were oxidized and exfoliated from natural graphite (~150 mm flakes) through improved Hummers¡¯ method.[1] BP nanosheets were synthesized using a liquid exfoliation method. Poly(propylene fumarate) was synthesized and used as the cross-linkable polymer for 3D printed scaffolds.[2] The 3D scaffold model was designed using computer-aided design (CAD) software (SolidWorks Corp., Concord, MA) and printed using the Viper si2 stereolithography system (3D Systems, Valencia, CA).[3] The ammonolyzed 3D-PPF-Amine scaffolds were coated with 2D materials by soaking in solution containing BP nanosheets, GO nanosheets or both BP and GO nanosheets. Sterilized functionalized 3D scaffolds were adhered to the bottom of 48-well tissue culture polystyrene (TCPS) plates and MC3T3 pre-osteoblast cells were seeded. Upon harvest, cells were fixed in paraformaldehyde (PFA, 4% solution) and stained with anti-vinculin-FITC antibody (Sigma-Aldrich Co., Milwaukee, WI) to label vinculin and rhodamine-phalloidin (RP, Cytoskeleton Inc, Denver, CO, USA) to stain cellular filaments. The immune-fluorescence stained cells on the functionalized 3D Scaffolds were immediately scoped on an inverted laser scanning confocal microscope (Carl Zeiss, Germany).

3. Results and Discussion: All 3D scaffolds after coating with both types of 2D materials showed good biocompatibilities (Fig. 1b). In addition, the increased surface area provided by GO sheets enhanced cell attachment at the initial stage. Afterward, cells were further stimulated by phosphate, which was continuously released from slow oxidation of BP nanosheets and played an important role in osteoblast differentiation. Through the use of 3D confocal imaging, unique interactions between cells and BP nanosheets were observed, including a stretched cell shape and development of filaments around the BP nanosheets, along with increased cell proliferation when compared with scaffolds incorporating only one of the 2D materials (Fig. 1c). All these results indicate that the incorporation of 2D BP and GO materials could effectively and synergistically stimulate cell proliferation and osteogenesis on 3D tissue scaffolds.

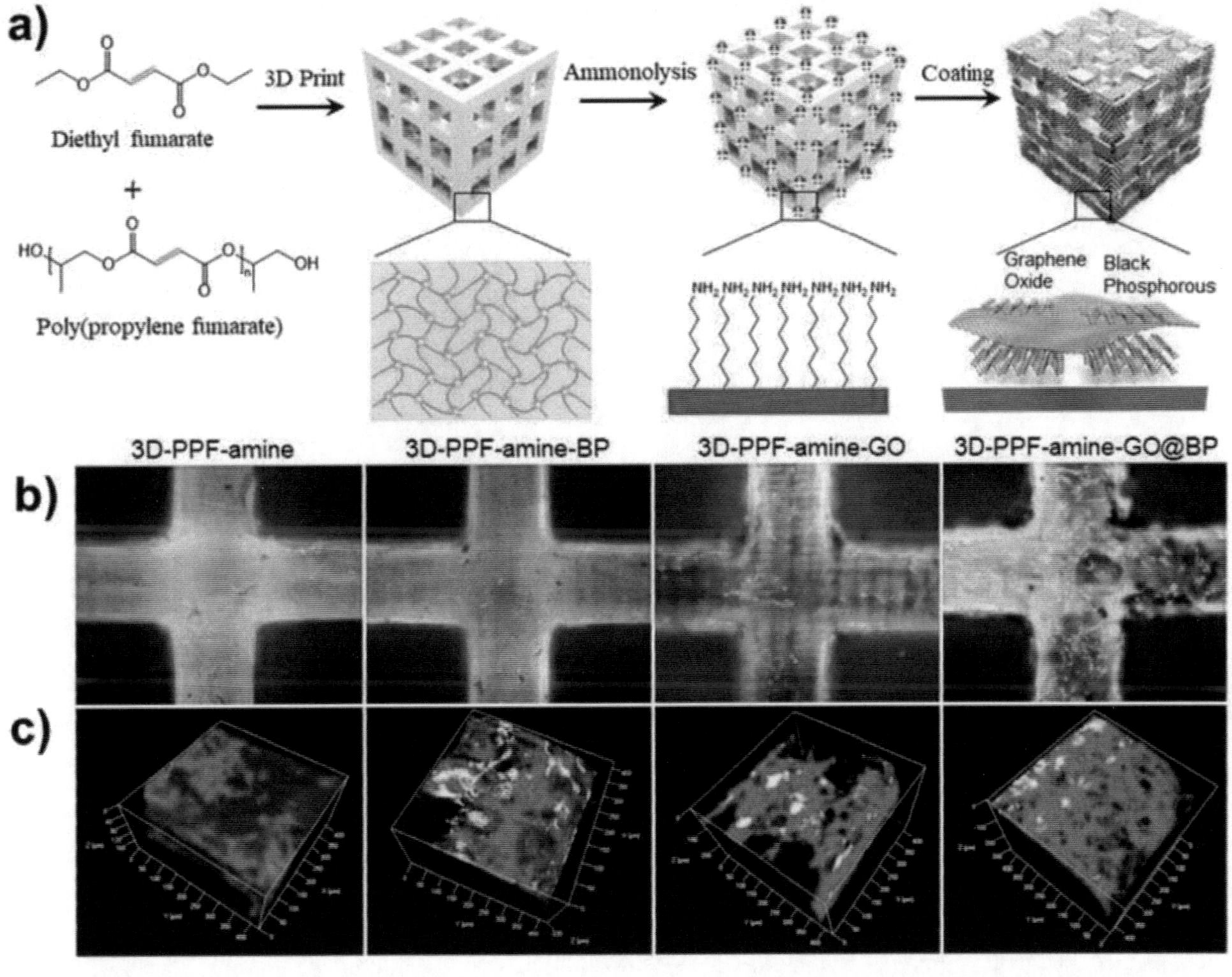

Fig. 1 a) Fabrication process of 3D printed scaffolds functionalized with 2D GO and BP nanosheets. b) Live/dead staining of ammonolyzed 3D scaffolds as well as scaffolds functionalized with BP, GO and GO@BP nanosheets. c) Immunofluorescence images of MC3T3 pre-osteoblasts growing on functionalized 3D scaffolds.

4. Conclusions: We fabricated a series of 3D-printed scaffolds functionalized with 2D materials for use in tissue engineering. Our results showed that the addition of 2D BP and GO nanosheets to scaffolds improved cell adhesion, proliferation, and differentiation.

Acknowledgments: This work was supported by National Institutes of Health grant R01 AR56212.

References

[1]. X. Liu *et al*, Journal of Materials Chemistry B 2016, 4 (43), 6930-6941.

[2]. X. Liu *et al*, RSC Advances 2015, 5 (27), 21301-21309.

[3] K. W. Lee *et al*, Biomacromolecules 2007, 8 (4), 1077-1084.

567121 Biological Response Precision-Controlled By Spherical Pores

Implanted biomaterials, after approximately one month, are observed in an avascular, collagenous capsule – this outcome is referred to as the foreign body reaction (FBR). This reaction starts with recruited macrophages that then fuse into giant cells leading to the avascular, fibrotic healing. Such healing can be undesirable for implanted sensors, electrodes, drug delivery systems and other devices. We have developed a number of strategies based on polymeric materials to address this FBR. One such strategy will form the focus of this talk. We have fabricated hydrogel materials with uniform, interconnected 30-40 micron pores by a sphere-template method -- upon implantation the pores are infused by macrophages. The macrophages within the pores orchestrate a different healing process from that seen with non-porous materials or materials with non-optimal pore sizes. Macrophages are biomechanically directed to a unique phenotype that encourages reconstructive healing (M2). Thus, the inflammatory process is exploited to generate improved healing – a new biocompatibility. Reconstructive healing has been noted subcutaneously, percutaneously and within heart muscle, cornea and bone. The surprising and desirable outcomes from these implanted porous structures led us to commercialize them. The path from laboratory curiosity to a product and medical device that is now used in humans will be discussed.

569596 Correlation between Fluid Shear Stress and Morphological Behaviour of Valvular Endothelial Cells

Valvular endothelial cells (VECs) line the outer layer of heart valves and withstand oscillatory and laminar shear forces along with other mechanical stimuli. Compared to vascular endothelial cells, VECs are understudied and as a consequence their exact physiological behaviour is unclear. Thus, endothelial dysfunction mechanisms leading to valve sclerosis or calcification are yet unknown. In vitro studies of VECs cultured under physiological mechanical stimulation can uncover mechanisms of VEC mechanobiology. Previous studies have shown that vascular endothelial cells align parallel to fluid flow while, in contrast, VECs align perpendicular to blood flow. Recent work from our group has demonstrated that this behaviour varies for different ranges of shear stress values. Here, we aim to show that VEC behaviour does not depend exclusively on fluid shear. There may be several other factors influencing such behaviours, such as confluence levels, time exposed to shear, as well as the stiffness of the supporting scaffold.

This work is based on the study of behaviour of VECs under different shear stress levels and different times of exposure. VECs are cultured on soft substrates on a custom-made parallel-plate chamber. This chamber is specially designed for cultures on soft scaffolds while applying fluid shear on cells in monolayer. Polydimethylsiloxane gels are prepared in the chamber and have their surfaces modified using sulfo-SANPAH as a linker for collagen type I. After collagen treatment, 10^4 VECs/cm^2 are seeded on the surface and allowed to adhere for 48 h without shear exposure. Upon cell adhesion, 2.6, 6.5, 13 and 20 dyne/cm^2 fluid shear are applied on the monolayer culture for either 24 h or 48 h. Cultures then undergo immunofluorescence processing prior to imaging and image analyses of cell aspect ratios, circularity, area, alignment, etc.

Our results demonstrate that with increasing fluid shear, VECs tend to orient in more diverse way, losing their traditional cuboidal morphology, as quantified by cell aspect ratio and circularity parameters. In addition, higher shear levels decrease VEC area, regardless of confluence level. In addition, we observed that, the threshold for cell alignment with shear is much higher than previously observed. High shear also increases VEC differentiation and mesenchymal transition, as validated by F-actin and interstitial cell phenotypes. These results increase the current understanding of VEC physiology and this knowledge is key for the future engineering of replacement heart valves.

578181 Metabolite-Based Modulation of Dendritic Cells for Developing Effective Immunotherapy

INTRODUCTION – Aging and diabetes associated chronic / slow healing cutaneous wounds have hallmark immune signatures, including *delayed* T-cell and macrophage infiltration, and increased pro-inflammatory immune cell infiltration for *extended periods* (compared to non-aged control). *If there were a method to modulate the local immune responses in a temporal manner then it would be possible to accelerate the wound healing of cutaneous wounds in aged and/or diabetic immune system.* Notably, metabolic pathways control immune cell functions, and modulating these pathways can determine overall immune system response. Therefore, in this work, polymers were generated with metabolite monomers that can deliver metabolites and directly modulate the function of immune cells.

RESULTS – This work for the first time (to the best of our knowledge) demonstrates utilization of metabolites to modulate immune-metabolic responses in wound healing. NMR and FTIR studies were performed to characterize the polymers. Moreover, oil in water emulsions were utilized to generate microparticles of these metabolite-based polymers. These microparticles are capable of encapsulating drugs and release them in a sustained manner (shown via representative drug-molecule rhodamine). The release kinetics of metabolites from these polymers were determined using NMR. Moreover, degradation kinetics of the polymers was determined using weight loss experiments. Importantly, it was demonstrated that these polymers could modulate the function of mouse bone marrow derived dendritic cells and allogenic T-cells in a mixed lymphocyte reaction. Lastly, we demonstrate that these polymers when applied in the presence of tegaderm were able to modulate the wound healing rate in BALB/c mice.

CONCLUSION - In summary, metabolite-based polymers were generated that were capable of controlling the pro- or anti-inflammatory responses of immune cells *in vitro* and *in vivo*. Notably, these polymers were able to lead to healing of cutaneous wounds in mice.

576630 Probing the Sensitivity of Point Charge Potentials in an Attempt to Improve Their Ability to Rank the Stability of Pharmaceutical Polymorphs

Temperature and entropic effects cannot be neglected when ranking the stability of organic polymorphs. Enormous effort has been put into structure prediction to seek out new crystalline phases for organic materials, with much of the stability ranking being an afterthought. Structure predictions have relied on lattice energy to determine the crystalline stability despite numerous experimental and computational studies that have shown that enantiotropic behavior is prevalent. In particular, one study has shown that 20% of a set of > 500 experimentally know polymorphs re-rank with temperature.

The bottleneck preventing free energy stability rankings in CSPs is largely due to the availability of accurate and efficient force fields. We have developed and tested molecular and lattice dynamic approaches to include temperature and entropic effects of enantiotropic polymorphs. For both approaches, we used point-charge potentials to compute the free energy of 12 systems. We found that the entropy differences using the point charge potential was the correct sign for 100% of the systems relative to experiment, but the enthalpic stability between polymorphs was only the correct sign for 58% of the systems. While more complex hamiltonians could be used to correct these issues the increase in computational expense becomes disfavorable, especially if added to a CSP.

We hypothesize that point charge potentials can be tuned with simple scaling factors improve the lattice energy, without negative implications on the entropic contributions. Initial results have shown that the entropic contributions to free energy stability are less sensitive to the potential used than the lattice energy. To better understand the tunability of point charge potentials we will look at how scaling the intermolecular interactions independently affect the enthalpy and entropy differences of the crystalline systems previously studied. The end goal would be to take the lattice energy rankings of a CSP performed with a highly level of electronic theory, tune the intermolecular interaction to minimize the error in the lattice energy stability, and then consider the effects of temperature and entropy. I propose to present the results tested to prove or disprove that hypothesis.

552602 Crystal Structure Prediction Applications - Going Beyond a Stable Form Selection

Crystal structure prediction (CSP) in pharmaceutical industry is typically performed to support commercial solid form selection of a freebase API. However, the variety of tasks of solid state chemistry in industry is much broader than that important application. The presentation will cover some virtual screening CSP-based applications such as poor solubility improvement during lead optimization and coformer screening for cocrystallization.

563344 Accurate and Efficient Lattice Energy Models for Industry-Oriented Crystal Structure Prediction

Many organic molecules of industrial interest exhibit crystalline polymorphism, forming multiple solid-state structures, each displaying different physico-chemical properties. Consequently, understanding the crystal energy landscape of a commercial drug or chemical can be of great importance, especially if the compound is delivered in solid form, as is the case with suspended agrochemicals or oral dosage forms for pharmaceuticals. The discovery of new polymorphs is thus desirable from an R&D perspective and can help to avoid potentially disastrous situations from an operational standpoint. These observations have motivated the development of methods for polymorph prediction that can now be applied to molecules of industrial relevance, provided that conformational flexibility is limited to a few degrees of freedom.[1].

Recent blind tests [2,3] organised by the CCDC have highlighted the growing capabilities of crystal structure prediction (CSP) for increasingly complex systems, but what is often considered a success in a blind test does not always provide sufficient information to complement experimental solid-form screening efforts in an effective way. In addition to these concerns, ensuring that CSP can be used in the early stages of product development requires that the crystal energy landscape can be mapped quickly and efficiently [4]. Despite the requirement for efficient methods, the use of computationally expensive periodic density functional theory (DFT) calculations has become a staple for many practitioners of CSP. Although this typically achieves greater accuracy, the computational cost may often be prohibitively large for the study of systems of industrial relevance in the early stages of product development.

In this talk, we investigate the role of parameter estimation and adaptive force fields in the development of bespoke *ab initio* potentials. Models developed in this way can approach the accuracy of periodic DFT calculations with a computational cost reduction typically around 6 orders of magnitude. This is partly achieved a) through the rigorous construction of training sets from theoretical calculations and b) by taking cues from successful developments in DFT dispersion corrections [5,6] over recent years to improve the description of intermolecular interactions. Overall, significant improvements in the accuracy of the predictions of the energy and geometry of organic solids can be obtained, over conventional force fields, increasing both the speed and reliability of CSP as applied in industrial applications.

[1] Nyman, Jonas, and Susan M. Reutzel-Edens. "Crystal structure prediction is changing from basic science to applied technology." *Faraday discussions* 211 (2018): 459-476.

[2] Bardwell, David A., et al. "Towards crystal structure prediction of complex organic compounds–a report on the fifth blind test." *Acta Crystallographica Section B: Structural Science*67.6 (2011): 535-551.

[3] Reilly, Anthony M., et al. "Report on the sixth blind test of organic crystal structure prediction methods." *Acta Crystallographica Section B: Structural Science, Crystal Engineering and Materials* 72.4 (2016): 439-459.

[4] Price, Sarah L., and Susan M. Reutzel-Edens. "The potential of computed crystal energy landscapes to aid solid-form development." *Drug Discovery Today* 21.6 (2016): 912-923.

[5] Grimme, Stefan, et al. "A consistent and accurate ab initio parametrization of density functional dispersion correction (DFT-D) for the 94 elements H-Pu." *The Journal of chemical physics* 132.15 (2010): 154104.

[6] Tkatchenko, Alexandre, and Matthias Scheffler. "Accurate molecular van der Waals interactions from ground-state electron density and free-atom reference data." *Physical review letters* 102.7 (2009): 073005.

567777 How Many Ritonavir Cases Are Still out There?

On rare occasions, a crystal structure that has been prepared for years is unexpectedly "superseded" by a thermodynamically more stable polymorph, rendering the initial crystal structure almost impossible to obtain—a so-called "disappearing polymorph" [1]. Famous cases include the drug compounds Ritonavir and Rotigotine: The thermodynamically more stable polymorph turned out to be substantially less soluble and required a costly reformulation before a product with comparable specifications could be manufactured again.

On a case-by-case basis, the probability of a kinetically hindered but thermodynamically stable polymorph having been missed in an experimental polymorph screen can be assessed by a computational crystal structure prediction study: the calculations do not suffer from kinetics and will in principle enumerate all possible crystal structures for a given compound; comparison against the experimental structures identifies any missed polymorphs. In a recent paper, we presented the results of 41 such computational crystal structure prediction studies to draw up statistics regarding the number of hidden Ritonavir cases [2]. The main finding of the paper was that between 15 and 45% of all experimental crystal structures are in fact thermodynamically unstable.

The large spread in our estimate was due to the computational error in the calculation of the relative free energies of the predicted polymorphs. Even though the computational error is estimated to be as small as 0.5 kcal/mol, about half the size of the gold standard referred to as "chemical accuracy", the experimental energy differences between polymorphs are also around the 0.5 kcal/mol mark, and to reduce the large spread in our estimate an even more accurate energy model was needed.

The shortcomings of our old energy model such as neglect of temperature, relatively poor accuracy of the exchange part of the functional used in the DFT calculations, and the assumption that Van der Waals interactions are pairwise additive are well documented, and in this contribution we will present the results of an energy model in which these shortcomings have been addressed [3]. With the availability of more accurate relative lattice energies, we are able to narrow down our estimate of how many experimental crystal structures represent a thermodynamically unstable polymorph at ambient conditions.

References

[1] D.-K. Bučar, R. W. Lancaster & J. Bernstein (2015), "Disappearing Polymorphs Revisited" *Angew. Chem., Int. Ed.* **54**, 6972-6993.

[2] M. A. Neumann & J. van de Streek (2018), "How Many Ritonavir Cases Are There Still out There?" *Faraday Discuss.* **211**, 441-458.

[3] J. Hoja, Hsin-Yu Ko, M. A. Neumann, R. Car, R. A. DiStasio Jr. & A. Tkatchenko (2019), "Reliable and Practical Computational Description of Molecular Crystal Polymorphs" *Sci. Adv.* **5**, eaau3338.

562227 A Comparison of Methods for Capturing Hydrate-Anhydrate Transition Thermodynamics

A key milestone in the small molecule drug development pipeline is the selection of a solid form for both research and commercial development. One of the main deliverables from the solid form selection process is to identify a solid form of the Active Pharmaceutical Ingredient (API) that is physically stable and unlikely to undergo a solid form change during a later stage of manufacturing, transportation, and storage. One significant physical transformation that can occur in a drug product is a conversion from an anhydrous to a hydrate form because a hydrate can have a lower aqueous solubility and dissolution rate by over an order of magnitude relative to the corresponding anhydrous form.

Computational models of hydrate-anhydrate transitions can reduce the risk of unintended transformations during late-stage development. Recent molecular modeling approaches have been developed to compute the stability of anhydrous polymorphs pairs over a range of experimentally relevant temperatures. Here we extend this work to the comparison of anhydrous forms to their corresponding hydrates over a range of temperature as well as ambient humidity levels. The stabilities of multiple different small molecule hydrate-anhydrate pairs are computed using a variety of classical potentials including OPLS, GAFF, and the polarizable AMOEBA force-field. Additionally, the interstitial waters are modeled with the TIP3P, TIP4P, and AMOEBA water models. Overall, we find that the estimated entropy differences between forms are significantly less sensitive to the force-field than the corresponding enthalpy differences. In addition, the entropy differences are in decent agreement with experimental measurements derived from coexistence points. This result suggests that quantitative prediction of the transition temperature could be achieved in conjunction with a sufficiently accurate enthalpy difference estimate at a single temperature.

583064 Genarris 2.0: A Random Structure Generator for Molecular Crystals

Genarris 2.0 is an open source Python package for performing configuration space screening by random structure generation for crystals of molecules with no bond-rotational degrees of freedom. The target unit cell volume is determined by estimating the molecular volume. Crystal structures are then generated in all space groups compatible with the requested number of molecules per cell (Z) with one molecule in the asymmetric unit (Z'=1), including those with special Wyckoff positions. To avoid generating unphysical structures, constraints are applied on the intermolecular distances, based on van der Waals radii. Special settings are applied for strong hydrogen bonds, which are automatically detected. Once an initial dataset of several thousand structures is generated, a smaller curated dataset may be selected based on user-defined quality and diversity criteria to serve as an initial guess for structure search algorithms or as a training set for machine learning. For clustering based on packing motif similarity Genarris uses the affinity propagation machine learning algorithm with either a relative coordinate descriptor (RCD) or a radial distribution function (RDF) representation.

583911 Crystal Structure Prediction in Lead Optimization Phase of Drug Discovery

Crystal structure predictions (CSP) has become a very valuable tool to explore crystal interactions in early discovery phase. At AstraZeneca we have developed a fast and accurate protocol that allows for applying CSP already in the lead optimization phase. The protocol will be described and compared to the most accurate protocols published. A few case studies will be presented to illustrate the speed and accuracy of the protocol.

560162 Development of a Continuous Hydrogenation Process for Pharmaceuticals Using a Trickle Bed Reactor

In past decade there has been significant interest in the development of continuous processes in the pharmaceutical industry. Key drivers for continuous processing is product quality improvement, yield and process robustness, footprint reduction of processing area and increased process safety. Hydrogenations are commonly used in synthesis of pharmaceutical intermediates and active pharmaceutical ingredients. In the present work we have demonstrated:

- Application of a laboratory scale trickle bed reactor for development and scale-up of a continuous hydrogenation process.
- Detailed description of catalyst packing procedures and startup of trickle bed reactors.
- Nitro to amine reduction for a pharmaceutical intermediate using a trickle bed reactor.
- The effect of operating parameters such as pressure, gas to liquid ratio, temperature catalyst loading and residence time of the substrate on product quality, is studied to develop a scalable and robust process.
- Development of a reactor scale kinetic model to predict conversion in the nitro-amine reduction group.

'Fit for purpose' model developed was very effective to efficiently identify optimal reaction conditions, assess risk to product CQAs, and increase overall robustness. The evaluation of these parameters led to a multi-faceted model capable of efficiently identifying optimal reaction conditions, assess risk to product quality attributes and increase robustness of the process.

Disclosures:

All authors are employees of AbbVie and may own AbbVie stock. AbbVie sponsored and funded the study; contributed to the design; participated in the collection, analysis, and interpretation of data, and in writing, reviewing, and approval of the final publication.

577249 Characterization of a Flow Hydrogenation System and Application to a Sensitive Reaction for Production of an Active Pharmaceutical Ingredient Intermediate

Hydrogenation is a chemical transformation frequently employed in the synthesis of active pharmaceutical ingredients (APIs). These reactions are traditionally performed in batch in stirred, pressure rated autoclaves. Such autoclaves can represent a significant initial capital investment and are expensive to clean and maintain. Operationally, they introduce safety concerns due to their inherent large hydrogen headspace. Further, these batch reactors often restrict the space of achievable operating conditions as a result of poor heat and mass transfer characteristics and limitations on maximum operating pressure.

Implementation of flow reactor technology offers solutions to many of the limitations of traditional batch hydrogenation. Flow reactors can be inexpensive to construct. They can be made from relatively low-cost stainless steel tubing and compression fittings capable of withstanding high pressures and temperatures. This allows a reactor to be dedicated to a single chemical transformation and may reduce or eliminate routine cleaning and maintenance. Flow reactors improve safety by minimizing the hydrogen headspace and increasing the surface-to-volume ratio available for heat removal. Of particular relevance to this presentation is the ability of flow reactors to maintain tight control of residence time and enable use of innovative process analytical technology (PAT).

This presentation will focus on characterization of a flow hydrogenation system and its application to a sensitive asymmetric hydrogenation for production of an API intermediate. Drivers for performing this reaction in flow will be discussed. Elements of both reaction and reactor design were investigated. The reaction was screened for a variety of temperatures, pressures, catalyst loadings, and solvents. Studies were performed to understand gas-liquid flow dynamics and residence time distribution (RTD).

572282 The Application of Flow NMR Toward the Development of a Kinetic Model for Continuous Grignard Formation

A kinetic model was developed to describe the formation of benzylmagnesium chloride from benzylchloride and magnesium metal. This work follows on previously reported kinetic studies by Eli Lilly and Co. (Changi, S.M., and S. Wong, *Org. Process Res. Dev.* 2016, 20, 525-539), although for a different substrate and the new studies include the kinetics of the Wurtz homocoupling impurity. Interestingly, the initial rate of homocoupling was found to be sensitive to the type of magnesium initiation employed. The significant species could be followed and quantified for each experiment using flow NMR when an internal standard was used. The impact of nuclei relaxation and absolute integral response, relative to flow rate through the NMR flow-loop, necessitated the use of correction factors for accurate quantitation of each species. The kinetic model was ultimately used to guide the operating conditions including feed strategies for a continuous stirred tank reactor (CSTR). The details of the flow NMR, the kinetic model development, sensitivities identified, and the process modeling results will be discussed.

568019 Application of a Dynamic Flow Reactor Platform for Data-Rich Experimentation of Complex Flow Reactions

Advancements in continuous processing technologies in parallel with the expansion of complex organic reactions enabled by flow chemistry have led to the increasing use of flow reactions throughout the pharmaceutical industry. However, to date, reaction development in flow is plagued by the inefficiencies associated with collecting data under steady-state conditions and data-rich experimentation options for characterizing flow reactions remain sparse. In the early stages of drug development, this challenge is compounded by limited availability of key raw materials and short development timelines. Transient reaction experiments provide a viable process development intensification option to lessen the resource burden and inefficiencies surrounding flow reaction process development; this approach has recently been successfully implemented to rapidly obtain kinetic profiling data [1-4].

This work focuses on the application of a recently developed automated continuous flow reactor, capable of well-defined transient experiments, to study two reaction cases where performing the chemical transformation in a flow reactor offers several benefits over a comparable batch operation: mixing sensitive and photochemical reactions. As these complex reaction systems are sensitive to reactor geometry and configuration, operating these flow reactions dynamically allows for the efficient screening of numerous reaction conditions in the specific reactor of interest. Furthermore, this added experimental efficiency can enable more facile testing of various reactor configurations (i.e., different static mixers, light sources). For example, using controlled transient flowrates and various pressure sensors, numerous mixing intensities can be screened in a single experiment and translated into trends relating reaction performance to mixing time. Similarly, any controlled transient variations in residence time, photocatalyst equivalents, and reaction concentration at various light intensities can potentially elucidate the relationship between photon equivalents, catalyst equivalents, and reaction concentration. Scale-up of these dynamic/transient flow test cases to kilo scale flow reactors may also be discussed.

[1] S. Mozharov, A. Nordon, D. Littlejohn, C. Wiles, P. Watts, P. Dallin, and J. M. Girkin, *J. Am. Chem. Soc.* 2011, 133, 3601-3608.

[2] C. A. Hone, N. Holmes, G. R. Akien, R. A. Bourne, and F. L. Muller, *React. Chem. Eng.* 2017, 2, 103-108.

[3] J. S. Moore and K. F. Jensen, *Angew. Chem., Int. Ed.* 2014, 53, 470-473

[4] K. C. Aroh and K. F. Jensen, *React. Chem. Eng.* 2018, 3, 94-101.

561125 Scale-up of a Continuous Extraction Process for an Equilibrium-Limited Reaction

Scale-Up of a Continuous Extraction Process for an Equilibrium-Limited Reaction

Eric G. Moschetta[*,1], Michael T. Tudesco[1], and Eric A. Voight[2]

[1]AbbVie, Inc. Process R&D, 1401 Sheridan Road North Chicago, IL 60064

[2]Discovery Chemistry and Technology, AbbVie, Inc., 1 North Waukegan Road, North Chicago, Illinois 60064

[*]corresponding author: eric.moschetta@abbvie.com

Continuous processing is an emerging paradigm the pharmaceutical industry because of the potential improvements in cycle time and the final quality and yield of the desired product. Flow chemistry is perhaps the most popular method for implementing continuous processing the pharmaceutical industry to date, but impacts in continuous separation and isolation of API have enormous potential to offer improvements in process efficiency. One especially fruitful area of continuous processing is liquid-liquid extraction. In the pharmaceutical industry, liquidliquid extraction is executed batch-wise on the manufacturing scale, almost exclusively. The execution of multiple batch extractions on the manufacturing scale is often inefficient, expensive, and labor intensive. Developing and implementing continuous liquid-liquid extraction processes represents significant reductions in processing time and solvent use. Additionally, continuous extraction offers flexibility in terms of the choice of contacting device and can be fit to the needs of the process in question, whether the separation involves handling emulsions, separating liquids of similar densities, etc. Therefore, it is critical to identify opportunities to develop and implement continuous liquid-liquid extractions for improving processing efficiency from the laboratory scale all the way to manufacturing.

We present the design and implementation of a fit-for-purpose continuous liquid-liquid extraction for the scale-up of an equilibrium-limited reaction. This fit-for-purpose approach is easily adaptable to a number of equilibrium reactions, can be customized to fit the needs of the process, and requires simple laboratory glassware and equipment to implement. We show how the process parameters were varied to reduce the overall amount of extraction solvent needed to drive the reaction to completion. Our method is ideally suited to situations where both material and time are limited. We implemented our continuous extraction for ABBV-168, a 2'-bromouridine for the treatment of Hepatitis C, and demonstrated scale-up in the laboratory up to 50 g, havingdeveloped the process on the 0.5-5 g scales.

All authors are employees of AbbVie and may own AbbVie stock. AbbVie sponsored and funded the study; contributed to the design; participated in the collection, analysis, and interpretation of data, and in writing, reviewing, and approval of the final publication.

558527 A Virtual Plant for Synthetic Continuous Manufacturing Via Integrated Systems-Based Modeling

The biopharmaceutical industry is moving towards continuous manufacturing to enable a smaller manufacturing footprint, improved environmental sustainability, improved control strategy, flexibility to meet changing demand and safer operations. Despite these benefits, the transition from batch to a continuous mode of operation significantly expands the number of parameters that require evaluation for optimal process design and process characterization, which directly translate to the need for a large and resource intensive experimental data set. These resource requirements can be reduced significantly by augmenting empirical design with modeling.

Here we present a 'Virtual Plant for Synthetic Continuous Manufacturing' via integrated systems-based modelling. As a part of this methodology, process characterization was executed to enable mechanistic understanding and modeling of the physical and chemical phenomena and to obtain quality data that are fed into model development. Those model components were integrated to form the end-to-end process model, with which *in silico* studies (sensitivity analysis, uncertainty propagation, etc.) are performed to identify the key process parameters. The systems-based model enables the subsequent development and testing of an integrated advanced process control strategy, which is envisioned to support regulatory filings for CM processes. The complete Virtual Plant enables identification and evaluation of other requirements and policies related to equipment, facility, automation, and operations. Development of a virtual plant for synthetic CM enables us to predict process performance and product quality in a real plant environment.

569212 Starting the Road to Commercialization: Pharmacy on Demand (POD) Ciprofloxacin

Unfortunately, both the cost and supply gaps of medicine are increasing. There are on average 200 active drug shortages nationally each year. Lengthy supply chains, limited numbers of manufacturers (commodity chemicals and drug product), and surges in demand are some of the more prominent causes of drug shortages. This lack of responsiveness, uncertainty, and diversity create pressure for drug suppliers, hospitals, and care givers to provide in time medication.

Through the lifetime of the Pharmacy on Demand (POD) initiative, researchers have worked toward addressing these chemical and economic inefficiencies associated with "typical" large-scale batch pharmaceutical manufacturing. Substantial advances in complex reaction telescoping, real time formulation, reaction engineering and pumping technologies have enabled the design, construction and implementation of a compact and reconfigurable platform. Tight control of process inputs and the implementation of in-line process analytical technologies enabled high volumes of APIs (Atropine, Diphenhydramine, Ciprofloxacin, Diazepam, Doxycycline, and Lidocaine to name but a few) to be manufactured over a four-week manufacturing campaign.

This production campaign served as a demonstration of these platforms ability to produce multiple API in one equipment train. However, to move towards the potential commercial realization of such platforms, it was thought prudent to select only one API, in this case Ciprofloxacin (a broad spectrum anti-biotic used to treat a wide spectrum of infections, most notably anthrax exposure) and target of filing an Abbreviated New Drug Application (ANDA) with the FDA.

The goal of this phase of the project was to create a platform/process capable of the end-to-end production of Ciprofloxacin HCl from commercially available starting materials to final oral solid dose. This platform/process had to be capable of meeting current Good Manufacturing Practices (cGMP) whilst producing material that would pass United States Pharmacopeia (USP) specifications for Ciprofloxacin HCl. Design of experiments, residence time distributions etc. were completed to aid in the process characterization, development, and optimization work required for the pharmaceutical development phase.

562790 The Effect of Centrifugal Force on the Mechanism of Wharton's Jelly Mesenchymal Stem Cells Transfection

A human mesenchymal stem cell (hMSC) is an adult stem cell characterized by: being an undifferentiated cell, having adapted to a specific growth environment, producing a large number of differentiated and functional progenies, and regenerating tissues after injury or lesion. All those characteristics make them a promising cell source for regenerative medicine, especially for their multipotent capacity which render them capable of differentiating into multiple lineages including cartilage, adipose, and bone. Although controlling the phenotype of MSCs is a central challenge in tissue engineering, gene therapy has become more attractive since with the introduction of exogenous genes into the cell, the latter is expected to produce its own necessary growth factors and cytokines for its differentiation. This approach would be preferable over the delivery of recombinant cytokines and growth factors, as it involves the administration of non-physiological concentrations with a short half-life, fast body clearance and a lower therapeutic effect in comparison to natural proteins.

In gene therapy, viral vectors such as retrovirus, lentivirus and adenovirus, have been used for the delivery of genes into cells via a process known as transduction. Even though it offers high gene delivery (nearly 99%) and stable gene expression, many limitations remain associated with viral vectors such as insertional mutagenesis, immunogenicity, limited DNA packaging capacity and cumbersome large-scale production. Non-viral gene carriers are promising alternatives for gene delivery and have the potential to address these limitations. Moreover, the transient expression associated with these systems can be more compatible with the natural wound healing processes. However, transfection has significantly lower gene entry rates compared to the viral transduction (approximately 10% to 30% in hMSC). This lower efficiency is due to the mechanism of transfection, which has been theorized to be affected by multiple factors regarding nanoparticle properties, including: size, charge, degradation, endosomal escape and arrival to the nucleus. These barriers depend directly on the chemical nature of the material that covers the gene and its non-specific interaction with all the pathways within the cell. Therefore, there is still need for more in-depth studies, with the aim of improving transfection techniques using this type of vectors or physical mechanisms.

One of the most promising non-viral gene carriers is polyethylenimine (PEI). This polymer, which has a great variety of molecular weights and degrees of branching, has been extensively studied *in vitro* ever since its initial use as gene carrier. PEI has an advantage over other polycationic polymers in that it acquires a remarkable DNA condensation capability and a potential ability to protect DNA from enzymatic degradation. In addition, the three different types of amines that it has are responsible for generating the buffering capacity over a wide range of pH values. This facilitates endosomal escape of carriers, which has been named the "proton-sponge" effect.

As discussed before, transfection presents a poor efficiency of gene uptake. However, we tested here the hypothesis that the efficiency of gene transfection is limited by a simple physical barrier: low concentration of gene (DNA) on the surface of the cell. Therefore, the present study evaluated the mechanism of centrifugation during the transfection process in order to increase the concentration of DNA on the cell surface, thus generating higher transfection efficiencies. The present study evaluated the mechanism of centrifugation during the transfection process in order to increase the concentration of DNA on the cell surface, which allowed higher transfection efficiencies. For this purpose, mesenchymal stem cells derived from umbilical cord, reporter genes such as GFP and luciferase and, branched PEI 25kDa were used to evaluate and optimize the transfection technique by assessing various centrifugal forces. We were able to standardize a transfection protocol in which the weight ratio and different concentrations of complex were evaluated to obtain transfection rates above 20% with a non-viral method.

569466 Aggregation of Mesenchymal Stem Cells Rejuvenates Functionality Following *in Vitro* Aging through Induction of the Integrated Stress Response

Mesenchymal stem cells (MSCs) have shown promising therapeutic potential due to their impressive ability to secrete inflammatory modulatory, angiogenic, and regenerative cytokines along with their ability to differentiate and aid in regeneration. Most work with MSCs though has been done in model systems requiring only a small amount of MSCs. Scaled-up expansion of isolated MSCs is carried out on stiff culture plastic which increases the surface modulus one thousand-fold [1]and limits cell-cell interactions. This type of culture results in *in vitro* aging characterized by increased senescence, decreased glycolytic metabolism, reduced autophagic activity, and loss of multipotent differentiation potential [2]. For these reasons understanding the mechanism of culture induced changes is critical for maintaining MSC phenotype during expansion and proper implementation of MSC therapy.

Protein homeostasis is critical for cells functional maintenance and retention of stemness. As cells age, they loss control over proteasome activity leading to the accumulation of misfolded proteins that would normally be handled by the integrated stress response (ISR). Normally, when misfolded proteins occur they are coated in chaperone proteins such as binding immunoglobin protein (BiP). BiP normally prevents activation of eukaryotic initiation factor 2 alpha kinase 3 (PERK), but as soluble BiP decreases PERK dimerizes/auto-phosphorylates and signals a reduction in protein initiation complex activity causing a cessation of protein synthesis. The misfolded protein is then ubiquinated and lysed. Proteasome autophagy is especially critical in maintaining cell stemness and fighting senescence. It has recently been shown that when autophagy was inhibited in naïve muscle-stem cells they took on the phenotype of *in vitro* aged cells, and when aged cells had autophagy restored they re-established their naïve phenotype [3]. Previously, it has also been shown that embryonic stem cells hyper-control their proteome to maintain stemness by limiting initiation, increasing folding through chaperone protein content, and increasing proteasome activity to degrade build-up of misfolded proteins and prevent defects from going into daughter cells.

It has previously been shown that aggregation of MSCs leads to heightened autophagy [3], and a reversal of metabolic phenotype back to aerobic glycolysis [4]. For this reason, we have explored what effects aggregation has on the proteome to determine what pathways are elevated by aggregation in bone-marrow derived MSCs (bMSCs), adipose derived MSCs (ASCs), and dermal fibroblasts (FB). Pathway analysis showed a decrease in the eIF2 pathway; however, western blot and ingenuity pathway analysis revealed that bMSC and ASCs had augmented eIF2α and eIF2γ. Specifically, eIF2α was significantly phosphorylated and eIF2γ was decreased showing heightened initiation control following aggregation. Since phosphorylation of eIF2α leads to the subsequent translation of activating transcription factor 4 (ATF) which in turn starts a cascade increasing autophagy. Measured levels of ATF in bMSCs and ASCs decreased following aggregation, while FB didn't shift. Downstream proteins DNA-damage inducible transcript 3 (CHOP) showed no shift though autophagy substrate microtubule associated light chain protein 3B (LC3B) showed a significant upregulation following aggregation in bMSCs and ASC. CHOP is also known to regulate growth arrest and DNA-damage inducible protein (GADD34), which is known to dephosphorylate eIF2α, showed lower levels in bMSC and ASC aggregates. This would lock the aggregated stem cells into a higher basal ISR state and recovering the *in vivo* homeostatic state. Further proteomic analysis revealed that MSCs and ASCs showed heightened proteasome control, while FB had higher chaperon content showing a distinct difference in stem and differentiated cells ISR mechanism.

This work has illustrated the importance of 3D culture in the form of aggregation for the maintenance of MSC stem cell properties. This was carried out in MSCs by induction of the ISR pathway. The ISR pathway *in vivo* allows cells to mitigate cellular stress, in particular stress associated with protein misfolding, which is a problem commonly encountered during *in vitro* aging. Aggregation of FB did not result in the same trend of ISR upregulation suggesting a loss of sensitivity following differentiation. Aggregation induce ISR in MSCs lead to heightened basal levels as the *in vivo* tempering protein GADD34 was not upregulated, this heighten basal ISR also lead to increased autophagy and increased stem protein levels. Taken together these results show a promising alternative to traditional expansion culture for MSCs which leads to loss of function. This will allow for more effective scale-up thereby increasing the therapeutic efficacy of MSC treatments.

References

[1] Z. Cesarz, J. L. Funnell, J. Guan and K. Tamama, "Soft Elasticity-Associated Signaling and Bone Morphogenic Protein 2 Are Key Regulators of Mesenchymal Stem Cell Spheroidal Aggregates," *Stem Cells and Development,* vol. 25, no. 8, pp. 622-35, 2016.

[2] A. Banfi, A. Muraglia, B. Dozin, M. Mastrogiacomo, R. Cancedda and R. Quarto, "Proliferation kinetics and differentiation potential of ex vivo expanded human bone marrow stromal cells: Implications for their use in cell therapy," *Experimental Hematology,* vol. 28, no. 6, pp. 707-15, 2000.

[3] L. Garcia-Prat, M. Martinez-Vicente, E. Perdiguero, L. Ortet, J. Rodriguez-Ubreva, E. Rebollo, V. Ruiz-Bonilla, S. Gutarra, E. Ballestar, A. L. Serrano, M. Sandri and P. Munoz-Canoves, "Autophagy maintains stemness by preventing senescence," *Nature,* vol. 529, no. 7584, 2016.

[4] R. Pennock, E. Bray, P. Pryor, S. James, P. McKeegan, R. Sturmey and P. Genever, "Human cell dedifferentiation in mesenchymal condensates through controlled autophagy," *Scientific Reports,* 2015.

[5] Y. Liu, N. Munoz, A.-C. Tsai, T. M. Logan and T. Ma, "Metabolic Reconfiguration Supports Reacquisition of Primitive Phenotype in Human Mesenchymal Stem Cell Aggregates," *Stem Cells,* vol. 35, no. 2, pp. 398-410, 2017.

579923 Three-Dimensional Culture of Human Adipose-Derived Stem Cells By Surface Modification Using Elastin-Based Copolymers

During their differentiation and maturation into adipocytes *in vitro*, human adipose derived stem cells (hASCs) begin to accumulate lipids during maturation and begin to become more buoyant over time. Such fat-laden cells eventually lift off of the culture surface and are lost during the next media change. Because of the loss of physiologically relevant spheroids, elastin-like polypeptide-polyelectrolyte coated surfaces were used to form spheroids for long term culture, in order to provide a proven, long term culture method for mature adipocytes *in vitro*. While there are several methods of forming spheroids, there is not a general consensus on which methods are the most useful for stem cell proliferation, differentiation, and maturation ability. This is also true for most cell types, including adipocytes. Overall, our research provides a proven cell culture model for future studies involving long-term culture of mature adipocytes derived from hASCs. In addition to using elastin-like polypeptide-polyelectrolyte coated surfaces, alternative methods such as hanging drop, ultra-low attachment plate culture, and suspension cultures were compared for their ability to form and retain productive hASC spheroids. It was found that there are differences in spheroid formation and maturation within different molecular weights of the same polyelectrolyte, polyethyleneimine, as well as different compositions of the same molecular weight. Though the selected alternative spheroid forming methods were comparable to the elastin-like polypeptide-polyelectrolyte coated surfaces in terms of adipogenic maturation, the coated surfaces had greater amounts of spheroid retention, which allows for more physiologically relevant spheroids to be acquired.

575197 Delivering Therapeutic Drugs to Rejuvenate Progenitor Cells

The controlled delivery of bioactive molecules to the cell surface can significantly widen the therapeutic window, as well as enhance therapeutic potential of progenitor cells. Here we describe a strategy to rejuvenate endothelial progenitor cells (EPCs) via the conjugation of adjuvant drug-loaded nanoparticles to the surface of progenitor cells. Cord blood EPCs isolated from patients with gestational diabetes exhibit lower vasculogenic potential and altered epigenetic regulation. By providing sustained pseudoautocrine stimulation, we can significantly rejuvenate the progenitor phenotypes and improve therapeutic potential of EPCs. Collectively, strategy to deliver therapeutic drugs to the cell surface represents a powerful tool that has a wide range of translational applications.

573531 Peptoid-Based Substrates for Human Neural Stem Cell Differentiation

Pluripotent human neural stem cells can repair damage in the brain due to stroke or neurodegeneration. Neural stem cells will differentiate into different neural cell types depending on substrate and cell-to-cell interactions. We propose that peptoid-based substrates could be utilized to promote neural stem cell growth and differentiation. Peptoids are biomimetic poly-N-substituted glycines that are resistant to degradation by proteases. Peptoids are similar in structure to peptides, with the side chains bound to the backbone nitrogens rather than the alpha carbons. These structures have potential uses in biomedical applications due to peptoid stability under physiological conditions, customizable sequence design, and structural similarity to naturally occurring peptides. The tunability of peptoids allows for the addition of a myriad of side chains that could be added to achieve the desired function. In this study, a peptoid that self assembles into 2-4 micron diameter microspheres and a peptoid capable of engaging in non-metal catalyzed click chemistry was synthesized. Human neural stem cells were plated onto the peptoid-microsphere substrate and compared to commercially available substrates to differentiate neural stem cells. The capability of the click chemistry peptoid to attach growth factors in physiological conditions was also tested.

575616 A Comprehensive Sensitivity Analysis for Risk Assessment of a Pharmaceutical Crystallization Process

In pharmaceutical industries, crystallization is one of the most critical unit operations. Performance of a crystallization process can be improved by effective use of state of the art science knowledge, engineering principles and tools. As emphasized by regulatory mechanism, the quality of a pharmaceutical product should be built-in or should be by design, rather than being tested. A comprehensive understanding of the process is essential in order to build the quality in both process and product. For this purpose, process systems engineering methods and tools can significantly support and speed up process understanding and development by means of mechanistic modeling, integrated process analytical technology (PAT) tools and control strategies, identification of uncertainties in the process and quantification of process risk assessment. Therefore, in this study we developed an *in silico* tool for a pharmaceutical crystallization process to perform a global uncertainty/sensitivity analysis for process risk assessment. Identification of operational variables with uncertainties in the process and analysis of how these process variables affect the performance of the crystallization process and outcome are critical for a risk based approach. Thus, the consistency of the process and probability of failing to meet desired quality requirements can be assessed. To this end, several uncertainties in the process design space of the crystallization process were identified. A comprehensive uncertainty and sensitivity analysis was performed to quantify the influence of these variabilities as well as to obtain importance ranking in order to help to assess the risks in the process. In conclusion, this study shows that *in silico* tools provide a fast, effective and cost-efficient platform to pharmaceutical industries to enhance their process and product developments through investigating design spaces, testing the feasibility of different strategies and assessing process risks.

Acknowledgment

We would like to thank the Danish Council for Independent Research (DFF) for financing the project with grant ID: DFF-6111600077B.

570306 Advancing from QbD to Operational Excellence in Continuous Pharmaceutical Manufacturing

The advent of Industry 4.0 (Smart Manufacturing) in the 2010s, is a manifestation of cultural and mindset change across multiple industries towards the acceptance of digital tools in process design and operations, resulting in highly automated and data-rich systems to realize lean manufacturing. These process intelligence tools leverage cross-industry advances in manufacturing methods, data analytics, fault management and implementation technologies, for sustainable process operations, safety, quality risk assessment, and enhanced customer relations. In pharmaceutical systems, such digitalization tools facilitate the implementation of continuous manufacturing and real-time release testing by tracking product quality in real-time, thereby enabling the release of life critical medication with reduced off-line quality testing compared to existing practices.

Continuous manufacturing of drug product is the outcome of targeted process intensification with systematic integration of product and process knowledge, instrumentation and automation systems, quality control protocols and real-time process management for a solids processing system handling pharmaceutical materials. Success in the technology necessitate appropriate infrastructure for Information and Operations Technology integration.

This poster discusses an integrated data-driven and model-based framework for robust process monitoring and maintenance management, as continuous manufacturing advances from technology development to adoption and implementation. The infrastructure for digital process operations implemented at Purdue University uses an analytically redundant sensor network and automation systems such as Emerson DeltaV, OSIsoft PI System and Applied Materials Smart Factory Rx for continuous tableting via direct compaction and dry granulation. A hierarchical systems integration architecture closely following ISA-95 is setup for the implementation of Quality-by-Design. Further, the poster discusses leveraging the systems for implementation of supervisory control and reporting for QbD implementation.

Further Reading:

1. S. Ganesh, G. Reklaitis et al. Leveraging the digital infrastructure for real-time operations management in continuous pharmaceutical manufacturing: Condition-based maintenance for sensor network robustness (in preparation).
2. M. Moreno, S. Ganesh, Y. Shah, Q. Su, M. Gonzalez, Z. Nagy, G. Reklaitis. Sensor Network Robustness using Model-based Data Reconciliation for Continuous Tablet Manufacturing. J Pharm Sci (2019).
3. Q. Su, S. Ganesh, M. Moreno, Y. Bommireddy, M. Gonzalez, G. Reklaitis, Z. Nagy. A perspective on Quality-by-Control (QbC) in pharmaceutical continuous manufacturing. Comput Chem Eng (2019).
4. Su Q, Bommireddy Y, Shah Y, et al. Data reconciliation in the Quality-by-Design (QbD) implementation of pharmaceutical continuous tablet manufacturing. International Journal of Pharmaceutics. (2019).

558769 Assessment of Blend Uniformity in a Continuous Tablet Manufacturing Process

Blend uniformity was monitored throughout a continuous manufacturing (CM) process by near infrared (NIR) spectroscopic measurements of flowing powder blends and compared to the drug concentration in the tablets. The NIR spectra were obtained through the chute after the blender and within the feed frame, while transmission spectra were obtained for the tablets. The CM process was performed with semi-fine acetaminophen as a cohesive active pharmaceutical ingredient (API) at a target level of 10.0% (w/w), silicified microcrystalline cellulose as the filler, and magnesium stearate as lubricant. The blender was operated at 250 rpm, where the system's best performance was expected, and 106 and 495 rpm where a lower mixing efficiency was expected. The variation in blender RPM increased the variation in drug concentration at the chute but not at the feed frame. The results demonstrated that the drug concentration of powder blends obtained in the chute during this study were not representative of the drug concentration in tablets. However, statistical results show that the drug concentration of tablets can be predicted, with great accuracy, from blends within the feed frame. This study demonstrated a mixing effect within the feed frame, which contribute to a 60% decrease in the relative standard deviation of the drug concentration, when compared to the chute. This study demonstrates that the feed frame is an ideal location for monitoring the drug concentration of powder blends for a continuous manufacturing system.

558998 Continuous Pharmaceutical Manufacturing of Uniform Crystals in Slug Flow

In the pharmaceutical and chemical industries, the continuous generation of crystals of target size distribution has the potential to improve efficiency for post-crystallization operations[1–7]. The control of crystallization processes can be challenging when undesirable phenomena such as particle attrition and breakage occur. Segmented/slug flow crystallization is one of the effective continuous processes demonstrated to allow enhanced control of organic crystal properties such as size and shape[2–7]. The slurry flow is combined with an air flow and fed to a tube to induce a multiphase hydrodynamic instability that spontaneously generates well-mixed slugs where the crystals continue to grow.

This presentation will elaborate: (1) the process design and strategies for continuous slug-flow crystallization that controls the crystallization phenomena; (2) the tuning of crystal properties without changing the equipment set-up[4]. Process intensification strategy is also discussed which makes the whole process more robust. Experimental validation confirms that the proposed crystallizer designs reduce production time and equipment cost evidently while suppressing secondary nucleation, attrition, and aggregation—dominant but undesired phenomena that worsen the ability to control the crystal properties.

References

(1) Sun, Z.; Ya, N.; Adams, R. C.; Fang, F. S. Particle Size Specifications for Solid Oral Dosage Forms: A Regulatory Perspective. *Am. Pharm. Rev.* **2010**, *13* (4).

(2) Jiang, M.; Zhu, Z.; Jimenez, E.; Papageorgiou, C. D.; Waetzig, J.; Hardy, A.; Langston, M.; Braatz, R. D. Continuous-Flow Tubular Crystallization in Slugs Spontaneously Induced by Hydrodynamics. *Cryst. Growth Des.* **2014**, *14* (2), 851–860.

(3) Eder, R. J. P.; Schrank, S.; Besenhard, M. O.; Roblegg, E.; Gruber-Woelfler, H.; Khinast, J. G. Continuous Sonocrystallization of Acetylsalicylic Acid (ASA): Control of Crystal Size. *Cryst. Growth Des.* **2012**, *12* (10), 4733–4738.

(4) Jiang, M.; Papageorgiou, C. D.; Waetzig, J.; Hardy, A.; Langston, M.; Braatz, R. D. Indirect Ultrasonication in Continuous Slug-Flow Crystallization. *Cryst. Growth Des.* **2015**, *15* (5), 2486–2492.

(5) Su, M.; Gao, Y. Air-Liquid Segmented Continuous Crystallization Process Optimization of the Flow Field, Growth Rate, and Size Distribution of Crystals. *Ind. Eng. Chem. Res.* **2018**, *57* (10), 3781–3791.

(6) Daniel Scott, C.; Labes, R.; Depardieu, M.; Battilocchio, C.; Davidson, M. G.; Ley, S. V.; Wilson, C. C.; Robertson, K. Integrated Plug Flow Synthesis and Crystallisation of Pyrazinamide. *React. Chem. Eng.* **2018**, *3* (5), 631–634.

(7) Lu, J.; Litster, J. D.; Nagy, Z. K. Nucleation Studies of Active Pharmaceutical Ingredients in an Air-Segmented Microfluidic Drop-Based Crystallizer. *Cryst. Growth Des.* **2015**, *15* (8), 3645–3651.

565067 Experimental Analysis of Influence Screw Configuration in View of Understanding Granulation Mechanism in Pharmaceutical Twin-Screw Melt Granulation

PURPOSE

For 15 years, there has been increasing interest in continuous manufacturing in pharmaceutical industry. In order to improve process efficiency and reduce costs, innovative continuous manufacturing processes, e.g. twin-screw melt granulation, are gaining interest. Twin-screw melt granulation is a particle enlargement process where a meltable solid binder is used to achieve agglomeration of pharmaceutical powder particles. This process has an extensive list of advantages e.g. the exclusion of a drying step in the production process compared to twin-screw wet granulation and the ability to formulate high-dosed formulations with up to 95% of API.

The aim of this study was to gain process understanding of the granulation mechanisms in continuous twin-screw melt granulation by evaluating the influence of different screw elements and process parameters for a miscible and an immiscible formulation on the granule properties.

METHODS

Soluplus was used as a thermoplastic binder in the 2 formulations selected for this study. Metoprolol tartrate (MPT) and Caffeine anhydrous (CAF) were chosen as model drugs due to their different miscibility[1]. The former is miscible, while the latter is immiscible with Soluplus. Physical mixtures (85% API/binder w/w) were processed using a twin-screw extruder (Prism Eurolab 16, Thermo Fisher Scientific) with a screw configuration containing 2 kneading zones of each five kneading elements and a screw mixing element at the end of the screws. A D-optimal screening design was performed with screw speed (150-500 rpm), material throughput (0.4kg/h-0.8kg/h), barrel temperature (60°-80°C for MPT and 90°-110°C for CAF), stagger angle (30°, 60° and 90°) and direction of the kneading zone (forward and reversed) as factors. Several granule and process properties were evaluated as responses like friability, granule size, shape, torque and porosity.

RESULTS

The stagger angle, direction of the kneading zone and the interaction of these two factors determine the formation of granules and their quality attributes most significantly. In addition, the formulation also has an influence on granule formation. Granule formation for MPT and SLP can already be observed with screw configurations with lower shear compared to CAF and SLP due to the difference in drug-polymer miscibility. The largest, strongest and more needle-shaped granules are obtained with a small stagger angle of kneading elements in reversed direction due to back-mixing and therefore more material build-up occurs at the kneading zones. Due to this material build-up, more densification and more melting of the binder takes place and thus stronger, more dense and larger granules can be obtained. For the reversed configuration, the material build-up occurs due to the little amount of space for the material to pass the kneading zone, which results in the material passing along the gap between the wall of the barrel and the edges of the kneading elements which results in shear elongation. How smaller the stagger angle in reversed configuration, how more shear elongation that takes place and how more needle-shaped granules are obtained. The needle-shaped granules become more spherical by adding a screw mixing element at the end of the screws which can break up the needle-shaped granules. For this screw configuration, shear elongation followed by breakage and layering can be proposed as the granulation mechanism for both formulations. In contrast, forward direction does not cause back-mixing and shear elongation. Therefore, breakage and layering occurs as granulation mechanism with this screw configuration.

CONCLUSION

This study demonstrates that screw configuration is the most influencing factor that determines granule and process properties. Depending on raw material properties, desired granule properties can be obtained by optimizing the screw configuration. The reversed configuration with the smallest stagger angle is the most robust configuration to obtain granules with the best properties for downstream processing.

REFERENCES

1. Monteyne, T. *et al.* The use of rheology to elucidate the granulation mechanisms of a miscible and immiscible system during continuous twin-screw melt granulation. *Int. J. Pharm.* **510,** 271–284 (2016).

570547 Agitator Impact on the Net Weight Signal of a Loss-in-Weight Feeder Operating at Low Mass Flow Rates

Continuous manufacturing is being adopted at an accelerated rate as a platform to produce new drug products and has been described as one of the most important tools modernizing the pharmaceutical industry.[1] Continuous loss-in-weight screw feeders are the key dispensing units for enabling the mass flow driven requirements of continuous drug product manufacture. These feeders most often operate in gravimetric mode where feedback from the net weight signal is used to update and control the screw speed. Continuous feeding performance evaluation methodologies [2,3] as well as the impact of raw material properties [4] on continuous feeding performance have been well identified and studied in the literature. It is recognized that the interaction of the feeder agitator and feeder screws has an impact on the mass flow feed rate during continuous feeding.[3] Understanding the impact of the agitator on the net weight signal (and hence screw speed control) is necessary for optimizing feeding performance.

In this work, the impact of agitator movement on the net weight signal during loss-in-weight feeding is studied. Specifically, the position of the agitator is monitored in real time to observe the impact of agitator motion on the net weight loss signal at low mass flow rates. It is identified that the agitator has a significant impact on the net weight signal. The frequency of the agitator speed co-varies with the frequency in the change of the net weight loss signal. The operating net weight of the feeder is a key factor determining the magnitude of variation in the changing net weight signal. This study provides direction for improving the net weight mass loss signal and reducing potential impact to mass flow feed rate variation at low mass flow rates by optimizing the operating net weight range of the feeder. Directions are also provided on modified agitator configurations for the raw material being fed and material mass flow rate requirements.

References

1 Gottlieb, S. & Woodcock, J. Statement from FDA Commissioner Scott Gottlieb, M.D., and Janet Woodcock, M.D., director of the Center for Drug Evaluation and Research on FDA's modern approach to advanced pharmaceutical manufacturing, https://www.fda.gov/NewsEvents/Newsroom/PressAnnouncements/ucm632043.htm (2019).

2 Engisch, W. E. & Muzzio, F. J. Feedrate deviations caused by hopper refill of loss-in-weight feeders. Powder Technology 283, 389-400, doi:https://doi.org/10.1016/j.powtec.2015.06.001 (2015).

3 Engisch, W. E. & Muzzio, F. J. Method for characterization of loss-in-weight feeder equipment. Powder Technology 228, 395-403, doi:https://doi.org/10.1016/j.powtec.2012.05.058 (2012).

4 Wang, Y., Li, T., Muzzio, F. J. & Glasser, B. J. Predicting feeder performance based on material flow properties. Powder Technology 308, 135-148, doi:https://doi.org/10.1016/j.powtec.2016.12.010 (2017).

556417 Temperature Mapping of Pharmaceutical Trickle Bed Reactor for Continuous Hydrogenation

Temperature Mapping of Pharmaceutical Trickle Bed Reactor for Continuous Hydrogenation

Huibo Sheng, Tomislav A. Ljubicic, Angel R. Diaz, David Pfisterer, Joel M. Hawkins and Jason Mustakis, Pfizer Global Research & Development, Pfizer Inc., Groton, CT

Trickle bed reactors (TBRs) are fixed bed reactors that widely used in chemical and petroleum industries for gas-liquid-solid three-phase reactions[1] and they have been extensively studied under the conditions for these industrial applications on large scale.[2] Pharmaceutical applications are on a much smaller scale, and the general flow dynamics together with the multiphase interactions can be very different. Therefore, an online monitoring method is highly desired in the development of pharmaceutical TBRs for continuous processes. Axial temperature distribution of trickle bed reactors is of great interest as it directly indicates the status of many pharmaceutical reactions such as hydrogenation of active pharmaceutical ingredients (APIs). The profile is also of great importance to a safe operation of trickle bed reactors to prevent the formation of local hot-spot that may lead to catalyst deactivation or reaction failure.[3]

Traditionally, a set of thermocouples are typically used to measure the axial temperature along the reactor.[4] This method, however, can only provide with very limited and location dependant temperature data. Recently, NMR imaging technique has been developed for temperature mapping of lab-scale fixed-bed reactor.[5, 6] However, this technique requires calibrating the chemical shifting at different temperatures. Thus, it is by nature chemical compound dependent as any catalyst or chemical change will require a re-evaluation of whether it is still applicable or a re-correlation of chemical shifting-temperature relationship. In this work, an optic fiber temperature sensor is employed to measure the axial temperature distribution of benzyl alcohol hydrogenation as a model reaction. Along with the different temperature profiles, the TBR operation associated catalyst activation, deactivation, and regeneration are also discussed. This advanced technology enables real time axial temperature profile monitoring of trickle bed reactors from lab to large-scale and can be applied to any reaction systems.

References

1. M. H. Al-Dahhan, F. Larachi, M. P. Dudukovic, A. J. I. Laurent and e. c. research, *Ind. Eng. Chem. Res.*, 1997, **36**, 3292-3314.
2. C. N. J. A. J. Satterfield, *AIChE Journal*, 1975, **21**, 209-228.
3. J. Hanika, *Chemical engineering science*, 1999, **54**, 4653-4659.
4. J. Hanika, K. Sporka, V. Ružicka and J. Hrstka, *The Chemical Engineering Journal*, 1976, **12**, 193-197.
5. I. V. Koptyug, A. V. Khomichev, A. A. Lysova and R. Z. Sagdeev, *Journal of the American Chemical Society*, 2008, **130**, 10452-10453.
6. L. F. Gladden, F. J. Abegão, C. P. Dunckley, D. J. Holland, M. H. Sankey and A. J. Sederman, *Catalysis Today*, 2010, **155**, 157-163.

579337 In-Line NIR Spectroscopy for Monitoring the Preparation of *Ginkgo Biloba* Extract Solid Dispersions By Hot-Melt Extrusion

Hot melt extrusion (HME) is regarded as an emerging but promising pharmaceutical technique to enhance drug solubility. The purpose of this study was to prepare the high drug-loading *Ginkgo biloba* extract solid dispersions (GBE-SD) using HME technique, achieving good dissolution profiles at the same time. Polyethylene glycol (PEG) 4000 was the only-used expicient, and various formulations with varying GBE concentrations (30%, 35% and 40%) were explored. Specific HME parameters setting (e.g. barrel temperature, rotation speed) corresponding to different GBE concentrations was based on the result of rheology testing and differential scanning calorimetry (DSC). Quercetin (QCT), kaempferol (KMF) and isorhamnetin (ISR) were chosen as representative substances to evaluate GBE-SD dissolution profiles via HPLC method. Compared to the marketed dripping pill (DP) formulation, the in vitro dissolution studies indicated that both GBE-SD and DP had good solution performance at 30 min (>85% labeled GBE amount). Near-infrared spectroscopy in the die head was used for monitoring the HME process and assessing extrusion stability, indicating molecular interactions between GBE and PEG4000 as well. Considering that marketed DP was struggling with the poor DP appearance when it came to high GBE loading, GBE-SD could improve GBE concentrations and thus enhanced patient compliance. All these studies show that GBE-SD prepared by HME is a creative and valuable formulation approach.

571987 Filtration Studies Combined with Mechanistic Modelling to Reliable API Process Understanding and Scale-up

Process development and manufacturing in Pharma Industry comprises several unit operations that need to be studied and optimized along a project lifecycle. One of the main stages present in all manufacturing processes is the isolation of the product or side-products in solid form. Among the different solid-liquid separation techniques that can be applicable, one of the most common is the filtration of the product suspension. This unit operation is normally carried out easily at laboratorial or small scale but often brings concerns when increasing scale, or changing equipment, mostly due to lack of process understanding. The combination studies of scale dependent and scale independent parameters allied with mechanistic modelling can bring real gains to the process, including optimization of process conditions, time saving and risk mitigation associated with scale-up.

Unlike the empirical or statistical models, the mechanistic model of a filtration relies on data from specific experiments related with the mathematical description of the physical processes during filtration. Thus, the model provides accurate answers to important questions raised when developing and scaling-up a filtration, namely the cake resistance and the compressibility index, and allows a deeper knowledge of the process.

Our methodology produces experimental data that will support the filtration development and the mechanistic model using DynoChem. Filtration data gathered from previous large-scale runs was used to validate the mechanistic model by direct comparison of key process variables.

We demonstrate the application of this methodology by showing case studies where specific experiments were performed to estimate the filtration scale-independent parameters and scale-up of the process. The examples will show the evolution of filtration processes with high cake resistance during development and the consequent results for the large scale runs, and the using of the models to support the scale-up of filtration processes with accurate predictions and consulting on the best strategies to optimize cycle time and reduce process risk.

561041 Filtration Process Modeling and Scale-up for Robust Drug Substance Manufacturing

During drug substance manufacturing, the products and intermediates are normally isolated from crystallization, filtration and drying processes. Filtration is a typical liquid-solid separation process that sometimes has operational challenges which may reduce process efficiency and/or impact product quality. In this presentation, a filtration process modeling and scale-up approach will be discussed. A lab scale column setup was used to characterize the slurry and wet cake. The obtained parameters such as specific cake resistance and compressibility index are then used to develop a model to predict the filtration performance for larger scale filtration equipment. The model was validated with the pilot scale operational data. The simulation results with typical filtration equipment, such as pressure filter pots and centrifuges, can be used to guide selection of the appropriate filtration equipment. For selected filtration equipment, the model can be utilized to explore the operational design space and identify the optimum operation conditions for minimum cycle time.

572332 Forget Throughput, It's about the Risk- Rethinking Dead-End Membrane Filtration Scale-up

Forget throughput, it's about the risk

- Rethinking Dead-End Membrane Filtration Scale-Up

Kelly Wei

Purification Process Sciences

AstraZeneca

One MedImmune Way

Gaithersburg, MD 20878

USA

In the biopharmaceutical industry, purification processes utilize chromatography and filtration methods to remove process and product related impurities. Membrane filters with pore sizes of 0.2 µm or 0.45 µm are often employed between unit operations to remove small particulates and mitigate bioburden risk in accordance with current good manufacturing practice (cGMP). For chromatography, substantial effort has been devoted to understanding mechanisms and scale-up principles which enables accurate prediction of process performance and product quality. Unfortunately, the same cannot be said for filtration. The current empirical practices used for sizing inline membrane filtration are often not predictive during scale up and technology transfer. This gap in understanding can lead to costly deviations at scale. Here, we challenge the conventional filter sizing approach based on throughput and propose a new risk analysis approach based on membrane fouling mechanism. By evaluating membrane fouling behavior during scale up, we determine the mechanism and develop the appropriate filtration strategy.

565620 Development of an Expanded Parametric PBE Using Experimental Data from a High Shear Wet Granulation Process

Wet granulation process is widely used in the pharmaceutical industry to increase the particle size distribution of a solid formulation as liquid binder is applied. Even though this process is considered as one of the most complex processes currently performed in the manufacturing process, it can help in improving the blend flow properties, to minimize cross contamination, and dust accumulation in the room. As the effect of the blade angle and its RPM, the massing time, and the addition of a hydrophobic agent to the initial formulation were studied, more insights of the final PSD of a high shear batch wet granulation process were obtained. The objective was to expand the PBE to include explicitly the operating parameters. This expanded model could be used to predict on real time the PSD behavior and control the granulation process. For the PBE development, the experimental data were fit using the GSolids program. The expanded PBE obtained was able predict the PSD behavior of particles below 1218 μm. To increase the particle size range of the PBE, a suitable agglomeration kernel is needed, which should consider the agglomeration of larger particles.

567926 A Chemical Engineering Approach to Modelling Drug Dissolution and Transport Phenomena in the Lower Gastrointestinal Tract

The human proximal colon offers opportunities for targeted drug delivery to treat local conditions such as ulcerative colitis and irritable bowel disease, as well as more systemic diseases. Whilst the colon has a lower surface area than the small intestine, in some cases the long residence time has certain advantages for treatments.

Understanding of how drug dissolution and absorption in the proximal colon might be achieved is hampered by a lack of *in vitro* tools which mimic the complex hydrodynamics of the colonic environment. The dynamic colon model (DCM), shown in Figure 1, has been developed to replicate the anatomy, motility patterns and physical pressures of the human proximal colon [1]. This biorelevant dissolution apparatus provides additional mechanistic insights on release alongside or in place of standard USP methods, which rely on simple stirred vessels and low viscosity buffer solutions which are not representative of conditions within the colon. More complex *in vitro* models have been developed for the colon, such as the three-stage tubular model and the TNO TIM-2 [2, 3]. These models operate with enhanced biorelevance in many areas such as the incorporation of fermentation reactors, but are compromised in terms of physical representation of biological features and biomechanical motility.

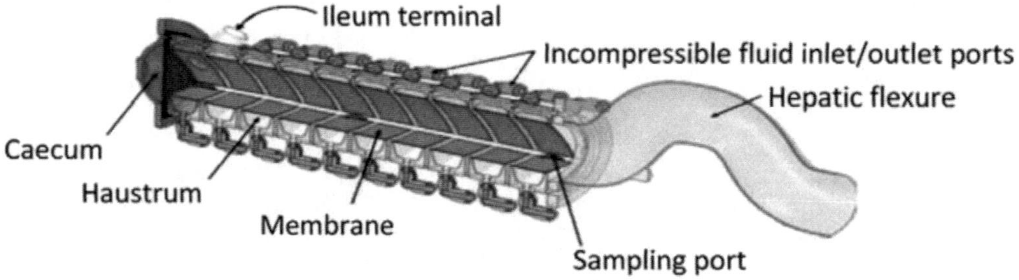

Figure 1: Schematic of the dynamic colon model (DCM) taken from Stamatopoulos, K. et al. 2016 with permission [1].

The schematic in Figure 1 shows how the DCM replicates the anatomy of the human proximal colon. The colon receives undigested food material (chyme) from the ileum through the ileocaecal valve. This is represented by the ileum terminal on the DCM, the injection point for dosage forms, dyes etc. The ileocaecal valve joins the colon at the top of the caecum – the pouch-like base of the proximal colon. The ascending colon, extending from the caecum to a sharp bend known as the hepatic flexure, is the favoured section for targeted drug delivery as the colonic contents have lowest viscosity in this region. The ascending colon has a segmented appearance comprising saccules called haustra. Contraction of smooth circular muscle causes a haustrum to distend whilst the subsequent haustrum is relaxed, driving the luminal contents forward at a rate controlled by a complexity of factors that maximise water absorption whilst avoiding over-solidification [4]. The DCM uses a computer controlled hydraulic system to mimic this circular muscle activity by inflating and deflating the silicone membrane of each haustrum, thus replicating motility patterns of the human proximal colon observed *in vivo*.

In this paper, an extensive analysis of the fluid dynamics with the DCM is discussed by comparison of results obtained from magnetic resonance imaging (MRI) and positron emission tomography (PET). These clinically relevant non-invasive imaging modalities visualise the mixing processes within the DCM from very distinct perspectives. A comparison will deliver further insights into fluid flow and how this can impact upon the disintegration, dissolution and dispersion of solid dosage forms within the colon. It is anticipated that the morphological MRI data will complement the functional information gathered by PET, and vice versa, in a way that progresses the current understanding and interpretation of each individual dataset. The coupling of MRI and PET techniques will enable discussion in the context of the synchronisation of colonic wall motion with distribution of drug/dye molecules throughout the DCM. This forms part of a wider project to establish an *in vitro in vivo* correlation (IVIVC) for dissolution testing and validate the DCM for use as an enhanced biorelevant method for measurement of dissolution of controlled release formulations, ultimately to enable improved dosage forms to be developed and tested.

MRI has the capacity to produce a high resolution map that tracks the dynamic morphology of the interior DCM environment during motility patterns. Figure 2 depicts MRI snapshots of the partially filled DCM for a single haustrum unit (Fig. 2 part 1) and the entire DCM tube comprising ten haustrum units (Fig. 2 part 2). The signature soft tissue contrast exhibited by MRI easily distinguishes between the water used to hydraulically control the motion of the haustra walls (d) and the water in the lumen (e) without the addition of contrasting agents. The critical advantage of MRI is the ability to synchronise wall motion with fluid flow and map the flow field induced within the DCM. Recent work has demonstrated that MRI of the human colon can be used to measure the motion of colonic contents as well as the colonic wall [5, 6]. MRI therefore provides the opportunity to demonstrate the similarities between the DCM and human colon in mixing and fluid flow.

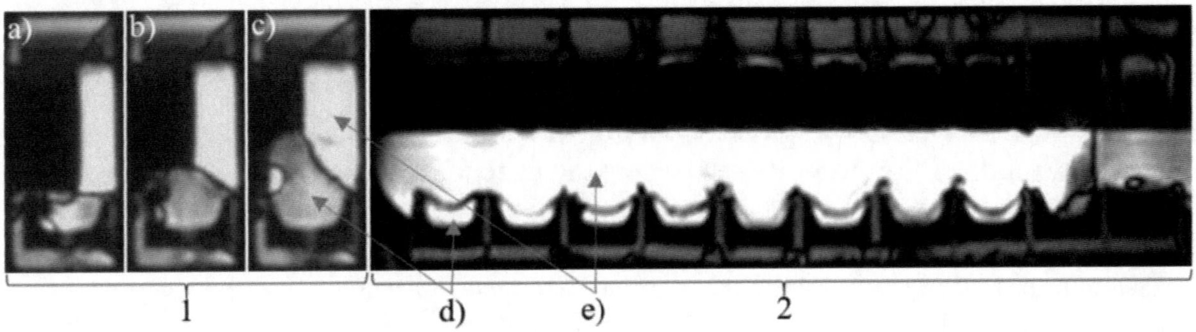

Figure 2: Magnetic resonance imaging (MRI) slices visualising snapshots of luminal fluid displacement in a single haustrum (1) induced by deflation (a) and inflation (intermediate at b and maximum at c) of the haustrum and the whole DCM (2). Water in the haustra used to hydraulically control wall motility and water in the lumen are indicated by d and e respectively. Images were captured at the Sir Peter Mansfield Imaging Centre, University of Nottingham.

Inside the DCM, PET offers the complementary ability to image the key underlying mass transport processes that are driven by the anatomical motility mapped by MRI; such as convection of a dynamically disintegrating extended release dosage form within the lumen. PET enables the local spatial concentrations of a tracer (analogous to a drug) to be mapped during the different stages of a motility pattern. For example, Stamatopoulos, et al. observed the accumulation of tracer molecules around the caecum and ileum terminal likely due to lack of propulsion forces at such early stages in development of an antegrade propagating wave [1]. Unlike the dynamically changing colonic environment, conditions were fixed in these studies to permit scrutiny of the interplay between a range of parameters such as fill level, media viscosity and motility pattern (occlusion degree, velocity of propagating wave, contraction/relaxation sequence, etc.). Media viscosity has been shown to influence the convective mass transport of tracer as can be seen in Figure 3 [1]. Increasing viscosity from 0.25 to 0.50 % (w/w) NaCMC (Fig. 3 Parts 1 and 2 respectively) decreases the uniformity of tracer distribution along the lumen. This is shown by lower intensity of the radioactive signal as x-axis position increases from 0-20 cm. Furthermore, the amount of tracer remaining around the caecum increases with viscosity, whilst a transition to plug flow is also observed in the distribution of the remaining tracer along the lumen following 5-10 antegrade propagating sequences (Fig 3. Parts 2c & 2d). As the wave reaches the end of the lumen, media flows up the wall of the rigid siphon without passing over the flexure, before gravity forces induce backflow. This backflow is stronger, closer to the siphon. For this reason, the higher viscosity media used to generate Figure 3 part 2 imparts more resistance to backflow, thus a more significant accumulation of tracer is observed around $x = 13 \pm 1.5$ cm. These phenomena have also been observed during *in vivo* scintigraphy studies [7, 8, 9].

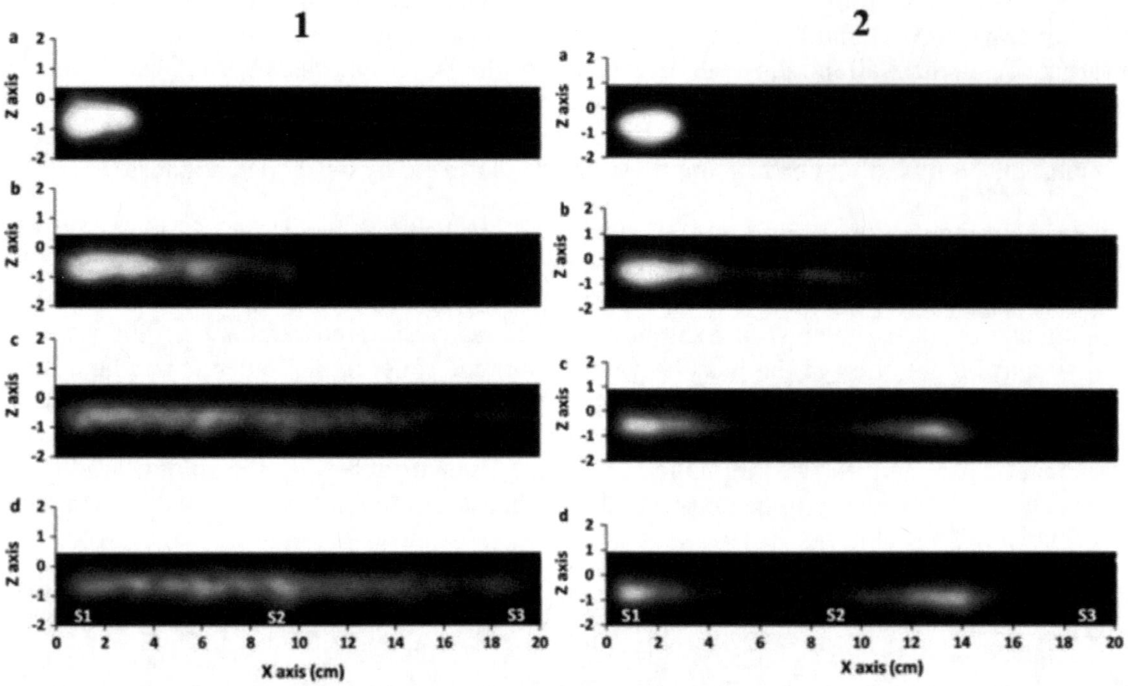

Figure 3: PET images showing the dispersion of a solution of radionucleotide ^{18}F along the x and z axis of the DCM lumen (where the caecum is located at $x = 0$ cm and the hepatic flexure at $x = 20$ cm) following (a) injection into the ileum terminal, (b) first antegrade wave, (c) fifth antegrade wave and (d) tenth antegrade wave using 0.25 % (w/w) and 0.50 % (w/w) NaCMC in 1 and 2 respectively. S1, S2 and S3 represent sampling points along the DCM. Images are taken from Stamatopoulos, K. et al. 2016 with permission [1].

Acknowledgements

Conner O'Farrell is sponsored by the EPSRC Centre for Doctoral Training in Formulation Engineering (EP/L015153/1), AstraZeneca AB R&D, Gothenburg. We acknowledge the Sir Peter Mansfield Imaging Centre at the University of Nottingham for their support in carrying out the MRI experiments.

References

1. Stamatopoulos, K., Development of a Biorelevant Dynamic Model of Human Proximal Colon: a Tool for Designing Colon-Specific Drug Delivery Systems, in School of Chemical Engineering. 2016, University of Birmingham: Birmingham.

2. Spratt, P., Nicolella, C. & Pyle, D. L., An Engineering Model of the Human Colon. Food and Bioproducts Processing, 2005. 83: 147-157.

3. Venema, K., The TNO In Vitro Model of the Colon (TIM-2). Verhoeckx, K., Cotter, P., Lópezexpósito, I., Kleiveland, C., Lea, T., Mackie, A., Requena, T., Swiatecka, D. & Wichers, H. (eds.) The Impact of Food Bioactives on Health: in vitro and ex vivo models. 2015, Cham: Springer International Publishing.

4. Sarna, S. K. & Shi, X-Z. Chapter 39 - Function and Regulation of Colonic Contractions in Health and Disease A2 - Johnson, Leonard R. Physiology of the Gastrointestinal Tract (Fourth Edition). 2006, Burlington: Academic Press.

5. Hoad, C.L., et al., Colon wall motility: comparison of novel quantitative semi-automatic measurements using cine MRI. Neurogastroenterology and Motility, 2016. 28(3): p. 327-335.

6. Pritchard, S.E., et al., Assessment of motion of colonic contents in the human colon using MRI tagging. Neurogastroenterology and Motility, 2017. 29(9).

7. Munjeri, O., Collett, J. H., Fell, J. T., Sharma, H. L., Smith, A-M., In vivo behaviour of hydrogel beads based on amidated pectins. Drug Deliv. J. Deliv.Targeting Therap. Agents, 1998. 5: p. 239-241

8. Cook, I. J., Furukawa, Y., Panagopoulos, V., Collins, P. J. & Dent, J., Relationships between spatial patterns of colonic pressure and individual movements of content. Am J Physiol Gastrointest Liver Physiol, 2000. 278: p. G329-41.

9. Dinning, P. G., Szczesniak, M. M. & Cook, I. J. 2008. Proximal colonic propagating pressure waves sequences and their relationship with movements of content in the proximal human colon. Neurogastroenterol Motil, 2000. 20: 512-20.

578896 Optimizing the Pharmaceutical Cleaning Process: Challenges and Opportunities

Optimising the pharmaceutical cleaning process: challenges and opportunities

-

Rabah Mouras[1,2], Gavin Walker[1,2] and Ahmad Albadarin[1,2]

[1] Department of Chemical Sciences, Bernal Institute, University of Limerick, Castletroy, Limerick, Ireland

[2] Pharmaceutical Manufacturing Technology Centre (PMTC), Bernal Institute, University of Limerick, Castletroy, Limerick, Ireland

Abstract:

Cleaning validation and verification is a regulatory requirement within the manufacturing of active pharmaceutical ingredients (APIs) and drug products. Thus production equipment in pharmaceutical plants ((reactors, dryers, pipes…) need to be periodically cleaned to eliminate cross-contamination between different products due to the adhesion of APIs, intermediates, by-products and raw materials. This process has been proven to be very difficult, time consuming and result in undesirable effects such as staining, and the manufacturers strive to improve the efficiency, cost, quality, and time required to design, qualify, and monitor the cleaning process. There is an urgent need to optimize the cleaning process that leads to a high competitiveness through reduced batches rejection, reduced laboratory costs, and improved process yields and reduced changeover time.

Our approach towards developing an efficient cleaning process is based on: ***(1)*** selection of best cleaning agents with the help of modelling and solubility predictions using conductor-like screening model for real solvents (COSMO-RS) that allows the screening for innumerable solvents and mixtures, ***(2)*** APIs and intermediates solubility measurement in selected cleaning fluids, ***(3)*** cleaning trials using metal or glass coupons in a dipping test apparatus to mimic the cleaning in plants, ***(4)*** ensuring the surfaces are clean by means of spectroscopic and analytical techniques and ***(5)*** developing rinse solution analysis using process analytical technology (PAT) methods.

Our approach allowed us to choose the most efficient cleaning fluid resulting in a better cleaning with a considerable reduction in time and usage of cleaning solvents.

In this presentation, theoretical and experimental data will be presented and discussed and challenges in the optimization and developing efficient cleaning process will be addressed.

Key words: Pharmaceutical cleaning Process, **Cleaning validation,** Active Pharmaceutical Ingredient (API), COSMO-RS, process analytical technology (PAT)

Acknowledgment: The research leading to these results has received funding from Enterprise Ireland Innovation Partnership Programme (IP-2018-0727).

579208 The Development of a 2-Step Buprenorphine Intermediate 5 Process

A 7-step process has been successfully reduced to a 2-step process in making a key buprenorphine intermediate with significant yield (> 20%) and quality improvements.

Old Process

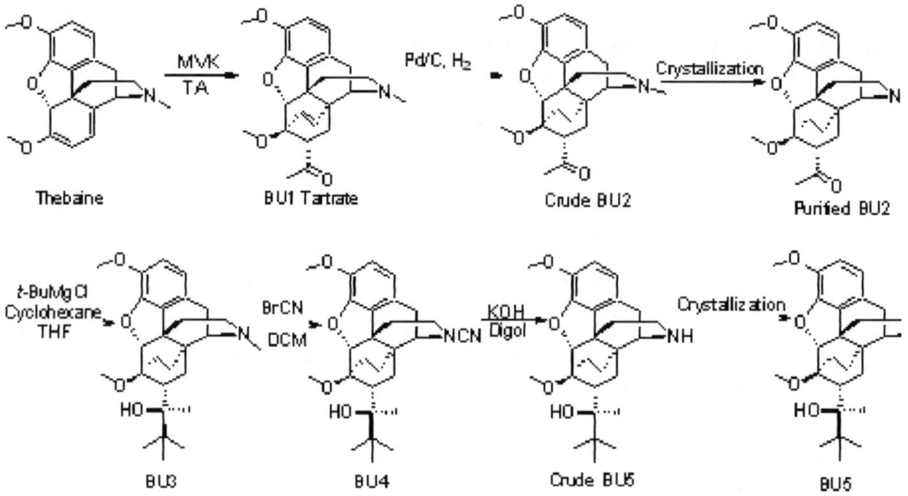

New Process

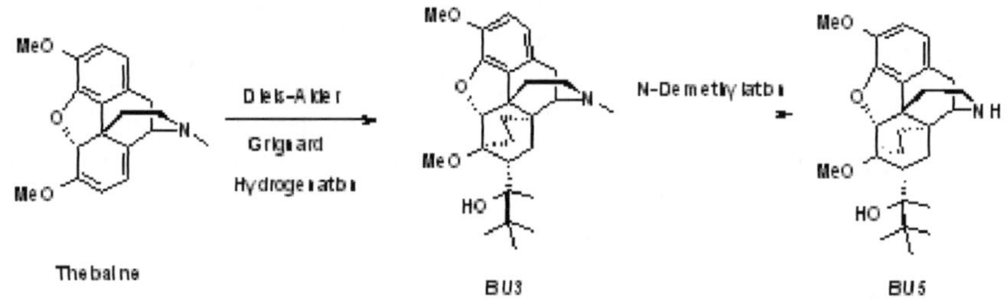

- - Abstract Text

572179 Arginine Interactions Resulting in Virus Inactivation

Therapeutic proteins treat a range of diseases including cancers, infections, autoimmunity/inflammation, and genetic disorders. To guarantee the effectiveness and safety of the final product, advancements in the manufacturing process are desired. For example, the low ph and/or surfactant treatment used to inactivate enveloped viruses during therapeutic protein manufacturing can cause aggregation or fragmentation of the final protein product. To maintain the stability and structure of proteins during viral clearance, it is favorable to keep pH levels above 4.0 or detergent concentrations low; however, established processes do not meet these criteria. One solution to increase protein stability and yields is to use the synergistic effects of arginine. Arginine can inactivate enveloped viruses at a less acidic pH (= 4) or lower temperature (30-40°C) than conventional methods. However, the mechanisms and optimal conditions for inactivation are not understood, and therefore, synergistic arginine viral inactivation is not widespread.

Optimal solution conditions for viral inactivation found in literature are high arginine concentrations (0.7-1M), a time of 60 minutes, and a synergistic factor of moderate temperature (40°C), low pH (pH 4), or Tris buffer (5 mM). However, full inactivation does not occur over all viruses tested. Viruses that are resistant to arginine inactivation contain membrane proteins with increased stability or a matrix protein that stabilizes the lipids bilayer. Therefore, increasing the stability of protein or lipids limits the effects of arginine. To determine if arginine inactivation relies on protein and/or lipid destabilization, herpes simplex virus-1 (HSV-1), bovine viral diarrhea virus (BVDV), equine arteritis viral (EAV), and pseudorabies virus (PRV) were subjected to arginine inactivation. Similar to literature, inactivation varied across different viruses. For example, HSV-1 inactivated to higher levels than BVDV across all solutions conditions. HSV-1 is a large enveloped virus with a diameter of 100-300 nm while BVDV has a diameter of 50-65 nm. The smaller diameter of BVDV increases lipid packing density, likely opposing membrane deformations and manipulations by arginine. The correlation between low viral inactivation and tightly packed lipids supports the mechanism of lipid destabilization by arginine. In addition to testing several viruses, a range of solution conditions were tested to optimize viral inactivation and determine arginine mechanisms. Agmatine, a derivative of arginine, was tested to understand the influence of charge on inactivation. Pyrenebutyrate, a hydrophobic cation, was added to arginine solutions to evaluate added hydrophobicity on inactivation. When the mechanisms are known, viral inactivation by arginine may be further enhanced by the addition of functional groups, charges, or molecules to completely inactivate all enveloped viruses.

572817 Once Weekly Oral Ivermectin for Prevention of Malaria Transmission in Zone IVb

In regions where malaria is endemic, there is an unmet need for malaria vector control. Insecticide treated nets (ITN) have provided significant reduction in malaria transmission [1] but are only effective when users are physically confined in the nets. Oral ivermectin administration has been proposed as a method to provide vector control outside of nets.

Malaria is most prevalent in sub-Saharan Africa, where temperature and humidity are high. ICH classifies the region as zone IVb and defines long term storage condition as 30°C with 75% relative humidity (RH) [2]. Moreover, US military has shown that shipping crates being transported to zone IVb region in Panama have recorded temperatures as high as 57°C. To provide ivermectin for prevention of malaria transmission in sub-Saharan Africa without cold chain storage, ivermectin oral medications should have stable performance after exposure to 60°C.

An oral ultra-long acting dosage form developed by Lyndra Therapeutics incorporates polycaprolactone (PCL) to deliver ivermectin for 14 days via gastric retention [3]. In this study, gastric resident ivermectin formulations were evaluated for their stability to high temperature excursions. Formulations were exposed to 60°C for hours to days and tested for performance, including physical shape retention, in vitro drug release, drug purity, and in vivo pharmacokinetics (PK) in a dog model.

1. WHO. World Malaria Report 2017. ISBN 978 92 4 156552 3.
2. WHO Technical Report Series, No. 1010, 2018 Annex 10. ISBN 978-92-4-121019-5
3. Kanasty et al. A pharmaceutical answer to nonadherence: Once weekly oral memantine for Alzheimer's disease. 2019. Journal of Controlled Release, doi https://doi.org/10.1016/j.jconrel.2019.03.022.

572419 Characterization of the Formulation-Process Interaction for Improved Thermal Bonding in the Manufacture of a Novel Ultra-Long Acting Oral Dosage Form

Lyndra Therapeutics is developing an ultra-long acting oral dosage form that reduces pill burden and helps improve patient adherence to medication. The performance requirements associated with achieving extended gastric residence and linear, multi-day drug release are significantly different than those of traditional solid oral dosage forms. Lyndra's dosage form meets these requirements through the incorporation of multiple specialized layers that are joined together using a unique thermal bonding process. In order to determine the optimal material formulations and process parameters for creating strong thermal bonds, we sought to better understand the behavior of the materials during the bonding process through thermal and rheological analysis.

The thermal bonding process consists of loading the selected layers into a nest in the desired configuration, applying radial pressure such that all interfaces make contact, and subjecting the exposed side of the components to heat. This process takes advantage of the use of the same base polymer throughout all components, promoting material compatibility at component interfaces. Strong thermal bonds are created when polymer chains are heated to the point at which they can flow across the joint interface and intermingle with chains from the adjacent component. Therefore, the melt viscosity, or resistance to flow, is an important material characteristic to analyze in the context of the Lyndra joining process.

The first step to characterizing the joining process was to understand how the key parameters impacted the materials in question. As a proof of concept, we chose to work with two adjacent layers that have substantially different polymer compositions. The temperature reached by each material during the heating process varies because each material has its own absorptive and conductive properties. The process temperatures were measured using thermocouples inserted directly into the materials. The temperatures of the layers were monitored under standard processing conditions, and alternative processing conditions were also characterized across a range of material formulations.

With the processing temperature space mapped out, the materials were evaluated using a capillary rheometer to determine the melt viscosities at relevant temperatures. Plasticizers were incorporated to modulate melt viscosity in order to determine the effect of viscosity on bond strength. Bond strength of the materials was evaluated using tensile testing to measure the force required to pull the components apart, and various dosage form configurations that incorporated relevant process and material changes were evaluated for in vivo durability in a beagle model. Through thermal and rheological analyses of the process we were able to reformulate the modular layers and choose the appropriate process parameters to improve bonding without negatively impacting the performance of each component.

572461 Development of pH -Responsive Disintegrating Matrices for Safe Intestinal Transit of Gastric Resident Dosage Forms

Multi-day oral drug delivery has recently been demonstrated in humans using a unique gastric resident dosage form pioneered by Lyndra Therapeutics. After oral administration in a standard sized capsule, the polymer-based dosage form unfolds in the stomach into a stellate geometry that is large enough to prevent passage through the pylorus and is able to withstand gastric emptying waves. Once the stellate has fully eluted its drug contents, it will exit the low-pH gastric environment and encounter the high-pH intestinal environment. To facilitate intestinal transit of dosage form components, enteric disintegrating matrix (DM) materials were developed that provide structure to the stellate at gastic pH but lose their physical integrity upon exposure to intestinal pH.

The enteric disintegrating matrix (DM) is an HPMCAS-based polymer blend that connects the drug-containing arms to the body of the dosage form. When exposed to high pH during intestinal transit, the enteric DM softens and can break, reducing the overall size of the dosage form. This promotes safe exit through the intestinal tract.

By varying the composition of the polymeric matrix, a formulation was developed that demonstrated structural integrity for 7 days in a simulated gastric environment and quickly lost structural integrity upon exposure to a simulated intestinal environment. Robust DMs were successfully manufactured by via hot-melt extrusion and evaluated by 3-Pt Bending, water uptake in gastric media, mass loss in intestinal media, and morphology via SEM. The materials were incorporated into dosage forms by thermal bonding, and the thermal bond strength was evaluated via tensile testing and dynamic mechanical testing.

575662 Solubility Enhancement of Hydrophobic Drug Molecules *Via* Amorphous Solid Dispersions and Flash Nanoprecipitation

Hydrophobic small molecule drugs present unique obstacles for oral administration. Their physical properties, namely poor aqueous solubility and large oil/water partition coefficient, often result in poor *in vivo* bioavailability. It is estimated that 60-70% of pipeline drug candidates may be classified as BCS Class II compounds, which exhibit solubility-limited *in vivo* absorption. Thus, there is an opportunity to improve the bioavailability of such compounds by enhancing their aqueous solubility. One strategy for solubility enhancement involves generating and stabilizing the amorphous form of the drug molecule in drug-polymer systems. Lacking the long-range order present in the crystalline form, the amorphous material represents a higher energy state compared to the crystalline state. This greater free energy increases the thermodynamic driving force for dissolution, resulting in improved aqueous solubility. Our research investigates two platform technologies for producing amorphous drug-polymer systems: Amorphous Solid Dispersions (ASDs) and Flash NanoPrecipitation (FNP). Here, we illustrate our study of these two technologies to enhance the aqueous solubility of hydrophobic drug molecules. We show that the amorphous form of the drug can be obtained as a result of the fast dynamics of the spray drying and FNP processes and that the amorphous drug is further stabilized in the solid form by the chosen amphiphilic polymers. We present characterization of our materials and an assessment of release performance and supersaturation.

577850 Comparison of Oral Drug Dissolution between Medium and Long Chain Unsaturated and Saturated Triglycerides: A Modeling-Based Approach

An increasing number of drugs emerging from discovery pipelines are lipophilic, poorly water-soluble drugs. When co-administered with food, these drugs can exhibit either a positive or a negative effect of food on absorption and bioavailability. Modeling such a food effect successfully would be very valuable, as it would allow for an improved understanding of food related pharmacokinetic effects and ensure better co-dosing or formulation decisions when advancing towards clinical trials. Lipids in food are particularly significant in affecting the bioavailability of poorly soluble drugs. Our group previously developed a model of the impact of lipids on oral drug dissolution and bioavailability[i], with these initial efforts focused on studying and predicting the effects of soybean oil, a mixture of many different lipid molecules, as a model lipid vehicle.

The goal of this study is to more accurately model lipolysis and drug dissolution during the dynamic digestion of structurally diverse lipids, representing the variety of lipids available in our foods. This model may then be combined with an absorption model to more accurately predict the bioavailability of drugs under fed conditions with different meal constituents. The modeling framework incorporates lipolysis kinetic parameters, drug solubility data, and partition parameters which are experimentally measured. Thus, *in vitro* experiments are employed to obtain kinetic digestion and partitioning modeling parameters that highlight the differences between medium and long chain, unsaturated and saturated, and solid and liquid triglycerides. These differences ultimately affect pharmacokinetics of lipophilic drug co-dosed with lipids. For our experiments, we employ Griseofulvin as a model drug known to have a positive food effect.

Experiments were performed using a simulated fed state intestinal solution consisting of a buffer, ($CaCL_2$), a bile salt (NaTDC), and phospholipid. The lipid is added to the fed state solution (after melting for lipids solid at room temperature) and sonicated to form microparticles, which are stabilized due to the amphiphilic character of the bile salt and phospholipid. The bile salt and the phospholipid also form bile micelles that assist in drug dissolution by allowing for drugs to partition into the micelles. Pancreatin extract is added to begin digestion. As digestion occurs, the digestion products, mainly monoglycerides and fatty acids, partition into the micelles and the micelles grow allowing for higher concentrations of drug to partition into them. In vitro digestion experiments with unsaturated lipid, triolein (long chain length: C18 - liquid) and saturated lipids, tristearin (long chain length: C18 - solid) and tricaprylin (medium chain length: C8 - liquid) were conducted. As expected, the medium chain triglyceride exhibits faster drug dissolution kinetics than longer chain triglycerides. Initial experiments also show that saturated triglyceride exhibit slower digestion and lower drug solubility than the unsaturated triglycerides. With all three lipids, a significant increase in drug solubility was observed when lipid was introduced into solution as compared to fed state solution without lipid present, in concurrence with the positive food effect observed with griseofulvin when co-dosed with lipids. The mechanistic mathematical model is currently under development and initial results show successful prediction of triolein lipolysis and its impact on drug dissolution and overall absorption.

[i] F. Buyukozturk, S. Di Maio, D. E. Budil, R. L. Carrier. Phram. Res. 2013, 30, 3131

557070 Enhancing the Long-Term Storage Stability of Amorphous Drug-Polyelectrolyte Nanoparticle Complex Via Incorporation of Hypromellose

Enhancing the Long-Term Storage Stability of Amorphous Drug-Polyelectrolyte Nanoparticle Complex via Incorporation of Hypromellose

By

Kunn Hadinoto and Chew Jia Wei

Associate Professor

School of Chemical and Biomedical Engineering

Nanyang Technological University

Singapore

Various solubility enhancement strategies of BCS Class II and IV drugs have been proposed, for example, by nanonization, amorphization, cyclodextrin inclusion, and complexation/conjugation with each of them come with its pros and cons. Recently, our group developed a new strategy of combined amorphization and nanonization via electrostatically driven drug-polyelectrolyte complexation to produce amorphous drug nanoparticle complex (or nanoplex in short) exhibiting solubility enhancement capability. This strategy is highly attractive owed to (1) its simple preparation involving only the mixing of drug and polyelectrolyte solutions under ambient condition, (2) high drug utilization rate without the need of solvent denoting its high mass and energy sustainability. The amorphous drug-polyelectrolyte nanoplex, however, exhibit poor physical stability during long-term storage in which the amorphous state crystallizes, hence the solubility enhancement is lost. In this talk, a new formulation strategy in which a well-known crystallization inhibitor, hypromellose, is incorporated into the drug-polyelectrolyte nanoplex is discussed using two model poorly-soluble drugs, i.e. curcumin and ciprofloxacin. The amorphous ternary nanoplex was formed based on electrostatic, hydrophobic, and hydrogen bond interactions among the drug, polyelectrolyte, and hypromellose molecules. The results showed that the incorporation of hypromellose successfully enhance the long-term stability of the drug nanoplex (> 12 months) while at the same time prolong the supersaturation generation in vitro, resulting in high and prolonged apparent drug solubility. The inclusion of hypromellose, nevertheless, resulted in a decrease in the drug payload and larger nanoparticle size with lower zeta potential due to the non-ionic nature of hypromellose, in contract to the charged nature of the polyelectrolytes. The variations in the drug payload and nanoplex characteristics, nevertheless, did not influence its dissolution characteristics or its storage stability. These results established the amorphous drug nanoplex as an effective bioavailability enhancement strategy of poorly soluble drugs.

561162 Understanding Polymorphic Phase Transformations of Acetaminophen in Polymer-Based Formulation Processes

In recent years, advanced polymer-based formulation processes (e.g. hot melt extrusion, HME) have emerge as enabling technologies for continuous solid dosage formulations of active pharmaceutical ingredients (APIs). However, aside from thermal stability of the individual components (polymer + API), polymorphic stability of the API is thought to be a requirement when HME is employed to produce crystalline solid dispersions (CSDs). Polymorphism enables molecules to exhibit multiple crystalline phases with inherently different physical properties and it is known to occur in up to 80% of APIs. Acetaminophen (ACM) is a widely used analgesic that, to date, exhibit five polymorphic forms from which only two are stable at ambient conditions. Form I (ACM I) is the most stable and commercially available form, however, presents low compressibility, which renders the tableting process cumbersome. The metastable form, ACM II, possess higher compressibility and solubility, but it is not readily accessible without using additives in the crystallization process prior to the formulation. The present study shows how ACM I can be transformed into ACM II during the formulation process without additives to obtain CSDs. For the proof of principle study, the CSDs were obtained by temperature-simulated hot melt extrusion (TS-HME) experiments using a differential scanning calorimeter (DSC). The physical mixtures of ACM and polyethylene glycol 10,000 and the solid dispersions thereof were characterized by powder X-ray diffraction, Fourier transform infrared spectroscopy, DSC, thermogravimetric analysis, and hot stage microscopy. The results demonstrate that employing this method can provide access to CSDs of ACM II with high polymorphic purity. The polymorphic phase transformation process from ACM form I into form II can be controlled by precisely defined heating and cooling profiles. Ultimately, this investigation helps to gain fundamental understanding of the processing needs of CSDs with the desired polymorphic form while contributing to process intensification efforts.

566603 Application of Artificial Neural Network As a Predictive Tool for Continuous Liposome Processing

Purpose:

Liposomes are lipid bilayer vesicles which are used as drug delivery systems, especially as part of anti-cancer and anti-fungal therapies. At UConn, a continuous liposomal processing system has been developed and implemented where lipids are dissolved in ethanol and mixed with an aqueous phase in a co-axial, co-flow arrangement. Predictive modeling of this process is essential due to the tedious nature of laboratory experiments and the challenging time consuming mechanistic computational modeling. In this work, an artificial neural network (ANN) was used as a predictive tool. We used experimental data results to train the ANN with input features such as the critical material attributes (CMA) and critical process parameters (CPP) that are known to affect the ANN output representing liposomal critical quality attributes (CQA). A major benefit of this work is that ANN has been shown to have predictive power by relating CMA and CPP to the CQAs of liposome formulation.

Methods:

Experiments have shown that the CMAs and CPPs that affect the liposome formation process are hydrocarbon tail length, percent cholesterol, temperature and flow rate of aqueous phase, type of buffer used. Natural and synthetic phospholipids were used for the hydrocarbon tail length studies. The liposomal formation process was modeled using ANN to predict CQAs (such as particle size, particle distribution index, and zeta potential). Multiple-input-multiple-output model (MIMO) and multiple-input-single-output (MISO) models were implemented using Levenberg- Marquardt (LM) algorithm. Initially 175 points were used to test the MIMO and MISO models. Later the natural phospholipid was added to make a total of 210 data points. Buffer was also added as one of the input features and the total data points were around 600

Results:

Mean Relative Errors (MREs) between the target values and predicted values were calculated for performance evaluation. In case of particle size, the training and testing MREs for MIMO were 32% and 33% and for MISO were 18% and 21%, respectively with 175 points. MISO model tested with 220 points showed the training and testing MREs as 12.01% and 12.35%, respectively. The incorporation of an additional input feature did not cause any significant change in the error as the training and testing MREs were found to be 11.55% and 12.65% respectively.

Conclusions:

Our findings indicate that the MISO model was superior than the MIMO model in terms of quantitative prediction of the CQAs. Our model is capable of predicting CQAs for natural as well as synthetic phospholipids. Ongoing work includes increasing the number of test points performing process optimization to further decrease the errors.

Acknowledgements:

FDA Grant# 1U01FD005773-01.

Disclaimer:

This article reflects the views of the authors and should not be construed to represent FDA's views or policies.

563574 Development of a Blistering Packaging Process with Nitrogen-Modified Headspace

A nitrogen overlay during blister packaging is used to reduce oxidative degradation of a drug product packaged in cold form foil blisters. The nitrogen purges ambient air to achieve = 2% oxygen in the blister headspace. An offline method to measure the oxygen content of individual blister cavities was developed to improve method robustness and minimize sample manipulation. Blister packaging process parameters that impact oxygen content were identified and operating ranges were established through engineering studies. In-line oxygen measurements were correlated to offline measurements of blister headspace oxygen content to establish in-process limits and minimize rejects.

576542 Dropwise Additive Manufacturing for Pharmaceuticals

In recent years additive manufacturing (AM) technology has attracted considerable interest for its potential applications to on-demand manufacturing of dosage forms with individually adjusted doses, which is a necessity for implementation of precision medicine, and for development of a robust manufacturing platforms for early drug development or continuous manufacturing. Commensurate with the level of interest is the variety of different technologies based on two-dimensional (2D) or three-dimensional printing (3D) techniques, all of which present the common advantages of precision and flexibility in dosing inherent to AM processes. However, differences exist in the printing material requirements and in the achievable production rates for the process, both of which must be considered when proposing implementations in the context of compounding in clinical/pharmacy settings or full-scale production.

Of the AM techniques, the system based on the drop-on-demand method (2D printing) for fluid deposition/dispensing has demonstrated the capacity to serve as flexible manufacturing platform for pharmaceutical dosage forms, with doses in amounts of 1-100 mg active ingredient produced from printing materials which ranged from drug solutions to drug-polymer melts and concentrated suspensions of micronized particles [1, 2, 3]. Most recently demonstrated was the development of the system for precision dispensing of micronized powders by processing them as particle suspensions, which, as a fluid-based method, avoids powder flow issues and may readily be adapted to new particulate materials using dimensional analysis [3, 4]. For dosage units which contain 100mg of active drug, production rates of up to 2000 doses per hour, with product RSD $< 2\%$, have been achieved with suspension processing [3, 4]. This system uses the drop-on-demand printing method to deposit drops on to a prepared substrate, wherein the content of each dose is determined by the number, volume and composition of drops; consequently, quick adjustments to dosage unit properties may be made by changing the deposited fluid volume, and/or composition.

In this work the capabilities of the dropwise additive manufacturing process for pharmaceuticals production are described, with particular emphasis given to dosage forms produced by printing of micro-particle suspensions. Applicability of the system to precision medicine is demonstrated through the production of oral solid dosage units from multiple drug compounds at low to high dosage strength levels; discussion of process monitoring tools for real-time prediction of dosage properties during production runs and characterization of drug product properties is included. In the context of process development, high speed photography of the drop formation dynamics is combined with an in-house image analysis tool to provide quantitative evaluation of performance metrics, such as process throughput and drop volume reproducibility, which are then related to particle properties in the printing materials. Presented together, these results demonstrate the capability of the dropwise AM system for dose production at multiple manufacturing scales.

References

1. Hirshfield, L., Giridhar, A., Taylor, L. S., Harris, M. T., & Reklaitis, G. V. (2014). Dropwise additive manufacturing of pharmaceutical products for solvent-based dosage forms. Journal of pharmaceutical sciences, 103(2), 496-506.
2. Içten, E., Giridhar, A., Taylor, L. S., Nagy, Z. K., & Reklaitis, G. V. (2015). Dropwise additive manufacturing of pharmaceutical products for melt-based dosage forms. Journal of pharmaceutical sciences, 104(5), 1641-1649.
3. Radcliffe, A. J., Hilden, J. L., Nagy, Z. K., & Reklaitis, G. V. (2019). Dropwise Additive Manufacturing of Pharmaceutical Products Using Particle Suspensions. Journal of pharmaceutical sciences, 108(2), 914-928.
4. Radcliffe, A. J., & Reklaitis, G. V. (2017). Dropwise additive manufacturing using particulate suspensions: feasible operating space and throughput rates. In Computer Aided Chemical Engineering (Vol. 40, pp. 1207-1212). Elsevier.

568930 Buckling of a Drying Colloidal Drop

It is well known that drying drops of colloidal dispersions undergo complex morphological transitions involving buckling of a particle-packed outer shell during drying. Although capillary stresses generated during drying have been identified as the cause for buckling, the exact conditions for buckling and its relation to the particle size, rigidity, and nature of packing have not been understood. Here, we derive explicit expressions for the critical capillary pressure for buckling of droplets based on the mechanical properties of the particle network formed during drying and the conditions under which buckling can be avoided. We anticipate our results to form the basis for design of formulations for the pharmaceutical, food and ceramics industry

573254 Engineering Quality and Manufacturability of a Trivalent Protein Subunit Vaccine

Global delivery of vaccines would benefit from ultra low-cost prices (less than 15 cents per dose) at intermediate scales of production (10-100 million annual doses). Currently, for vaccines comprising recombinant proteins, the cost of manufacturing remains higher than these targets. Protein subunit vaccines are increasing in prevalence due to enhanced safety and understanding of epitopes and immune response. Extensive characterization has enabled recent work in rational design and scaffolding of vaccines. While promising for function and efficacy, current designed therapeutics are subject to the same manufacturing challenges as biopharmaceuticals. Thus, the key drivers of cost include volumetric productivity, quality, and product stability. Here, we present an approach that uses bioinformatics to improve the expression, quality, and stability of a subunit vaccine for rotavirus gastroenteritis.

Currently, three serotypes of a truncated VP8 rotavirus subunit protein are in clinical development. We expressed each serotype in the alternative yeast host *Komagataella phaffii* (*Pichia pastoris*), which boasts high growth rates, simple genetic manipulation, and is Generally Recognized as Safe (GRAS) by the FDA. The small genome size and fast iteration time for expression of new products (~4 weeks) enabled a bioinformatics-based approach for rational product engineering. With minimal, targeted sequence modifications we successfully mitigated different manufacturing challenges encountered with each serotype, including glycosylation, aggregation, truncation, and low expression. Finally, we were able to reduce process cycles and cost by co-expression of high quality serotypes. This case study motivates manufacturability as a key component of protein therapeutic development.

575121 Design of Experiments Reveals Critical Parameters for Bulk Freeze-Thaw of Lactate Dehydrogenase

Purpose: Nearly half of biotherapeutics bulk drug substance are stored frozen. Storage in a frozen state provides operational flexibility for fill-finish of drug products and extends the shelf life of protein solutions. However, the freezing and thawing (F/T) processes themselves can destabilize the complex structure of proteins. Formulations and processes, therefore, must be designed to be protective against potential F/T-induced stresses to proteins. The overall goal of this research project was to investigate the impact of F/T process parameters on the stability of a model protein under pharmaceutically relevant conditions at a large scale.

Methods: A response surface model was designed and the results were evaluated using JMP® 13 (SAS Institute, Cary, NC) to investigate how F/T process parameters affect the lactate dehydrogenase (LDH) protein stability in 1-L polycarbonate containers. The LDH protein solutions in histidine buffer were frozen at -20°C, -50°C, or -80°C and thawed at room temperature after 24 hours. The F/T cycles applied either single or full loading configuration, i.e., a total of nine bottles arranged in a 3x3 array. The following process parameters were applied: solution fill volume (50%, 70%, and 90%), F/T loading distance (1 cm, 5 cm, and 10 cm), and the thawing method (with or without forced air flow). The F/T rate was monitored using type-T thermocouples inserted at different axial and radial locations inside the solutions. After each F/T cycle, a sample withdrawn from the homogenized bulk was analyzed for protein concentration, LDH tetramer purity, and LDH enzymatic activity.

Results: Analysis from 40 experimental runs indicate that increasing the loading distance and lowering the solution fill volume to 50% significantly increased the freezing rate with 95% confidence interval. Furthermore, decreasing the solution fill volume and applying a forced air flow at 102 fpm during thawing significantly increased the thawing rate. The F/T runs performed at faster F/T rates also correlated with improved protein stability. F/T cycles performed at faster rates resulted in enhanced retention of the protein enzymatic activity and native structures percentage. For the faster F/T processes completed in less than 6.0 hours of freezing and 8.7 hours of thawing, the percentage of LDH initial activity was greater or equal to 97.4%, correlating with approximately 99.9% of LDH native structures. However, longer F/T processes requiring more than 9.0 hours of freezing and 25.5 hours of thawing resulted in a poor protein stability: less than 70.7% of the LDH initial activity and approximately 98.2% of LDH native structures.

Conclusions: This study confirms the feasibility of a faster freezing rate and forced-air thawing procedure to enhance the stability of proteins exposed to F/T unit operations. Moreover, the data obtained provided a mechanistic understanding of possible stresses arising during the process and their ultimate effect on the protein stability. Overall, the data obtained shed light on how F/T process parameters impact on bulk protein stability and can be applied to establish F/T process guidelines.

578338 Multimodal Membrane Adsorbers for Clearance of Minute Virus of Mice

In recent years, several membrane adsorbers with high binding capacity have evolved for the removal of contaminants for the polishing steps during downstream purification of biologics. In particular, multimodal membrane adsorbers with tunable pore sizes and ligand properties have generated tremendous interest due to their many possible applications. Here the removal of Minute Virus of Mice (MVM), host cell proteins (HCPs) and DNA using multimodal membrane adsorber Puri™-MQ was studied. The membrane performance was evaluated over a range of feed and elution conditions. The effects of pH and HCP concentration on MVM clearance were investigated. Comparison with other membranes at similar conditions demonstrates that multimodal membranes are highly effective for virus clearance.

555716 Advances in Spray Freeze-Drying for Uniform Spherical Bulk Intermediate Product

Advances in Spray Freeze-Drying for Uniform Spherical Bulk Intermediate Product

Atmospheric spray freezing and rotational dynamic vacuum freeze drying is a new technology that is now being introduced for pharmaceutical aseptic processing of parenteral products. This process, which is designed to improve on traditional lyophilization techniques produces microspheres which can be held as bulk material and filled on demand into primary packages and devices. The spheres may be formed in a selectable size range from 300 – 1000 μm.

The freeze-dried spheres are homogeneous, free flowing and dust free. This physical form allows for significantly increased flexibility in the supply chain, either by the improved stability and handling of a free flowing solid for transporting bulk active, or by dosing and batch size flexibility when filling finished product. Producing a combination product becomes as simple as filling the necessary types of spheres into the primary packaging. Devices as well as vials may be filled, opening the possibility of new product types.

Further product innovation is possible due to the range of capability of the technique. Higher concentration formulations are suitable to process, while still offering fast reconstitution properties. The additional concentration may enable subcutaneous administration. The concept also allows for new formulation strategies by opening the range of solid dosage form technologies like coating and multiparticulate dosage forms for freeze dried products.

The process is performed in a fully contained process line which includes the generation of frozen micropellets and a dynamic rotary freeze drying process. This arrangement greatly increases the surface area available for heat and mass transfer leading to process efficiencies.

560182 Controversial Influence of Gravitational Force on the Crystallisation of Lysozyme

The manufacturing of biopharmaceuticals [1] in a cost effective and reliable route is a major industrial challenge due to the complex nature of the protein. Biocrystallisation [2,3] as an alternative to conventional downstream separations, offers route to efficient biopharmaceutical manufacturing, enabling higher throughput, lowering the product costs, and an increase in product quality and stability [4,5]. There are literatures reported the controversial results about the gravity, some reported micro-gravity enhanced the crystallisation of protein, and, however, some reported hyper-gravity accelerated the protein crystallisation.

In this work, the influence of gravity induced by centrifugation is investigated for variation in rotation speeds and time. This gravity effect is investigated for lysozyme crystallisation solution (70-90 mg/ml) with NaCl solution (90-110 mg/mL). The results obtained show no obvious influence of gravity force (< 100 g) and short duration (< 5 mins) of centrifugation on the crystallisation process. The results also answered the controversial results reported in literatures. For a continuously long time centrifugation, the high gravity (> 11000 g) hindered the nucleation process. However, the high gravity accelerated the crystal growth process, therefore, obtaining larger crystals comparing with the low gravity.The continuously manufacturing by combination of centrifugation and crystallisation can be an efficient platform for purifying macro-biomolecules.

Acknowledgements

We acknowledge the EPSRC (EP/N015916/1) for funding.

References

1. Narhi, M. and K. Nordstrom, Manufacturing, regulatory and commercial challenges of biopharmaceuticals production: a Finnish perspective. European journal of pharmaceutics and biopharmaceutics, 2005. 59: 397-405.
2. Curcio, E., et al., Membrane crystallization of lysozyme under forced solution flow. Journal of membrane science, 2005. 257: 134-143.
3. Harrison, R.G., et al., Bioseparations science and engineering. 2015: Topics in Chemical Engineering.
4. Basu, S.K., et al., Protein crystals for the delivery of biopharmaceuticals. Expert opinion on biological therapy, 2004. 4: 301-317.
5. Yang, H., et al., Workflow of continuous Protein Crystallisation: The Case of Lysozyme Crystal Growth and Design. 2019. 19: 983-991

565209 Application of CFD Modeling for the Development and Transfer of Mixing and Compounding Processes with Single Use Equipment

AIChE Annual Meeting 2019

Pharmaceutical Discovery, Development and Manufacturing Forum

26006 Advances in Large Molecule Processes - Oral Session –

Application of CFD Modeling for the development and transfer of mixing and compounding processes with single use equipment

Mario Hirth[1*], Kushal Sinha[2], Nandkishor Nere[2], Martin Bultmann[1]

*presenter: Mario.hirth@abbvie.com

[1]AbbVie Deutschland GmbH&Co KG, Knollstraße 50, 67061 Ludwigshafen

[2]AbbVie, 1401 Sheridan Road, North Chicago, Il 60064

Mixing is one of the most important unit operations in drug-product processes for large biological molecules as it assurers homogeneity of parenteral drug products in pooling and for additional compounding steps. In order to achieve successful scale-up and technology transfer from development to commercial scale of mixing process, it is crucial to understand the hydrodynamics as a function of equipment and drug substance. Single-use mixing systems are becoming an important option in manufacturing of large molecules for conventional mixing vessels due to the many advantages that come with disposable systems like reduced cross-contamination risk and shorter batch changeover times. The main challenge when transferring large molecule processes from stainless-steel to single-use vessels is insufficient knowledge of the equipment related performance. Computational Fluid Dynamics (CFD) is a well-established and powerful tool to characterize the mixing performance in single-use systems.

In this study, CFD-modeling was applied to study and predict the hydrodynamics of different types and sizes for single-use mixing system applicable from development to manufacturing scale. Understanding the fluid dynamics by CFD in those tanks is a crucial factor to understand the blending, intensity and duration of shear exposure. We present here the results from 5 different single-use mixing systems with a working volume of 50 L, 100 L and 200 L. Flow pattern and mixing performance were predicted as a function of viscosity, fill volume and agitation speed. Tracer simulation was applied to indicate how process parameters affect the mixing duration in the tank. A bolus tracer was injected and the distribution was monitored at several locations to calculate blend time and to identify poor mixing zones for each single-use system. The results can be used to recommend ideal mixing conditions up to batch sizes from 6 L to 200 L. In addition, lab trials were conducted to verify the CFD model. Results show that the CFD approach generates a profound understanding what happens inside the different reactor scales. As a result scale up and tech transfer of mixing and compounding processes get more robust and offer the possibility to decrease the amount of effort and API for mixing validation.

All authors are employees of AbbVie and may own AbbVie stock. AbbVie sponsored and funded the study; contributed to the design; participated in the collection, analysis, and interpretation of data, and in writing, reviewing, and approval of the final publication.

565297 Bio-Separation Via Selective Protein Crystallisation Facilitated By Porous Silica Templates: A Case Study Using Mesoporous Nanoparticles in Lysozyme-Thaumatin Mixture

Bio-separation via selective protein crystallisation facilitated by porous silica templates: A case study using mesoporous nanoparticles in lysozyme-thaumatin model binary protein mixture

Xiaoyu Li [1], Wenqian Chen [1], and Jerry Y. Y. Heng[1] *

1. Department of Chemical Engineering, Imperial College London, South Kensington Campus, London SW7 2AZ, United Kingdom

* Corresponding Author: Jerry Y. Y. Heng. E-mail: jerry.heng@imperial.ac.uk

Abstract

Recent progress of advanced biotechnology has enhanced the development of commercial biopharmaceutical products. Over 246 approved biotechnology drugs and vaccines had been developed with cumulative revenues reaching $140 billion [1]. To meet the growing market demand of protein-based biopharmaceuticals, a cost effective and reliable downstream route becomes a major industrial challenge in therapeutic protein manufacture. Crystallisation has been used for the purification of more than 95% small-molecule pharmaceuticals in various conventional chemical industries [2]. Crystallisation of proteins as downstream separation and isolation steps is considered to be more efficient and cost-effective compared to conventional purification techniques such as chromatography [3-5]. Additionally, crystalline proteins have a higher purity and stability which can benefit formulation, storage, and drug delivery steps [3, 6]. Even so, control of protein crystallisation remains challenging after decades of investigation. To address this problem, recent researches showed heterogeneous nucleants with designed properties have the potential to facilitate selective protein crystallisation and thus lead to better control of the crystallisation processes [7, 8].

Here, we report a demonstration of selective crystallisation of target protein from a model binary protein mixture achieved by addition of mesoporous silica nanoparticles (NPs) as heterogeneous nucleants. In this work, mixtures were prepared using model proteins: lysozyme and thaumatin. Protein mixtures with a composition range of 0 – 100 mg/mL lysozyme and 0 – 100 mg/mL thaumatin were crystallised using Hanging-drop vapour-diffusion (HDVD) crystallisation technique. $>10^3$ of 2-4 μl hanging drops containing mixture of protein and precipitant solutions were set-up. The crystallisation processes were monitored under optical microscope for 1-2 weeks. Protein crystals of these proteins were distinguishable from their habits (shapes). Both model proteins can be crystallised out by adding the same precipitant solution. Diagrams were produced to show the operation windows of selective crystallisation of target protein from the mixtures. Further, mesoporous silica NPs such as SBA-15 were used in the crystallisation experiments and experiments revealed that certain silica nanoparticles were able to promote lysozyme crystallisation while hinder thaumatin crystallisation in the range of this study. Therefore the mesoporous NPs were able to shift the operating window of target protein crystallisation. Furthermore, scaled-up –mL batch crystallisation were conducted to prove the results revealed by HDVD experiments. This work demonstrates that mesoporous silica facilitated crystallisation possesses the potential of separating a target protein from a complex mixture environment which can contribute to the future development of downstream bio-separation processes.

References

1. Demain, A.L., REVIEWS: The business of biotechnology. Industrial Biotechnology, 2007. 3(3): p. 269-283.

2. Yang, H. and J.H. ter Horst, Crystal nucleation of small organic molecules, in New Perspectives on Mineral Nucleation and Growth. 2017, Springer. p. 317-337.

3. Basu, S.K., et al., Protein crystals for the delivery of biopharmaceuticals. Expert opinion on biological therapy, 2004. 4(3): p. 301-317.

4. Parambil, J.V., et al., Effects of Oscillatory Flow on the Nucleation and Crystallization of Insulin. Crystal Growth & Design, 2011. 11(10): p. 4353-4359.

5. Delmas, T., M.M. Roberts, and J.Y. Heng, Nucleation and crystallization of lysozyme: Role of substrate surface chemistry and topography. Journal of Adhesion Science and Technology, 2011. 25(4-5): p. 357-366.

6. Harrison, R.G., et al., Bioseparations science and engineering. 2015: Topics in Chemical Engineering.

7. Shah, U.V., et al., Heterogeneous nucleants for crystallogenesis and bioseparation. Current Opinion in Chemical Engineering, 2015. 8: p. 69-75.

8. Shah, U.V., D.R. Williams, and J.Y.Y. Heng, Selective Crystallization of Proteins Using Engineered Nanonucleants. Crystal Growth & Design, 2012. 12(3): p. 1362-1369.

559305 Data Rich Experimentation and Machine-Assisted Process Development for Rapid Reaction Optimization

Identifying the optimal conditions for complex reactions is a challenging task in drug substance development and can be hindered by material availability, program timelines, limited data, and experimental bias. However, the combination of parallel reactor technology with automated sampling strategies and machine learning approaches can greatly accelerate the optimization procedure with minimal user intervention. To demonstrate these improvements, we applied these technologies to optimize two case study reactions – the copper catalyzed methoxylation of an aryl halide and the multihalogenation of an oligosaccharide. In each case study, reactions were conducted in a reactor block that enabled the execution and sampling of 10 reactions simultaneously. Data from these reaction profiles were inputted into a direct search optimization algorithm, which provided the conditions for the next round of 10 experiments. In this manner, the reactions were optimized in one week using 40 experiments and a minimal amount of material. Data analytics and kinetic modeling provided mechanistic understanding and further process insights using the same data set obtained from the optimization procedure.

564157 Optimal Time Sampling Strategy in Pharmaceutical Reactions for the Estimation of Accurate Drsm Models

Through the available robotic experimental devices, one is able to perform multiple experiments and sample the reaction mixture at multiple time instants. Such time resolved data have been successfully modeled through the recently introduced Dynamic Response Surface Methodology (DRSM). The development of the DRSM method has gone through phases, from the initial approach where the model parameters were estimated with linear stepwise regression[1], to the more complete and accurate method incorporating an exponential transformation of time[2], constrained regression[3] and non-linear parameter estimation of one of the critical parameters.

This presentation addresses the optimal time instants to take compositional samples that lead to the most accurate model. We first analyze two industrial practices. The most traditional approach is to take samples equidistantly in time. A second approach is to double the sampling interval as time increases, since most of the compositional changes happen at initial times.

We propose a novel sampling strategy, which is equidistant in an exponentially transformed time, denoted by θ. The proposed approach follows two steps. First, one estimates the time constant, t_c, characterizing how fast or slow the evolution of the chemical transformations will be represented by the DRSM model. This can be estimated from prior process knowledge, from historical data, or through a single new experiment. Second, depending on the number of measurements one is willing to take, the proposed equidistant in θ strategy is translated into the sampling times in through the relationship.

We will show that this strategy is the most desirable one with respect to the following interrelated aspects: (1) it makes the non-diagonal elements of the Fisher information matrix as small as possible and reduces the cross-correlation between model parameters; (2) it maximizes the determinant of the Fisher information matrix; and (3) it reduces the prediction interval of the estimated responses. Another approach for determining the sampling times is to define and solve a set of non-linear equations requiring that the non-diagonal entries of the Fisher information matrix are zero. Nevertheless, solving such a problem can be time-consuming when the number of samples is large and offers only small incremental improvements in the accuracy of the model. Because all versions of the DRSM modeling approach use shifted Legendre polynomials, we will also demonstrate their advantage over the simple power series expansions, in t or θ, in providing superior correlation properties and higher accuracy models.

Reference:

1. Klebanov N, Georgakis C. Dynamic Response Surface Models: A Data-Driven Approach for the Analysis of Time-Varying Process Outputs. *Ind Eng Chem Res.* 2016;55(14):4022-4034.
2. Wang Z, Georgakis C. New Dynamic Response Surface Methodology for Modeling Nonlinear Processes over Semi-infinite Time Horizons. *Ind Eng Chem Res.* 2017;56(38):10770-10782.
3. Dong Y, Georgakis C, Mustakis J, et al. A Constrained Version of the Dynamic Response Surface Methodology. *Ind Eng Chem Res.* 2019;(submitted).

574545 Expedite Particle Engineering Process Development Using Data Science

Expedite particle engineering process development using data science

T. Porfirio[1,2*], Inês Matos[1], J. Vicente[1], V. Semião[2]

[1] Hovione Farmacência SA, Estrada do Paço do Lumiar, 1649-038 Lisbon, Portugal; *tporfirio@hovione.com

[2]LAETA, IDMEC, Mechanical Engineering Department, Instituto Superior Técnico, Universidade de Lisboa, Av. Rovisco Pais 1, 1049-001 Lisbon, Portugal

The scale-up of particle engineering processes such jet milling, spray congealing or spray drying, is often considered a resource intensive challenge requiring large quantities of expensive drugs. The possibility of reducing the experimental burden at large-scale of such development by using well-structured data set from the prior knowledge is a very appealing proposition since it not only reduces the development costs but also the time and material resources. The proposed work uses a hybrid model to link the operating conditions and process descriptors to the product quality attributes such particle size and bulk density.

Firstly, the operating conditions are used as input to a first principles model comprising the key phenomena of the process – thermodynamic, atomization and congealing kinetics in the case of spray congealing. Process descriptors are calculated using these models. Further, both process descriptors and operating conditions are used as input of a black box model. Different data science algorithms were evaluated as black box: neural networks, random forest, K-nearest neighbors, support vector machines, and generalized linear models.

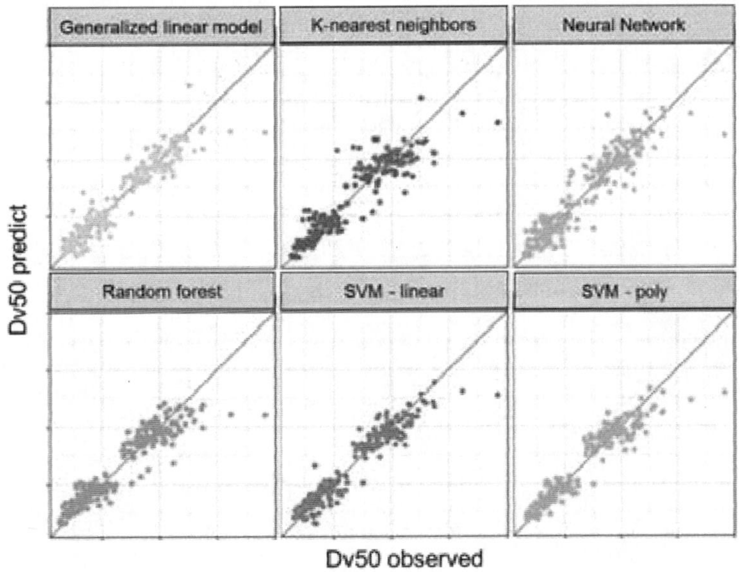

Figure 1 – Observed vs predict of a hybrid model a product quality attribute

Comprehensive case studies of application in jet milling, spray congealing or spray drying scale-up are explored. Figure 1 shows an example of the output of the hybrid model. The results of the data science algorithms are compared with PLS approach. Both data science algorithms and PLS can successfully predict the particle size from the selected input variables.

Using this approach, the development risk could be reduced by using a structured tool where the prior experience relates to suitable models for the prediction of the right descriptors. From this study was concluded that despite the many influences of process and formulation variables, trends and correlations could be founded between scale-independent parameters and the product quality attributes. The established models can be used to facilitate and expedite the scale-up development for new products in which lab-scale correlations could increase the degree of prediction.

569183 Data-Driven CLD-to-Psd Model for in-Line Crystal Size Monitoring

Recently, a data-driven model architecture was proposed for the correlation of CLD to PSD data [1,2]. The model quantifies the particle size distribution (PSD) based on solids' concentration data and the corresponding chord length distributions (CLD) measured inline using FBRM [1] and has been shown to be very robust at estimating various particle morphologies, concentrations and size ranges typically encountered in pharmaceutical crystallization processes. The model architecture consists of three distinct steps: first, the CLD data is compressed into moments, or singular values that describe the entire distribution. Second, advanced regression models convert these moments into a small set of PSD percentiles (also moments). Finally, these PSD percentiles are mapped into an alternate composite distribution using a set of parametrized functions called generating functions [1]. In this work, further progress on steps 2 and 3 is reported which is aimed at increasing model robustness, particularly when only a limited amount of data is available. Furthermore, model performance has been assessed for process monitoring under real-life conditions, including multi-modal size distributions. The technical issues of dataset generation and process monitoring are discussed as well.

[1] Irizarry R, Chen A, Crawford R, Codan L, Schoell J. Data-driven model and model paradigm to predict 1D and 2D particle size distribution from measured Chord-length distribution. Chem Eng Sci. 2017; 164: 202-218.

[2] Schoell J, Irizarry R , Sirota E, Mengel C, Codan L, Cote A. Determining particle size distributions from chord length measurements for different particle morphologies. AIChE J. Vol 65.

568812 Kernel Mean Embedding of Distributions: A New Data-Driven Model Approach for Modelling a Continuous Pharmaceutical Twin-Screw Granulation Process

Traditional pharmaceutical solid oral dosage processes comprise of a series of batch unit operations. More recently, a transition is developing from batch unit processes to continuous manufacturing application, to cope with the inefficiencies and high cost involved in process development & scale up. Twin-screw wet granulation is a developing pharmaceutical continuous process that is being assessed for its performance in solid dosage manufacturing. In this research, the twin-screw wet granulation unit is a unit operation of the ConsiGmaTM-25 continuous powder-to-tablet process line from GEA Pharma Systems. However, since these continuous processes are still under development in the pharmaceutical industry, detailed process knowledge and understanding is still evolving. Application of models, either mechanistic of data-driven, can help in filling knowledge gaps by thoroughly investigating gathered detailed experimental data and unravelling the underlying mechanisms.

Previous work of the authors focussed on population balance modelling of the continuous pharmaceutical twin-screw wet granulation unit [1]. In that work, a novel two compartmental population balance model was calibrated and validated using measurements of particle size after the wetting zone and at the end of the granulator. The population balance model used in that work can be described as a mechanistic model: the physical phenomena of aggregation and breakage of particles are explicitly described to allow a transformation of one particle size distribution into another.

In this work, the following question to is posed: can we predict particle size distributions without any assumption on the governing mechanistic phenomena in the system? In other words, can a model learn how to transform one distribution into another using only data-driven techniques?

The model under study is a Hilbert space embedding of distribution, or in short, a kernel mean embedding. The main idea behind this framework is to map distributions into a high-dimensional reproducing kernel Hilbert space (RKHS). It can be viewed as a generalization of the original "feature map" common to support vector machines and other kernel methods. Rather than manipulating the individual observations, a distribution of points (in this case a particle distribution) is represented as a mean in the RKHS. Under mild conditions, this mean retains all information on the moments of the original distribution. Kernel mean embedding has found application in fields ranging from kernel machines and probabilistic modelling to statistical inference, causal discovery, and deep learning. The embedding of distributions enables us to apply RKHS methods to probability measures which prompts a wide range of applications such as learning on distributional data [2].

This learning on distributional data is the core of this work. The kernel mean embedding framework allows for a direct feature mapping from the process variables (machine settings such as liquid to solid ratio and mass flow rate as well as blend properties such as the concentration of the active pharmaceutical ingredient and water binding capacity) to the particle size distributions at the end of the granulator. This reveals an upside of the data driven approach: it is no longer needed to provide an extra layer to the mechanistic model to link it with the process variables. The learning of distributional data can be extended to learning on conditional distributions, i.e. modelling several process manipulations on the particle distributions in series. This is learning the feature mapping between one (intermediate) distribution, e.g. the particle size distribution in the wetting zone, and the another distributions, e.g. the particle size distribution at the outlet of the granulator. Similar to the population balance model, a compartmental model can be constructed, describing the distribution transformations in every stage of the granulator.

Further, it is interesting to investigate the relation between the data-driven model in this work and the mechanistic population balance model in the previous work. The calibrated feature mapping could potentially provide insight in the physical phenomena governing in the twin-screw wet granulator by linking it with aggregation and breakage descriptions of the population balance model.

The results of this work are promising: using the kernel mean embedding, particle size distributions at the end of the granulator can be predicted. The model is fast to compute and does not require much computational resources to train. The speed of the model allows for future applications such as real time soft-sensors, model based control and optimal experimental design of the twin-screw wet granulator.

[1] Van Hauwermeiren, D., Verstraeten, M., Doshi, P., am Ende, M. T., Turnbull, N., Lee, K., ... Nopens, I. (2018). On the modelling of granule size distributions in twin-screw wet granulation: Calibration of a novel compartmental population balance model. *Powder Technology*, *341*, 116–125. https://doi.org/10.1016/j.powtec.2018.05.025

[2] Muandet, K., Fukumizu, K., Sriperumbudur, B., & Schölkopf, B. (2016). Kernel Mean Embedding of Distributions: A Review and Beyond. https://doi.org/10.1561/2200000060

553725 Breaking Force and Disintegration Prediction of Tablet Formulations Using Machine Learning

Conventional, quality-by-test methods for determining tablet breaking force and disintegration time usually involve destructive tests, which consume significant amount of time, labor and provide limited information. Recent advances in material characterization, statistical analysis and machine learning have provided multiple tools that have the potential to develop nondestructive, fast and accurate approaches in drug product development. In this talk, a methodology to predict the tablet breaking force and disintegration time of tablet formulations using nondestructive ultrasonics and machine learning tools will be presented. The inputs to the model include intrinsic properties of drug formulation as well as extrinsic process variables influencing the tablet during manufacturing. The model has been applied to predict tablet breaking force and disintegration time of tablets using small quantities of API and prototype formulation designs. The novel approach presented in this study is a step forward toward rational design of a robust drug product based on insight into the performance of common materials during formulation and process development. It may also help reduce development time, API usage and facilitate the development of a robust drug product.

569031 Development of Data-Driven and Hybrid Models for Continuous Pharmaceutical Manufacturing Lines Under Industry 4.0 Framework

Continuous manufacturing has been a critical research area in the pharmaceutical industry for the past decade[1]. With intensive collaboration among regulatory agencies, leading industrial players and academia, a large number of data-driven and first principle models have been developed to guide the design and simulation process[2]. Data-driven models attempt to capture the system behavior using only input and output data from the manufacturing line[3]. Even though they are accurate for specific datasets, the models are highly dependent on the choice of training data, and their description of process physical phenomena is not explicit[4]. First principle based models simulate the process with known mechanistic equations, but it is hard to model the complex behavior of the line completely[5]. Recent development in Industry 4.0 has led to a focus in cyber-physical system, internet of things, and smart manufacturing[6, 7], and combined with the quality by design initiative from the FDA, the volume of data generated for the entire manufacturing cycle increases dramatically, allowing for a better application of the generated data. As a data science tool, machine learning can be used for analyzing the large data set, and the resulting data-driven models can be used to enhance first principle models for the development of a hybrid model. This similar approach has been done in other areas like catalytic reactor modeling[8, 9] and HVAC system analysis[10].

As one of the three main routes of continuous pharmaceutical manufacturing, the continuous direct compaction (DC) line has mechanistic and semi-empirical models developed[2, 11]. In this study, an integrated data collection and data analysis framework is constructed for the DC line using Industry 4.0 concepts to facilitate the development of data-driven and hybrid models. After the implementation of the Industry 4.0 framework, both process and analytical data of the DC line can be centralized into a cloud-based platform, where users can retrieve and visualize all relevant manufacturing information in the system. With a large amount of data collected, various machine learning techniques like Support Vector Machine (SVM) and Artificial Neural Network (ANN) are applied to infer data-driven models of each unit operation and the entire flowsheet. The data-driven models developed can then be used as a gap-assessment tool to improve the previously developed mechanistic and semi-empirical models, leading to hybrid models. The wealth of data in the system are also used to test the applicability of both types of models. The standalone data-driven models can provide quick input-output correspondence in the absence of mechanistic models, but on top of it, the hybrid models can offer a layer of system understanding. The highlight of this work is to apply a data-driven approach to supplement the mechanistic models and to form hybrid models, which can be used to improve the predictability of the continuous pharmaceutical manufacturing system.

1. Ierapetritou, M.; Muzzio, F.; Reklaitis, G., Perspectives on the continuous manufacturing of powder-based pharmaceutical processes. *AIChE Journal* **2016,** *62* (6), 1846-1862.
2. Wang, Z.; Escotet-Espinoza, M. S.; Ierapetritou, M., Process analysis and optimization of continuous pharmaceutical manufacturing using flowsheet models. *Computers & Chemical Engineering* **2017,** *107*, 77-91.
3. Boukouvala, F.; Muzzio, F. J.; Ierapetritou, M. G., Dynamic Data-Driven Modeling of Pharmaceutical Processes. *Industrial & Engineering Chemistry Research* **2011,** *50* (11), 6743-6754.
4. Rogers, A.; Hashemi, A.; Ierapetritou, M., Modeling of Particulate Processes for the Continuous Manufacture of Solid-Based Pharmaceutical Dosage Forms. *Processes* **2013,** *1* (2), 67-127.
5. Boukouvala, F.; Muzzio, F. J.; Ierapetritou, M. G., Predictive modeling of pharmaceutical processes with missing and noisy data. *AIChE Journal* **2010,** *56* (11), 2860-2872.
6. Liao, Y.; Deschamps, F.; Loures, E. d. F. R.; Ramos, L. F. P., Past, present and future of Industry 4.0 - a systematic literature review and research agenda proposal. *International Journal of Production Research* **2017,** *55* (12), 3609-3629.
7. Zhong, R. Y.; Xu, X.; Klotz, E.; Newman, S. T., Intelligent Manufacturing in the Context of

Industry 4.0: A Review. *Engineering* **2017,** *3* (5), 616-630.
8. Azarpour, A.; N.G. Borhani, T.; R. Wan Alwi, S.; A. Manan, Z.; I. Abdul Mutalib, M., A generic hybrid model development for process analysis of industrial fixed-bed catalytic reactors. *Chemical Engineering Research and Design* **2017,** *117*, 149-167.
9. Luo, N.; Du, W.; Ye, Z.; Qian, F., Development of a Hybrid Model for Industrial Ethylene Oxide Reactor. *Industrial & Engineering Chemistry Research* **2012,** *51* (19), 6926-6932.
10. Park, S.; Ahn, K. U.; Hwang, S.; Choi, S.; Park, C. S., Machine learning vs. hybrid machine learning model for optimal operation of a chiller. *Science and Technology for the Built Environment* **2018**, 1-12.
11. Boukouvala, F.; Niotis, V.; Ramachandran, R.; Muzzio, F. J.; Ierapetritou, M. G., An integrated approach for dynamic flowsheet modeling and sensitivity analysis of a continuous tablet manufacturing process. *Computers & Chemical Engineering* **2012,** *42*, 30-47.

568217 Determination and Understanding of Dynamic Lead Lag between in-Line NIR Tablet Press Feed Frame and Off-Line NIR Tablet Measurements

INTRODUCTION

Process Analytical Technology (PAT) is crucial for the understanding, monitoring and control of (continuous) manufacturing processes and essential to move towards a Quality by Design (QbD) approach for pharmaceutical drug product development [1]. Different studies have evaluated near infrared (NIR) spectroscopy for blend potency determination inside the feed frame of a rotary tablet press and compared these results with off-line tablet potency measurements [2,3,4]. The final blend first passes the NIR probe before it undergoes additional mixing inside the feeding chamber and being filled into the dies of the tablet press. As a result, the in-line NIR response is always ahead of time and more intensive compared to the signal of the material inside the tablet cores (measured by offline analysis (e.g. NIR)). This phenomenon is defined as lead-lag. Although a time-shift or lead-lag has been observed between the in-line and off-line NIR response before [2], the effect of tablet press process parameters on the observed lead-lag was not investigated and no proposal was made on how to practically use the in-line NIR potency measurements for tablet diversion. Therefore, the aim of this study was to gain an in-depth understanding of the lead-lag between in-line NIR feed frame and off-line NIR tablet measurements and to propose an approach to support the use of in-line NIR feed frame measurements for product diversion and real-time release.

MATERIALS AND METHODS Materials

For this study, test formulations containing sodium saccharine as a surrogate API were used..

Blends containing different amounts of sodium saccharine were prepared using a three-dimensional mixer (Inversina, Bioengineering AG, Wald, Switzerland).

Methods

All experiments were performed using a Modul^{TM} P high speed rotary tablet press (GEA Pharma Systems, GEA Process Engineering, Halle, Belgium) equipped with ten 10 mm flat faced punches. All blends were compressed to a constant tablet weight of 375 mg. The NIR spectra were collected in reflection mode with a diode array spectrometer (SentroPAT FO, Sentronic, Dresden, Germany) positioned inside the feeding chamber of the feed frame, in very close proximity to the die filling station, thus monitoring blend potency immediately before the compression step [5]. Spectra were collected with an integration time of 7 ms, an averaging number of 200 spectra and a measurement cycle of 3s. A spectral range of 1100 – 2200 nm was used with a spectral resolution of 1 nm. The NIR spectra of the tablets were collected off-line by a Fourier transform near-infrared (FT-NIR) spectrometer (Antaris^{TM} II, Thermo Fisher Scientific, USA) in transmittance mode covering the range of 7501.74 to 11998.9 cm^{-1}, with a resolution of 8 cm^{-1} and 64 scans per spectra. In-line and off-line NIR quantitative models were developed based on measurements of blends and tablets, respectively, referenced using different known sodium saccharine concentrations. Paddle and turret speeds were varied according to a full factorial design of experiments (DoE), for both model calibration and validation. The NIR models were developed using the SIMCA software (Version 15, Umetrics, Umeå, Sweden).

In order to measure the lead-lag, two series of experiments were conducted. The first series of experiments were conducted by performing blend composition step-changes (90% - 110%) of sodium saccharine during tablet compression and, the second, by adding a spike of pure sodium saccharine into the feed chute above the tablet press feed frame. To enhance the understanding of the influence of process parameters during tablet compression on the observed lead-lag; turret speed, paddle speeds, paddle speed ratio, overfill level and feed frame design were varied according to a DoE. Other equipment parameters such as paddle wheel design, position of the NIR probe and amount of punches were kept constant.

Concentration versus time profiles were plotted and lead-lag was determined.

RESULTS AND DISCUSSION

An off-line univariate linear regression NIR model using the wavelength at 8801.52 cm^{-1} and an in-line PLS model using the spectral range between 1600 and 1750 nm were obtained. The two models had an RMSEP % of 0.40 and 0.44 respectively.

From the step change experiments, it was observed that the lead-lag decreases at higher paddle speeds due to the similar responses in blend potency between NIR and compressed tablets. This phenomenon can be explained by the higher mixing intensity inside the feed frame. Similarly, the lead-lag was observed to follow the same behaviour at higher paddle speeds, when conducting spike experiments. The turret speed and overfill level were found not having a significant influence on the lead-lag within the experimental process ranges studied.

An approach was developed to use current experimental findings for supporting future use of in-line NIR for in-process control (IPC), product diversion and real-time release.

REFERENCES

[1] Wahl, P. R., Fruhmann, G., Sacher, S., Straka, G., Sowinski, S., & Khinast, J. G. (2014). PAT for tableting: Inline monitoring of API and excipients via NIR spectroscopy. European Journal of Pharmaceutics and Biopharmaceutics, 87(2), 271–278.

[2] Pauli, V., Roggo, Y., Pellegatti, L., Nguyen Trung, N. Q., Elbaz, F., Ensslin, S., ... Krumme, M. (2019). Process analytical technology for continuous manufacturing tableting processing: A case study. Journal of Pharmaceutical and Biomedical Analysis, 162, 101–111.

[3] Sierra-Vega, N. O., Román-Ospino, A., Scicolone, J., Muzzio, F. J., Romañach, R. J., & Méndez, R. (2019). Assessment of blend uniformity in a continuous tablet manufacturing process. International Journal of Pharmaceutics, 560, 322–333.

[4] Šašić, S., Blackwood, D., Liu, A., Ward, H. W., & Clarke, H. (2015). Detailed analysis of the online near-infrared spectra of pharmaceutical blend in a rotary tablet press feed frame. Journal of Pharmaceutical and Biomedical Analysis, 103, 73–79.

[5] De Leersnyder, F., Peeters, E., Djalabi, H., Vanhoorne, V., Van Snick, B., Hong, K., ... De Beer, T. (2018). Development and validation of an in-line NIR spectroscopic method for continuous blend potency determination in the feed frame of a tablet press. Journal of Pharmaceutical and Biomedical Analysis, 151, 274–283.

568340 Digitalization in Active Process Control of Pharmaceutical Continuous Manufacturing of Oral Solid Dosage

Digitalization in Active Process Control of Pharmaceutical Continuous Manufacturing of Oral Solid Dosage

Qinglin Su, Alessandra E. Anderson-Lewis, Sudarshan Ganesh, Yasasvi Bommireddy, Marcial Gonzalez,
Gintaras V. Reklaitis, Zoltan K. Nagy

Davidson School of Chemical Engineering, West Lafayette, Indiana, United States, 47907

School of Mechanical Engineering, West Lafayette, Indiana, United States, 47907

Abstract: The emerging technology of continuous manufacturing in the pharmaceutical industry requires the use of active process control strategies in order to respond to disturbances or variations in raw material properties or process parameters and thus to maintain a state of control of in-process material or product critical quality attributes (CQAs). The development of these control strategies has been identified as a need in the recent United States Food and Drug Administration¡¯s (US FDA) draft guidance for industry: Quality Considerations for Continuous Manufacturing [[1]]. Purdue University, together with Rutgers University, had been funded by US FDA through the project of ¡°Industry 4.0 Implementation in Continuous Pharmaceutical Manufacturing¡±, to investigate the digitalization of active process control, as shown in **Figure 1**. One of the specific aims of our work is to develop nonlinear model predictive control (NMPC) strategies based on high degree understanding of process dynamics in the product and process for the continuous manufacturing of oral solid dosage. This advanced control strategy is capable of real-time prediction of the transition of controlled variables and optimization of the control moves in the manipulated variables within a receding time horizon, so as to achieving digital process operation. The actual implementation of NMPC in pharmaceutical continuous manufacturing lines is still in its infant stage: not much progress has been as yet reported [[2]]. The sluggish low-level control loops were regarded as a source of medium-level common risk in our recent study [[3]]. Hence, a second aim of our work is to monitor the process control system itself, using performance criteria or metrics of control design analysis to identify and diagnose the root causes of degrading control performance [[4]], especially, in the lower-level control loops of the control hierarchy. Instead of the overall control performance evaluation based on implicit final product qualities, each individual process control loop performance can be explicitly and digitally diagnosed. The proposed digitalized active process control strategy is demonstrated on the Purdue University Pharmaceutical Continuous Manufacturing pilot plant for oral solid dosage using the direct compaction route. Specifically, a nonlinear hybrid model based on classical Kawakita model for powder compressibility is used for NMPC implementation in the continuous tablet press unit operation. The real-time model parameter estimation for Kawakita model on the basis of joint data reconciliation and parameter estimation for sensor network was recently published in our group [[5]]. In addition, case studies of using PiControl Apromon for digitalized control performance monitoring and OSIsoft Pi System for digitalized process performance visualization are discussed. Finally, perspectives on process model maintenance and control performance monitoring will be given.

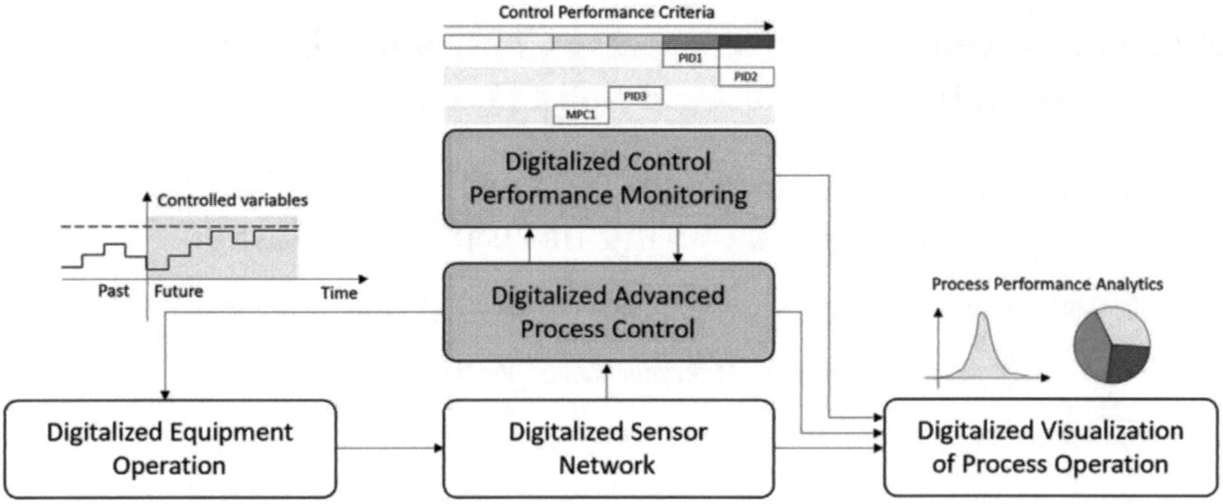

Figure 1. A schematic diagram of the digitalization in active process control.

References

1. United States Food and Drug Administration. *Quality Considerations for Continuous Manufacturing - Guidance for Industry (Draft Guidance)* 2019.

2. Mesbah A, Paulson JA, Lakerveld R, Braatz RD. Model predictive control of an integrated continuous pharmaceutical manufacturing pilot plant. *Organic Process Research & Development*. 2017;21:844-854.

3. Su Q, Ganesh S, Moreno M, et al. A perspective on Quality-by-Control (QbC) in pharmaceutical continuous manufacturing. *Computers and Chemical Engineering*. 2019;125:216-231.

4. Su Q, Moreno M, Giridhar A, Reklaitis GV, Nagy ZK. A systematic framework for process control design and risk analysis in continuous pharmaceutical solid-dosage manufacturing. *Journal of Pharmaceutical Innovation*. 2017;12:327-346.

5. Su Q, Bommireddy Y, Shah Y, et al. Data reconciliation in the Quality-by-Design (QbD) implementation of pharmaceutical continuous tablet manufacturing. *International Journal of Pharmaceutics*. 2019;563:259-272.

576091 Real-Time Monitoring and Model-Based Prediction of Product Properties in a Dropwise Additive Manufacturing Process for Pharmaceuticals

In recent years additive manufacturing (AM) technology has attracted considerable interest for its potential applications to on-demand manufacturing of dosage forms with individually adjusted doses, which is a necessity for implementation of precision medicine at the societal level, and for development of robust manufacturing platforms for early drug development or continuous manufacturing. Commensurate with the level of interest is the variety of different technologies based on two-dimensional (2D) or three-dimensional printing (3D) techniques, all of which present the common advantages of precision and flexibility in dosing inherent to AM processes. However, differences exist in the printing material requirements and in the achievable production rates for the process; both must be considered when proposing implementations in the various contexts (compounding in clinical or pharmacy setting; full-scale production). In the former, stringent requirements imply that the system would be suited to a narrow range of dosage forms, while the latter is an eminently practical consideration – if the production time for a single dose is on the order of minutes, then the system may struggle to meet demand.

Of the AM techniques, the system based on the drop-on-demand method (2D printing) for fluid deposition/dispensing has demonstrated the capacity to serve as flexible manufacturing platform for pharmaceutical dosage forms, with doses in amounts of 1-100 mg active ingredient produced from printing materials which ranged from drug solutions to drug-polymer melts and concentrated suspensions of micronized particles [1, 2, 3]. Most recently demonstrated was the development of the system for precision dispensing of micronized powders by processing them as particle suspensions, which, as a fluid-based method, avoids powder flow issues and may readily be adapted to new particulate materials using dimensional analysis [3, 4]. For dosage units which contain 100mg of active drug, production rates of up to 2000 doses per hour, with product RSD $< 2\%$, have been achieved with suspension processing (3, 4). This system uses the drop-on-demand (DoD) printing method to deposit drops on to a prepared substrate, wherein the content of each dose is determined by the number, volume and composition of drops; consequently, quick adjustments to dosage unit properties may be made by changing the deposited fluid volume, and/or composition.

From a process perspective, the DoD method presents a unique opportunity for monitoring of dose uniformity through the online measurement of each drop added to a given substrate; this enables model-based estimation of the drug product properties (e.g. active ingredient and/or excipient amount), which when combined with statistical modeling, can be used for real-time quality control purposes. To achieve this, one must have real-time measurements of drop volume and concentration in the printing fluid just upstream of the drop ejection point. As demonstrated in previous works, drop volume measurements may be obtained using an online image acquisition system and image analysis [1, 2], with corrections necessitated by particle effects made using a Bayesian statistical model [5]. In-line concentration measurement is complicated by the small scale of the process, which precludes the use of some sensors, and by the variety of process fluids for which the sensor must have the capacity to accurately measure. For pure fluids there are few limitations imposed by physical constraints, but for more interesting printing materials such as concentrated particulate suspensions, choices for sensing technology are limited to ultrasound velocity (correlation with particle fraction through modified Urick equation) or Raman spectroscopy.

In this work we investigate these potential methods for in-line concentration monitoring for particle suspensions within the dropwise additive manufacturing system. Combining the real-time drop volume and inline concentration measurements, we are able to make model-based predictions for properties of dosage units produced during manufacturing runs for doses of variable potency; comparison of model predictions with offline analysis of the dosage units is then presented. Through the use of a statistical approach for model-based prediction, and, consequently, quantification of uncertainty in the dose content, additional modeling work enables making a decision as to whether each dose should be released to the patient, formulated based on minimization of the expected loss in the Bayesian context. In this way, an important step towards the realization of precision medicine is made, namely, model-based real-time release of doses (at least, clinical/compounding settings).

References

1. Hirshfield, L., Giridhar, A., Taylor, L. S., Harris, M. T., & Reklaitis, G. V. (2014). Dropwise additive manufacturing of pharmaceutical products for solvent-based dosage forms. Journal of pharmaceutical sciences, 103(2), 496-506.
2. Içten, E., Giridhar, A., Taylor, L. S., Nagy, Z. K., & Reklaitis, G. V. (2015). Dropwise additive manufacturing of pharmaceutical products for melt-based dosage forms. Journal of pharmaceutical sciences, 104(5), 1641-1649.
3. Radcliffe, A. J., Hilden, J. L., Nagy, Z. K., & Reklaitis, G. V. (2019). Dropwise Additive Manufacturing of Pharmaceutical Products Using Particle Suspensions. Journal of pharmaceutical sciences, 108(2), 914-928.
4. Radcliffe, A. J., & Reklaitis, G. V. (2017). Dropwise additive manufacturing using particulate suspensions: feasible operating space and throughput rates. In Computer Aided Chemical Engineering (Vol. 40, pp. 1207-1212). Elsevier.
5. Radcliffe, A. J., & Reklaitis, G. V. (2018). Bayesian estimation of product attributes from on-line measurements in a dropwise additive manufacturing system. In *Computer Aided Chemical Engineering* (Vol. 43, pp. 1243-1248). Elsevier.

578800 Process Control Strategy Development for a Challenging Drug Powder Blending Process Via Real-Time Process Analytical Technology (PAT) Monitoring and Prediction

Powder blending is one of the key unit operations, for which the performance and outcome of the unit operation could significantly impact the critical quality attributes of oral solid dosage forms such as, assay, content uniformity (CU) or uniformity of unit dose. The significance of powder blending becomes more profound for cohesive drugs and narrow therapeutic index (NTI) drugs. The following three aspects will be covered in this presentation. 1) A laboratory case study with Near-Infrared Spectroscopy (NIRS) real time process monitoring and chemometrics model prediction for a high-shear granulator powder blending process consisting of the NTI drug, Phenytoin Sodium (10% w/w) and two common pharmaceutical excipients. 2) Proper choice of the data-preprocessing algorithms to facilitate scale-up using available publicly available information. 3) Some common practices of mitigating risks to blending uniformity and content uniformity as applied to batch processing and continuous processing. This presentation will demonstrate that from both public health and process engineering perspectives, real time process monitoring, and prediction can play a vital role for understanding, controlling, and risk mitigation of a challenging drug powder blending process.

565140 Advanced Control Strategies for a Continuous Tableting Line Via Hot Melt Extrusion

From a processing point-of-view the shift to continuous manufacturing mode in pharmaceutical industry is supported by the fact that most common unit operations are continuous machinery by design, such as feeding, tableting or granulation. In-line monitoring tools to supply real-time process data and to trigger control actions are available. One remaining challenge is the establishment of real-time control concepts that enable a fully automated, yet still flexible process, capable to cope with raw material variability and process events.

In this study a feedback control strategy for a continuous tablet manufacturing line has been developed. The line consists of a hot melt extrusion and pelletizing system to produce pellets of API and polymer. The pellets are transported to a direct compaction line, blended with a pre-mix of excipients and compacted to tablets. A PAT strategy to monitor critical quality attributes of intermediates and product was implemented. Different control loops covering quality and mass flow rate were designed. PID controllers were applied for API concentration in the melt and the blend, strand temperature and strand diameter corresponding to pellet thickness. Model predictive control was used to coordinate the mass flow rates of the different unit operations. The presentation highlights an assessment of different control concepts for the line, in which the tablet press, the extruder or none act as master unit.

565007 Exploring the Robustness of Terahertz-Based at-Line Porosity Measurements for Realising a Non-Destructive Dissolution Assay of Tablets during Manufacturing

INTRODUCTION

To achieve a continuous downstream manufacturing process with in-process quality monitoring and real-time release testing capabilities, different process analytical technologies (PATs) such as laser scattering methods, near infrared (NIR) spectroscopy, and Raman spectroscopy can be deployed to, for example, measure the particle size of raw materials and powder blends as well as to detect the content uniformity of powder blends and tablets. However, the use of the above-mentioned PAT methods to directly measure and quantify the bulk properties of tablets, e.g. porosity, that govern the mass transport processes and mechanical changes in tablets during disintegration and dissolution, are limited. For example, NIR methods to quantify tablet hardness typically probes surface properties only and these measurements are not necessarily representative of the bulk tablet property, hence the need for suitable analytical technologies, such as terahertz time-domain spectroscopy (THz-TDS), that can reliably measure the porosity in transmission of tablets [1, 2]. The total porosity, f, is a more meaningful metric to measure in order to predict the dissolution performance of the tablet as it can be directly linked to the mass transport of liquid into the tablet during disintegration through physical models whilst also being predictive of the mechanical strength of the tablet (f $1/\rho$, with ρ density).

We recently developed a terahertz spectroscopy method to directly measure tablet porosity without the need for any chemometric model [1-4]. This study showcases the robustness of this method for measuring tablet porosity for a range of geometries and formulations, specifically several batches of flat and biconvex tablets with either ibuprofen or indomethacin as the active pharmaceutical ingredient (API). The study further investigates the effect of debossing on the extracted terahertz porosity. To explore the suitability of the method for possible in-line application we determined the interplay between signal averaging and error in porosity in order to determine the minimum measurement time using present commercial terahertz spectroscopy equipment. Finally, we investigate the correlation between porosity measured by terahertz spectroscopy and the mean dissolution time of immediate release tablets.

METHOD AND MATERIALS

Several batches of flat and biconvex tablets with different porosities within the range of 6 - 25% were directly compressed using a compaction simulator (HB50, Huxley Bertram Engineering Ltd, UK). The production-scale tablet press (Fette 2090 B) compression profile, with a compression speed of 60 rpm, was used for compressing all the batches. The targeted porosity for each batch was achieved by keeping the weight of all tablets at 400 mg and adjusting their thickness. The formulation used was composed of microcrystalline cellulose (39.1% w/w), lactose anhydrous Supertab21AN (46.9% w/w), croscarmellose sodium (3.0 %w/w), Magnesium stearate (1.0% w/w), and ibuprofen/indomethacin (10.0% w/w). The proportions of the various powders were weighed and properly mixed using a Turbula mixer, T2F (Willy A. Bachofen AG, Switzerland). The blending process lasted for 11 mins with a speed of 32 rpm. Terahertz time-domain measurements of all the batches were acquired using a TeraPulse 4000 (TeraView Ltd., Cambridge, UK). The porosity was extracted from the measured terahertz effective refractive index using Bruggeman's effective medium theory [1,5].

RESULTS AND CONCLUSIONS

The comparison between the terahertz porosity values measured from the debossed tablets and those without debossing showed that the presence of the debossing has a negligible effect on the results. Both flat-faced and biconvex tablets exhibited an excellent linear correlation between the nominal and the terahertz porosity. Our results show the robustness of the terahertz approach for tablet porosity measurement over a wide range of geometries. We also observed that the root mean squared error (RMSE) of the linear correlation between the nominal and measured porosity stayed < 0.65% even when reducing the signal averaging to less than one second (for comparison one minute acquisition time results in a RMSE of around 0.4%). The equipment available today therefore would be suitable for at-line porosity measurements with a potential throughput of 1 tablet per second and future improvements to the terahertz equipment appear to make on-/in-line applications possible. Finally, an excellent correlation was observed between the porosity and the mean dissolution time, which highlights the suitability of this method for non-destructive prediction of the dissolution behaviour of pharmaceutical tablets during manufacturing.

ACKNOWLEDGEMENTS: The authors would like to thank Innovate UK (Project 5804) for the funding provided.

REFERENCES

1. Bawuah, P.; Pierotic Mendia A.; Silfsten, P.; Pääkkönen P.; Ervasti, T.; Ketolainen, J.; Zeitler, J.A. and Peiponen, K.E. Detection of porosity of pharmaceutical compacts by terahertz radiation transmission and light reflection measurement techniques. Int. J. Pharm. 465, 70-76 (2014).
2. Bawuah, P.; Silfsten, P.; Ervasti, T.; Ketolainen, J.; Zeitler, J.A. and Peiponen, K.E. Non-contact weight measurement of flat-faced pharmaceutical tablets using terahertz transmission pulse delay measurements. J. Pharm. 476, 16-22 (2014).
3. Markl, D.; Strobel, A.; Schlossnikl, R.; Bøtker, J.; Bawuah, P.; Ridgway, C.; Rantanen, J.; Rades, T.; Gane, P.; Peiponen, K.-E. and Zeitler, J. A. Characterisation of pore structures of pharmaceutical tablets: A review. J. Pharm. 538, 188-214 (2018).
4. Markl, D.; Sauerwein, J.; Goodwin, D.J.; van den Ban, S. and Zeitler, J. A. Non-destructive determination of disintegration time and dissolution in immediate release tablets by terahertz transmission measurements. Res. 34, 1012-1022 (2017).
5. Markl, D.; Wang, P.: Ridgway, C.; Karttunen, A-P.; Chakraborty, M.; Bawuah, P.; Pääkkönen, P.; Gane, P.; Ketolainen, J.; Peiponen, K.-E. and Zeitler, J. A. Characterization of the Pore Structure of Functionalized Calcium Carbonate Tablets by Terahertz Time-Domain Spectroscopy and X-Ray Computed Microtomography. Pharm. Sci. 106, 1586-1595 (2017).

555745 Commercial Control Strategy Development for a High Shear Wet Granulation Process

During development of an oral drug product commercial-scale manufacturing process using high shear wet granulation, challenges were encountered due to a change in the drug substance manufacturing process and resulting API properties. These challenges required a redesign of the manufacturing process and implementation of a control strategy to ensure/demonstrate equivalency of the granules, that when processed downstream, achieved the desired drug product critical quality attributes. Appropriate PAT tools were used to facilitate development at the commercial scale. The final control strategy uses real-time data monitoring, complex recipe control and formal in-process control (IPC) requirements.

All authors are employees of AbbVie and may own AbbVie stock. AbbVie sponsored and funded the study; contributed to the design; participated in the collection, analysis, and interpretation of data, and in writing, reviewing, and approval of the final publication.

581347 Chemical Engineering Innovations for a Renewable Economy

Chemical engineering evolved in the twentieth century with rise in the use of fossil resources and myriad of products derived from then. An array of these products required great innovations and made huge impact on the human race, and such examples include urea, transportation fuels, polymers, water purification chemicals, etc. Going forward as we make transition from fossil resources to renewable resources such as solar energy, wind and sustainably available biomass, we again have great opportunities for innovation that would not only make great impact but redefine chemical engineering. My talk will focus on my own experiences with challenges and opportunities presented to us as chemical engineers. I will share an overall perspective on the use of biomass as sustainable source of carbon, harvesting solar energy in conjunction with growing food without any land constraint, and creating solution processed solar cells that utilize chemical engineering principles along with knowledge from other engineering disciplines.

581350 Engineering Materials for Clinical Application

Dr. Schmidt's research has resulted in numerous patents that are licensed to companies, including AxoGen, Alafair Biosciences, Smith and Nephew, and Siluria Technologies. Comments for this panel will focus on peripheral nerve applications and one key success that is the foundation for the Avance Nerve Graft from AxoGen. In addition, along the way to uncovering strategies for nerve regeneration, Dr. Schmidt and her team also discovered a biomaterials processing approach that led to a separate application of protecting tendon's after surgery, resulting in the VersaWrap Tendon Protector from Alafair Biosciences, a start-up company in Austin, Texas.

Nerve Repair: Dr. Schmidt's technology for processing decellularized nerve tissue was licensed to AxoGen in Alachua, Florida (http://www.axogeninc.com) for the AVANCE™ nerve graft. Many tens of thousands (~100,000) of AxoGen's AVANCE™ grafts have been implanted in patients at hundreds of hospitals globally.

Post-surgical Wound Management: Using a novel in situ crystallization processing technique, Dr. Schmidt's team created mechanically robust hyaluronic acid-based hydrogel films that can be easily handled, sutured, and delivered laparoscopically. This research is the basis of Alafair Biosciences, a start-up company in Austin, TX (http://www.alafairbiosciences.com), which is run by Dr. Schmidt's PhD graduate and a neurosurgeon collaborator. Alafair's first product, VersaWrap™ Tendon Protector, has been implanted in more than 500 patients since December 2017.

564074 Continuous Synthesis, Crystallization, and Isolation of Amoxicillin

Continuous synthesis, crystallization, and isolation of amoxicillin

Matthew A. McDonald[*], Ronald W. Rousseau[*], Martha A. Grover[*], and Andreas S. Bommarius[*]

[*]*School of Chemical & Biomolecular Engineering, Georgia Institute of Technology, Atlanta, USA*

Keywords: Continuous manufacturing, ß-lactam antibiotics, reactive crystallization, penicillin G acylase

Amoxicillin is the highest-volume antibiotic in the world; production is in excess of 10,000 tons per annum. State-of-the-art production facilities use penicillin G acylase (PGA) immobilized on beads to first produce crude amoxicillin from the precursors 6-aminopenicillanic acid (6-APA) and 4-hydroxy phenylglycine methyl ester (4-HPGME), then purify the amoxicillin by crystallization(s), before finally readying it for formulation. Some of these steps have recently been made continuous, but the process is still characterized by a series of distinct batch operations. In this study, amoxicillin trihydrate, ready for formulation, is produced continuously in a single reactive crystallization[1]. The combination of reaction and crystallization has been shown to increase the yield of the PGA-catalyzed reaction and improve productivity of the entire process [2]. Process analytical technology is used to monitor the process and operate a segregation line, allowing aberrant material to be segregated from the product line without significant loss of material. The potential for model based feedback and feed-forward control is also discussed.

[1] McDonald, M. A., Bommarius, A. S., Rousseau, R. W., and Grover, M. A. (2019) Continuous reactive crystallization of ß-lactam antibiotics catalyzed by penicillin G acylase. Part I: Model development, *Computers & Chemical Engineering*, 123, 331-343

[2] McDonald, M. A., Bommarius, A. S., Grover, M. A., and Rousseau, R. W. (2019) Continuous reactive crystallization of ß-lactam antibiotics catalyzed by penicillin G acylase. Part II: Case study on ampicillin and product purity, *Computers & Chemical Engineering*, ACCEPTED

565997 Continuous Centrifugal Extraction for Intermediate Recovery from a Biocatalysis Reaction

Continuous processing, or flow chemistry, has started to transform the way pharmaceuticals are made over the past decade. The flexibility, reduced footprint, enhanced control strategies, and process intensification capabilities offered by continuous unit operations are a convincing case for transitioning certain processes from batch to flow. Everything from reactions, liquid-liquid extraction (LLE), purification, crystallization, and tableting processes are being developed in flow with the goal of achieving new chemistries not possible in batch as well as increasing economic savings and safety. Pfizer is working to develop continuous processes through a concerted effort at all levels across our organization, from R&D through full-scale commercial production.

Recent work that utilized continuous LLE coupled with phase separation enabled via centrifugal extraction will be presented. Performing reactions in flow is not sufficient if the subsequent work-up steps are done in batch. This is particularly true for gravity separation when phases can take days or weeks to separate, or when stable emulsions can form a thick rag layer at the liquid-liquid interface. This work will describe the use of a lab-scale centrifugal extractor to separate a post-enzymatic reaction mixture consisting of 2.5 wt% enzyme/protein, water, MTBE, and reaction intermediate.

This process was chosen since this reaction mixture was prone to forming emulsions and took several days to separate via gravity separation. The batch process also involved several protein pre-treatment steps, including low pH denaturation, addition of celite or carbon filter aid, and coarse filtration, before performing the extraction and separation at an elevated pH. The intermediate was sensitive to both low and high pH conditions, which could only be mitigated via sequestering the intermediate into the organic phase, preferably without significant agitation.

Three processing routes were trialed, two which eliminated one or more pre-treatment steps and one which maintained the same pre-treatment as the batch route. Successful extraction and phase separation was demonstrated for all three routes, including the most challenging route which included no pre-treatment. The results for the most successful route were scaled up from our lab unit, which had a 2.2 mL bowl volume (up to 33.3 mL/min throughput), to a larger pilot plant sized unit with a bowl volume of 80 mL (up to 1000 mL/min throughput). A discussion on the scale-up parameters and throughput capabilities of this technology will also be discussed.

571441 Continuous Mixing/Washing of an API Suspension in a Precipitating Environment

Continuous manufacturing has already been widely implemented in secondary pharmaceutical production in recent years. Efforts in primary manufacturing have also been made, but the linkage of the synthesis process with drug product manufacturing is still mostly completed in a batch process. Washing, filtration and drying are crucial process steps linking these production routes. Special consideration must be given to the washing and drying of crystals, which have been carefully engineered during crystallization of the active pharmaceutical ingredient (API). Particle size and morphology can easily be altered during a washing or drying step. Hence, it is very important to avoid any undesired side reactions during these processes. Continuous washing is challenging and only a few processes have been developed. In the continuous processing of API suspensions, both equipment fouling and the quality of the finished product are a challenge.

This work focuses on the application of ultrasound (US) to improve the performance of a continuous washing process in a precipitating environment. Washing in this continuous regime is carried out by dilution/mixing of the suspension with an anti-solvent, created by a filtration step to once again concentrate the suspension. These steps are repeated until the desired purity is reached. Optimized process parameters include process performance in terms of process time until blockage, avoiding changes in particle size distribution, as well as total amount of dry product produced in a single run.Ibuprofen 25 and GranuLac® 230 were used as representative model substances in different mass fractions. The corresponding anti-solvent was used as washing agent. The process of mixing the liquid feeds was carried out in an US-chamber (Hielscher Ultrasonics GmbH, Teltow, Germany) under constant US input, at various US-levels.

This work shows that disturbance free process times and efficiency can be increased by using a process chamber with ultrasound. Furthermore, it was possible to avoid blockage or fouling of the process, enabling blockage free test runs of up to 24 hours. Particle size of the product was not changed and no agglomerates were formed. This facilitates a large step forward in the effort to link primary and secondary manufacturing lines.

566762 Multistage, Multi-Zones Antisolvent-Cooling Crystallisation of a Proprietary API: Comparison of Two Continuous Crystalliser

The study aimed to establish parallel antisolvent-cooling crystallisation of a proprietary active pharmaceutical ingredient (API) in (5-stages antisolvent additions and three temperature zones) continuous oscillatory baffled crystalliser (COBC) and mixed-suspension mixed-product removal (MSMPR) crystalliser with 1000 and 180 mL(s) for each stage respectively. Each stage comprises of known solvents-API compositions based on the provided solubility profile. Process platforms used were off shelf NITECH-COBC and CMAC-MSMPR multistage crystalliser. The crystallisation solvent mixture is a blend of two solvents namely – ethanol, tetrahydrofuran (THF) and water as antisolvent. The throughput in both systems was based on the matched residence time of approximately 6 hr. The three temperature zones of 40 °C (stages 1 – 3), 30 °C (stage 4) and 20 °C (stage 5) were spread across the stages as indicated. While this worked effectively with MSMPR, a linearised gradient was adopted for the COBC to minimise fouling at interstage nodes – a non-optimised condition. Temperatures, particle counts, chord length distributions, feed, seed and antisolvent flow rates, and blockages were consistently monitored throughout the continuous operations.

At matched residence times for the two crystallisers, the particle size distribution obtained from the COBC was bigger having enhanced flow properties, less agglomeration propensity, higher yield and lower surface area compared to MSMPR. In addition, slightly reduced impurity rejection was observed in the COBC; meanwhile, the MSMPR resulted in lesser yield but higher impurity rejection. In terms of platform operations, more interventions were required to keep the COBC running with significant impact on process steady-state and potential shutdown. Fouling and encrustation were observed in both systems but could be reduced if processing conditions are optimised for COBC. In conclusion, the COBC which was primarily built for flexibility and relatively new in terms of applications resulted in better crystal attributes and higher yield compared to the MSMPR platform which is more established in terms of existing knowledge and operations, and more plant-ready. Optimisation of the process conditions in the COBC is strongly encouraged coupled with the redesign of the seed transfer line to prevent blockages.

575961 Continuous Crystallization Using Mixed Suspension Mixed Product Removal (MSMPR) Crystallizers and Oscillatory Baffled Crystallizers (OBC)

Emerging technologies, such as continuous manufacturing (CM), have been trending in the pharmaceutical industry in the past few years[1-2]. Continuous crystallization is an important step to achieve an end-to-end CM process. Currently, there are two popular approaches for continuous crystallization processes: sequential mixed suspension mixed product removal (MSMPR) crystallizers[3], and plug-flow oscillatory baffled crystallizer (OBC)[4]. Combined with our experience in developing continuous crystallization processes, we will discuss our understanding and lessons learned regarding the two systems from the perspective of risk assessment, process dynamics, implementation of PAT tools/methods, and the related control strategies. *This presentation only reflects the views of the authors and should not be construed to represent FDA's views or policies

References:

1. Lee, S. L.; O'Connor, T. F.; Yang, X.; Cruz, C. N.; Chatterjee, S.; Madurawe, R. D.; Moore, C. M. V.; Yu, L. X.; Woodcock, J., Modernizing Pharmaceutical Manufacturing: from Batch to Continuous Production. *J Pharm Innov* **2015**, 1-9.
2. Adamo, A.; Beingessner, R. L.; Behnam, M.; Chen, J.; Jamison, T. F.; Jensen, K. F.; Monbaliu, J.-C. M.; Myerson, A. S.; Revalor, E. M.; Snead, D. R., On-demand continuous-flow production of pharmaceuticals in a compact, reconfigurable system. *Science* **2016,** *352* (6281), 61-67.
3. Acevedo, D.; Peña, R.; Yang, Y.; Barton, A.; Firth, P.; Nagy, Z. K., Evaluation of mixed suspension mixed product removal crystallization processes coupled with a continuous filtration system. *Chemical Engineering and Processing: Process Intensification* **2016,** *108*, 212-219.
4. Lawson, S.; Steele, G.; Shering, P.; Zhao, L.; Laird, I.; Ni, X. W., Continuous Crystallization of Pharmaceuticals Using a Continuous Oscillatory Baffled Crystallizer. *Org. Proc. Res. Devel.* **2009,** *13*, 1357-1363.

576354 Risk Assessment of Encrustation in Continuous Pharmaceutical Crystallization Process – a Case Study

Encrustation is a risk factor that can cause product and process failure in continuous crystallization processes. Various risk mitigation strategies proposed in the literature, such as coating of crystallizer walls, use of additives to control encrustation kinetics, and periodic steady state operation show promising results in delaying or preventing encrustation.[1-3] Due to increased interest in the use of continuous crystallization in industrial applications, it is important to understand this risk factor further. Recent developments on dynamic models, mechanisms and risk factors of encrustation in continuous crystallization processes will be discussed. Methodologies for detection, mitigation, or control of encrustation will be evaluated. Cooling crystallization of carbamazepine will be used as a case study to evaluate detection methodologies (visual, process analytical technology, etc.). Various design and control strategies to mitigate the encrustation risk are also proposed. Appropriate control strategies should be implemented during continuous crystallization to avoid the impact of encrustation on drug substance quality.

1. Wu, T.; Sun, Y.; Li, N.; de Villiers, M. M.; Yu, L., Inhibiting surface crystallization of amorphous indomethacin by nanocoating. *Langmuir* **2007,** *23* (9), 5148-5153.
2. Powell, K. A.; Saleemi, A. N.; Rielly, C. D.; Nagy, Z. K., Monitoring Continuous Crystallization of Paracetamol in the Presence of an Additive Using an Integrated PAT Array and Multivariate Methods. *Organic Process Research & Development* **2016,** *20* (3), 626-636.
3. Koswara, A.; Nagy, Z. K., On-Off Feedback Control of Plug-Flow Crystallization-A Case of Quality-by-Control in Continuous Manufacturing. *IEEE Life Sciences Letters*

571017 Enhanced Dissolution Performance of Amorphous Solid Dispersions By Size Reduction

Amorphous solid dispersions (ASDs) are widely used in pharmaceutical industries in order to enhance the drug absorbance, solubility, dissolution rate. Although the ASDs have higher solubility, the corresponding dissolution rate naturally depends on the nature of the polymeric matrix and particle size. Unfortunately, particle size reduction promotes recrystallization rate, because it increases the available surface areas of the ASDs to environment, both during storage and during exposure to dissolution medium. The effect of particle size on the dissolution behavior of amorphous griseofulvin (GF) particles was investigated using 0-50% Kollidon® VA 64 as a polymeric crystallization inhibitor. Results show that both the final dissolved GF concentration, and the dissolution rate of GF amorphous solid dispersions (ASDs) are inversely proportional to the particle size. The solution concentrations for the smallest (45-75 μm) size group with different polymer loading at 2 hours were 1.5 to 7 and 1.3 to 4 times higher than those for the largest (250-355 μm) group under non-supersaturating and supersaturating conditions, respectively. The dissolution rate of GF ASDs with 50% polymer loading for the finest group was 2.7 times higher than for the largest group under supersaturating conditions. The individual rates of dissolution and recrystallization were assessed through surface concentration and Avrami recrystallization rate kinetics. That indicated that particle size reduction enhanced ASD drug loading by reducing the amount of polymer necessary as finest size ASDs initially dissolve faster negating their higher recrystallization rate. Kollidon® VA 64 at 30% loading was sufficient to inhibit the GF recrystallization. Thus the combination of particle size reduction and recrystallization inhibition is effective for improving the dissolution behavior of GF ASDs.

574108 Modeling Particle Formation in Spray Drying of Amorphous Solid Dispersions

In this presentation we present a mathematical model of the various concurrent physical phenomena that a polymer solution with an Active Pharmaceutical Ingredient (API) undergo prior to forming an amorphous solid dispersion particle, such as atomization, co-solvent evaporative dynamics, propensity for phase separation and shell formation.

The numerical model is benchmarked against single-droplet drying data obtained with a suspended droplet apparatus indicating that the model captures, to a first approximation, the dominant physical phenomena (see Figure 1).

The motivation of the work is to support the development of Amorphous Solid Dispersion powders of poorly water soluble APIs in hydrophilic polymeric matrices which is a commonly used strategy to increase the solubility and dissolution rate of oral-dosage forms. However, successful development of amorphous solid dispersions for optimal performance and stability depends on both the formulation (e.g. drug load, polymer system and surfactants) and on the manufacturing technology/process conditions. Optimizing the formulation and the process requires extensive laboratory work, which is often compromised by time and cost constraints. Therefore, the ability to perform virtual experiments is highly advantageous and cost-effective.

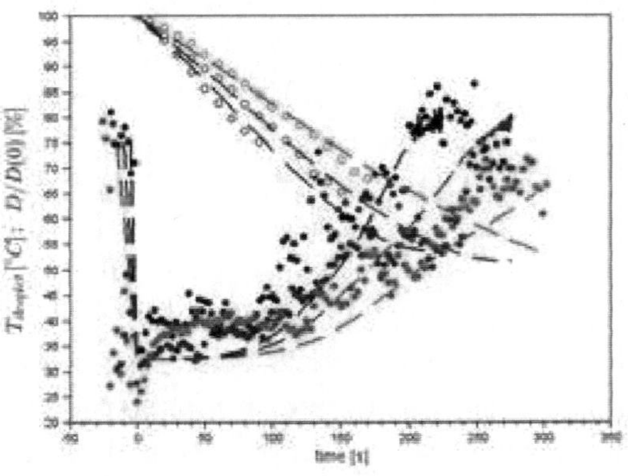

Figure 1: Model prediction (dashed-line) benchmarked against experimental data (filled circles represent Temperature measurements, whereas hollow circles represent droplet size).

*Vulcanus in Japan-Europe program for supporting Mr. Masahiro Nakai's internship

573159 Analysis of the Microstructure of Particles Obtained By Evaporating Acoustically Levitated Single Droplets Using X-Ray Computed Tomography

Analysis of the microstructure of particles obtained by evaporating acoustically levitated single droplets using X-ray computed tomography

Hassan Abdullahi [1], Christopher L. Burcham [2] Thomas Vetter [1*]

[1] *School of Chemical Engineering and Analytical Science, University of Manchester, Manchester, UK;*

[2] *Eli Lilly & Company, Indianapolis, IN, USA*

thomas.vetter@Manchester.ac.uk

The flowability and tabletability of powders is an important consideration in the manufacturing of pharmaceutical products. These properties can be successfully tuned by crystallising particles of a desired morphology, or, more flexibly, by engineering the microstructure of the individual particles in a spray drying process it can be tuned by altering the formulation and process conditions [1][2]. Therefore, spray drying offers a unique opportunity to design tailor made particles in a single step process involving simultaneous droplet drying and particle formation. However, carrying out an in-depth investigation of particle morphology and microstructure directly on the scale of a typical spray dryer (that contains millions of droplets at any time) is infeasible. Instead, single droplet studies are often used to understand how drying conditions affect the microstructure of the resulting dried particles [3][4]. In the present work, an acoustic levitation approach is used to suspend single droplets in an atmosphere of controlled relative saturation and temperature (Figure 1). Through a camera and automated image analysis, we study the droplet drying dynamics, as well as the macroscopic morphology of the resulting particle. We have further applied Raman spectroscopy and scanning electron microscopy to study the solid state form, as well as the surface characteristics. By applying X-ray computed tomography, we have elucidated the internal microstructure of the formed particles and have uncovered a fascinating variety of structures, depending on the mixture dried (we have investigated combinations of different solutes, solvents, and polymers acting as excipients), as well as the processing conditions (temperature and relative saturation of nitrogen used as drying gas). The results can be correlated with formulation and process conditions to produce drying maps for different systems.

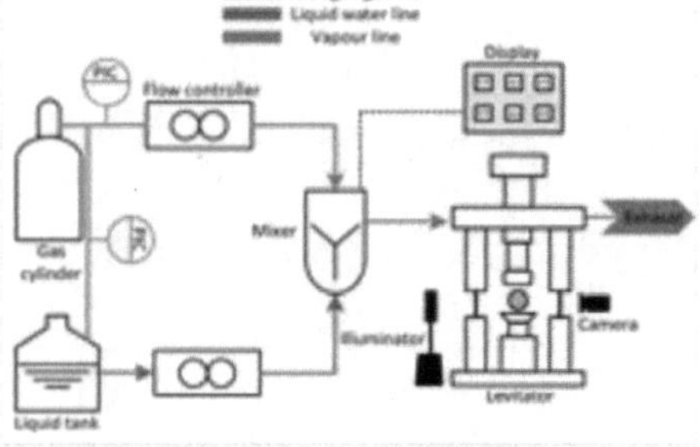

Figure 1: Suspension of a droplet in a controlled atmosphere using an acoustic levitator.

The droplet drying history during evaporation is shown in Figure 2a for an aqueous formulation of mannitol (4.5 wt%) and PVP (0.9 wt%). In the first drying stage, the diameter of the droplet decreases as solvent is evaporated from the droplet surface. In the second drying stage, a shell forms and evaporation occurs through interstices in the shell. The evaporation rate hence reduces significantly due to the added resistance of the crust. When presented as a plot of normalised squared diameter D^2/D_0^2 against time, the two drying regimes can be clearly identified. We have performed such characterisation for a variety of process conditions and the particles obtained can be subsequently analysed to provide useful information on particle properties such as size, shape, porosity, roughness and crystal structure. Figure 2b and 2c shows exemplary images of particles obtained using SEM and micro X-ray computed tomography. By investigating the droplet drying dynamics and resulting particle microstructure, for a variety of mixtures and operating conditions enables us to gain deep insight into the process behaviour and the properties of the resulting particles.

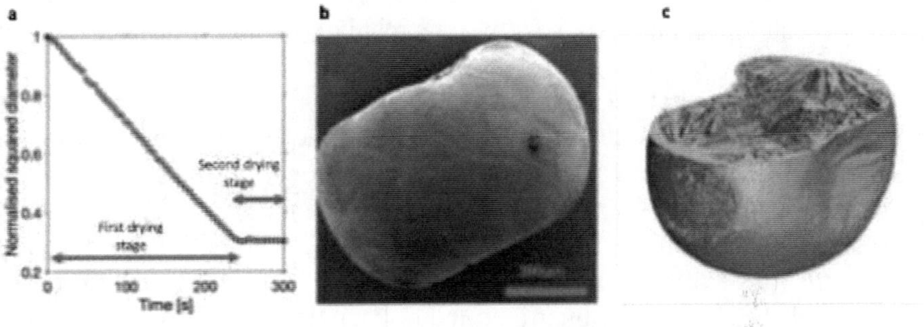

Figure 2: **a)** Droplet drying history and particle obtained from drying an aqueous formulation containing 0.9%w/w polymer, 4.5%w/w mannitol at 60C and 0% relative humidity **b)** SEM image **c)** cut-away surface from micro X-ray tomography.

References

[1] Carver, K. M., & Snyder, R. C. (2012). Ind. Eng. Chem. Res, *51*(48), 15720.

[2] Mšnckedieck, M., Kamplade, J., Fakner, P., Urbanetz, N. A. (2017). *Drying Technology*, *35*(15), 1843.

[3] Osman, A., Goehring, L., Patti, A., Stitt, H., & Shokri, N. (2017). Ind. Eng. Chem. Res, *56*(37), 10506.

[4] Sugiyama, Y., Larsen, R. J., Kim, J., & Weitz, D. A. (2006). Polymer Suspensions, (8), 6024.

572842 Decoupling Temperature and the Kinetics of Dye Conversion Via Scalable Nano-Encapsulation in a Thermal-Responsive Media

Chemical treatment processes in which substrates are exposed and interact with chemicals require quality control to ensure process specifications such as temperature, dosage, and duration are satisfied. One approach for quality control is the use of colorimetric dyes which undergo chemical reactions in response to an applied treatment, thereby shifting their absorbance spectrum. When incorporated in a process run alongside samples, operators can observe these post-treatment color changes as qualitative success metrics. While this approach is simple to implement, the reliance on the kinetics of dye conversion presents challenges including dye aging during storage which can lead to false positive results as well as the inability to decouple the effects of dosage and temperature on the final result.

We seek to overcome this disadvantage by decoupling the kinetics of dye conversion from temperature through the implementation of a wax matrix surrounding dye nanocapsules. The wax matrix serves as a barrier which melts and exposes dye-encapsulated nanoparticles at a specific melting temperature, commensurate with that of the treatment process. To form dye nanocapsules directly in wax, the versatility of flash nanoprecipitation (FNP), a scalable nanoparticle fabrication technique, is expanded. In this work, an organic solution of dye and polymeric stabilizer is impinged against a molten wax to create turbulent mixing, yielding the super-saturation, precipitation, and subsequent two-level encapsulation of the dye. Characterization of the resultant nanoparticle-imbued waxes is presented as well as the response of encapsulated dye to treatments at temperatures above and below the melting point of various wax matrices. As a result, the ability of these novel materials to thermally trigger exposure and initiate the kinetic response of the dye is demonstrated. Nanoparticle-imbued waxes are shown to be amenable to further post-processing including emulsification, spray prilling, and printing.

569768 Evaporative Heat Transfer Model for QbD of Agitated Filter Bed Drying

Drying of an active pharmaceutical ingredient (API) is an important unit operation in the pharmaceutical industry as it is used to control residual solvent content, in addition to controlling appropriate physical properties. Drying operation is often time consuming and may turn out to be a bottleneck in the overall process cycle time. Many APIs are potent compounds that require filtration and drying to be done in a single contained vessel, known as an agitated filter dryer (AFD), to minimize operator exposure.

Scale-up/tech-transfer of drying process is quite challenging as different AFDs have different surfaces that could transfer heat that ends up impacting effective heat transfer coefficients across the bed as we scale the process. Understanding of bed temperature and residual solvent across the filter cake is critical for reliable and robust scale-up of AFD drying protocol.

In this work, we present results from a novel 3D heat-transfer model that allows us to understand the temperature evolution across the filter cake considering the equipment variability at scale. We also present experimental techniques that can be employed to extract relevant model parameters. In the end, we would provide the decision tree to guide efficient design and reliable scale-up of agitated filter drying.

Disclosures:

All authors are AbbVie employees and may own AbbVie stock. AbbVie sponsored and funded the study; contributed to the design; participated in the collection, analysis, and interpretation of data, and in writing, reviewing, and approval of the final abstract.

560684 Novel Co-Processing Technology to Generate Drug Substances Amenable to Direct Compression

Many active pharmaceutical ingredients (APIs) display poor powder properties and cannot be directly compressed into intact tablets with sufficient strength. The desired powder properties are often difficult to achieve through conventional particle engineering approaches, such as particle size and habit modification during crystallization. Co-processing of API with excipients can significantly improve the functional properties to overcome these challenges. In this talk, we will present a novel co-processing technology that we developed at Bristol-Myers Squibb to improve powder properties in which polymer is precipitated and coats the crystalline particles resulting in discrete, nearly spherical agglomerates. We identified critical process parameters and demonstrated scalability up to 50 kg scale in the pilot plant using theophylline. Furthermore the co-processed theophylline generated under various process conditions were formulated into a blend at 50-90% loading and successfully processed by direct compression.

567918 Kinetic Parameter Estimation from Multiple Spectroscopic Datasets with or without Unwanted Spectral Contributions Using Kipet

Accurately determining reaction kinetics is an important step in the development of any commercial process. In many industries, lab-scale experiments are conducted on systems where little is known in advance regarding the reaction mechanism. These experiments can be costly to run and therefore it is imperative to maximize the use of information gathered during the experiment. A common, non-intrusive method of gathering data during reaction experiments is through the use of spectroscopic instruments that collect data based on absorption of the mixture over time in the ultraviolet visible (UV-vis), Raman, near-infrared (NIR), and infrared wavelength regions. Analyzing data from spectroscopic measurements from such devices is extremely challenging for a number of reasons. Firstly, the data often contains considerable noise and it is common for there to be unwanted spectral contributions from a number of time-variant and time invariant sources. Examples of time invariant contributions can include such effects as baseline shift, distortion, or the presence of absorbing species that do not form a part of the model or that are inert, interfering species. Time variant contributions could be due to scattering based on solids in the mixture or spectral drift. Unwanted contributions are often treated by preprocessing the data using one of a number of different techniques. However, none of the preprocessing strategies guarantee removal of these phenomena, and the choice of preprocessing strategy is not obvious. Another challenge for analyzing spectroscopic data is that the datasets can be extremely large, especially when attempting to estimate parameters that are common across datasets. This is often needed in laboratory applications where chemists attempt to estimate reaction behavior at different conditions.

In a breakthrough approach, Chen et al. (2016) presented a unified framework based on maximum likelihood principles, nonlinear optimization techniques, and orthogonal collocation methods to solve the systems of differential equations. The approach first estimates the variances related to both system variable noise and measurement noise through an iterative optimization-based algorithm and then simultaneously determines the concentration profiles, kinetic parameters, and individual species' absorbance profiles using the nonlinear programming solver, IPOPT. Using the sensitivity properties at the optimal solution, the technique is also able to obtain the covariance matrix and confidence intervals of the estimated parameters. This approach was recently developed into an open-source Python-based package called KIPET (Kinetic Parameter Estimation Toolkit) (Short et al., 2019). KIPET makes use of Pyomo (Hart et al., 2017) as the algebraic modeling language, however users need not be experts in Python nor Pyomo to make use of the many parameter estimation tools therein.

In this work we develop a novel extension to the framework presented by Chen, et al. (2016, 2019) where multiple experimental datasets are simultaneously used to determine the reaction kinetic parameters, concentration profiles in each experimental run, as well as individual species absorbances. In certain experimental setups, new reactants and reactions are introduced and in other examples, different process conditions with different feed times and temperatures are used. In this way, local parameters are introduced in certain datasets, with global parameters linking each experimental dataset. In addition to simultaneously estimating both local parameters and global parameters, we also introduce the ability to determine unwanted contributions that are either time-dependent or time independent. The new approach is demonstrated through a set of case studies that show the utility of performing parameter estimation across datasets as well as the effects of explicitly including unwanted contributions into the model.

568711 Using First-Principles and Data-Driven Models to Guide Inference of Pharmaceutical Reaction Data

Mathematical modeling of chemical reaction kinetics has been proven to aid the development of new reactions and processes. Chemical kinetic modeling is well-established principle in chemical engineering that uses fundamental knowledge of the reaction mechanism to predict conversion data. In the pharmaceutical drug development, elementary-type kinetics are hardly common, due to the nature of the complex organic reaction mixtures and low-level impurities. Thus, statistical modeling plays an important role in understanding the relationship between parameters and reaction profiles. Emerging new technologies, such as high-throughput platforms and auto-samplers, enable greater data collection to enrich our understanding of chemical reactions. As a result, statistical analysis has shifted from conventional end-point analysis to modeling the entire reaction profile using more advanced statistical models, such as the dynamic response surface model and Gaussian process regression model.

In this presentation, a rigorous and general modeling workflow is described on the application of kinetic models and statistical models to the same set of dynamic reaction data. In particular, a semi-parametric empirical model was chosen. An industrial case study is presented, as showcase of the performance and robustness of the two modeling approaches and their impacts, side by side, on parameter effect estimation, reaction robustness range finding, and reaction optimization and operation window prediction. Emphasis is placed on the impact of data and sampling on the model inference. New and innovative visualization techniques are shared in this article for efficient data and model result interpretation.

569597 Integrating Statistical Analysis and Kinetic Modeling to Increase Learning from High-Throughput Experiments

High-throughput experimentation is widely used in pharmaceutical process development to assess the impacts of process conditions on reaction performance. At BMS, a highly automated workflow is used to execute experimental designs. While statistical analysis performed on the data typically focuses on key responses such as yield, purity, and impurity levels at the reaction endpoint, the automated sampling platforms used to run these reactions also provide a significant amount of kinetic data.

This presentation describes the use of templated analysis scripts to complete statistical analysis of reaction endpoints, plot kinetic data for visualization, and export experimental conditions and data to enable kinetic parameter estimation in DynoChem. The web-based template minimizes the need for manual reformatting of the datasets to accommodate different software platforms, and allows flexibility to customize the analysis to address the requirements of each reaction studied. Case studies highlighting the use of this approach to gain additional insight into reaction performance and to inform process development will be discussed.

573803 Model Parameter Uncertainty Revisited

Emergence of the QbD initiative and design space concepts more than a decade ago led to productive discussions, research and publications about how to incorporate uncertainty when using mathematical models derived from experimental data.

Peterson (1) helped to elucidate how probabilistic tools could be used to define a design space using the concept of reliability. Albrecht (2) showed how Markov Chain Monte Carlo methods produced more realistic kinetic parameter confidence regions than conventional maximum likelihood approaches when measurement errors were biased. Garcia Munoz et al (3) sampled the model parameter covariance matrix to construct a probabilistic design space. The impact of uncertainty on process responses was estimated using asymptotic confidence and prediction bands in Figueroa et al (4). Best practices in this field are still evolving, including interesting techniques presented this year in (5).

There have been relatively few published examples where several techniques for modeling uncertainty have been applied and compared for the complex reaction kinetic models typical of pharma organic synthesis. We have therefore applied a range of techniques for quantifying uncertainty in these cases.

In this presentation we compare parameter estimates obtained with rapid methods such as maximum likelihood and sampling the parameter covariance matrix with more CPU intensive techniques such as parametric bootstrapping to estimate the posterior distribution of the parameters.

We propagate uncertainty with each method into simulations of multiple responses and compare rapidly calculated asymptotic confidence bands with sampling the covariance matrix and parametric bootstrapping.

Finally we calculate and compare design spaces created using each method to estimate the joint probability of meeting specific process goals.

Other than differences in CPU time, the techniques lead to interesting variations in parameter values and predicted responses, some of which are important for setting realistic expectations for future measured data and model accuracy in general. The results from each technique also vary somewhat depending on some relatively subtle choices and assumptions made in their application.

We have applied the same techniques to other common parameter estimation models for applications such as filtration and heat transfer and will include a brief discussion of these results.

References

1. Peterson, J. J. J. Biopharm. Stat. 2008, 18, 959–975.
2. Albrecht, Jacob. (2013). Estimating reaction model parameter uncertainty with Markov Chain Monte Carlo. Computers & Chemical Engineering. 48. 14–28. 10.1016/j.compchemeng.2012.07.011.
3. Garcia-Munoz, Salvador & Luciani, Carla & Vaidyaraman, Shankar & Seibert, Kevin. (2015). Definition of Design Spaces Using Mechanistic Models and Geometric Projections of Probability Maps. Organic Process Research & Development. 19. 15071609505300, 10.1021/acs.oprd.5b00158.
4. Isabel Figueroa, Shankar Vaidyaraman, and Shekhar Viswanath, Model-Based Scale-up and Design Space Determination for a Batch Reactive Distillation with a Dean–Stark Trap, Organic Process Research & Development 2013 17 (10), 1300-1310, DOI: 10.1021/op4001127
5. Laky, Daniel & Xu, Shu & S. Rodriguez, Jose & Vaidyaraman, Shankar & García Muñoz, Salvador & Laird, Carl. (2019). An Optimization-Based Framework to Define the Probabilistic Design Space of Pharmaceutical Processes with Model Uncertainty. Processes. 7. 96. 10.3390/pr7020096.

573809 Reaction Kinetic Modeling and Process Optimization of Multi-Enzyme Biocatalytic Cascade Syntheses

The application of biocatalysis for the synthesis of new molecules has been significantly enabled in the last decade by advancements in protein engineering. The presence of biocatalysis in pharmaceutical chemistry is rapidly growing due to its uniquely attractive features for highly selective, mild, safe, and sustainable synthesis. Traditional techniques for pharmaceutical reaction development are met with mixed success as they are applied to these systems. As enzyme evolution proceeds alongside traditional process development, potency and selectivity remain elusive properties which threaten reproducibility and continuity across historical data sets. The need for richer information in shorter timescales is further demanded by challenges with material stability, compatibility, and scale.

In this presentation, the challenges and successes of biocatalytic reaction engineering will be summarized with several case studies. Complex enzymatic mechanisms are examined, including for multi-substrate multi-enzyme cascades. Reaction kinetic models are built using DynoChem and are validated against experimental data. The regression of kinetic parameters in these models allows for the demonstration of a statistically robust process understanding; and in some cases, these models allow for useful process prediction and optimization. Often, the model-building process has proven to be useful for identifying gaps in process understanding, which is useful to inform further development and risk mitigation.

This discussion will also include a particular emphasis on the value of pairing PAT and benchtop automation capability with experimental design methodologies to maximize information content in these fast-paced and complex development projects.

577365 Understanding of a Continuous Barbier-Type Grignard Process Via Mechanistic Modeling

As part of the efforts to advance continuous manufacturing (CM) capabilities of small molecules at Amgen, an end-to-end mechanistic process model with in-depth understanding of each unit operation was developed to leverage in silico simulations to increase process understanding, improve process design, and synthesize control strategies. A heterogenous Barbier-type Grignard reaction was among the many unit operations characterized and modeled in the CM process.

Due to their heterogeneous nature and the hazards associated with processing highly exothermic mixtures, Grignard reactions can be challenging to operate at production scale. This presentation focuses on the development of a mechanistic model for this challenging class of reactions and the utilization of the developed model to investigate and identify operating conditions that satisfy process constraints for continuous manufacturing. Batch experiments were designed and executed to provide the kinetics data for estimating the kinetic parameters. The model was also compared with existing fit-for-purpose models in the literature.

567096 Coarse-Scale PDEs from Microscopic Observations Via Machine Learning

Complex spatiotemporal dynamics of physicochemical processes are often modeled on a microscopic level (through molecular dynamic models, agent-based models, lattice-based models) based on first principles. Some of these processes can also be successfully modeled at the macroscopic level using partial differential equations (PDE) for the right macroscopic observables (not molecules and their velocities, say, but rather concentration and momentum fields). However, deriving the macroscopic PDE that effectively models the microscopic physics (the so-called "closure problem") requires deep understanding/intuition about the complex system of interest; the discovery of macroscopic governing equations is often a difficult and time-consuming process. Recent developments of data-driven approaches via machine learning algorithms provide an alternative way to effectively extract the macroscopic nonlinear PDEs approximating the underlying microscopic observations, without prior knowledge/hypotheses.

In this work, we introduce a novel framework to identify unavailable coarse-scale PDEs from microscopic observations through machine learning algorithms. Specifically, using Gaussian processes, neural networks, and/or diffusion maps, the proposed framework uncovers the relation between the relevant macroscopic space fields and their time evolution (the right-hand-side of the explicitly unavailable macroscopic PDE). This framework will be illustrated through the data-driven discovery of the macroscopic, concentration-level PDE resulting from a fine-scale, Lattice Boltzmann model of a reaction/transport process. Long-term macroscopic prediction is facilitated by simulation of the coarse-grained PDE identified from data. The different features, as well as the pros and cons of our three different machine learning approaches for performing this task (Gaussian Processes, neural networks, and geometric harmonics based on diffusion maps) will also be discussed.

561318 Machine Learning Methods in Process Control

Process Systems Engineering (PSE) has been commissioned to merge mathematics and chemical engineering from about 50 years ago onwards [1]. Since the beginning of this effort, different mathematical approaches were used for tackling chemical engineering problems, and many of these approaches were materialized in real applications in the computational era. Artificial Intelligence (AI) is one particular approach that has been persistent along the decades with many ups and downs [2]. Among many problems that are found in AI during its initial phase are time-consuming and computationally expensive methodologies, and the impossibility of implementing neural networks with many layers [2]. Since the beginning of this century, many breakthroughs have been made in the area of machine learning and deep learning [3], which led to numerous important applications in human visual pattern recognition and natural language processing. Along with the rapid development of machine learning algorithms and computing resources/platforms, many free and open-source software libraries for machine learning applications have been created, which contribute to the broader use of machine learning techniques in classical engineering fields in addition to computer science and engineering.

In this work, feedforward neural networks (FNN) and recurrent neural networks (RNN) are first employed to model steady-state input-output nonlinear relationship and dynamical nonlinear systems, respectively. Moreover, advanced recurrent neural network architectures such as Long short-term memory (LSTM) [4] and Gated recurrent units (GRUs) that can overcome the issue of vanishing gradient are also applied in dynamical nonlinear system modeling. Additionally, since most of the systems in process control engineering are controlled in a dynamic and online manner, which requires the real-time calculation and implementation of control actions, parallel computing of ensemble regression models on a high-performance computing cluster is introduced to assign calculation tasks of multiple machine learning models to distributed processors to improve computational efficiency and prediction accuracy. To further improve the performance of machine learning models, online adaptation and training are employed using real-time data sets collected from multiple sensors to reduce modeling error and account for model uncertainties.

Due to the strong approximation capabilities and user-friendly, open-source machine learning libraries developed nowadays, machine learning models have great potential for process control and operations, for example, process fault detection, cybersecurity, control and real-time optimization. Specifically, we demonstrate that machine learning modeling method is able to approximate a broad class of nonlinear systems with a sufficiently small modeling error, and therefore, the closed-loop system under the controller that incorporates machine learning models is demonstrated to achieve desired stability and performance. The effectiveness of the proposed implementation of machine learning model in process control and the enhancement of computational efficiency via parallel computing are finally demonstrated through chemical process examples.

[1] Ramkrishna, D., and Neal R. A. Mathematics in chemical engineering: A 50 year introspection. AIChE Journal, 2004, 50: 7-23.

[2] Venkatasubramanian, V. The promise of artificial intelligence in chemical engineering: Is it here, finally?. AIChE Journal, 2019, 65: 466-478.

[3] Schmidhuber, J. Deep learning in neural networks: An overview. Neural networks, 2015, 61: 85-117.

[4] Wang, Y. "A new concept using LSTM Neural Networks for dynamic system identification." In Proceedings of American Control Conference (ACC), pp. 5324-5329. IEEE, 2017.

570184 Application of Deep Neural Networks for Artifact Removal from Sensor Data

Chemical processes inherently contain elements of stochastic noise and disturbances that affect their normal operation. Extensive research efforts have been devoted to filtering noises and attenuating disturbances to improve the process system operation [1]. Artifacts, however, represents a challenge in process operations that necessitates further investigation. Artifacts are considered as disturbances in the measured signal not originating from the process itself. Artifacts may arise due to the nature of the measurement devices or the susceptibility of the measurements to ambient or surrounding conditions. In contrast to the process noise, artifacts can have power spectral densities and characteristics similar to the signal being measured, thus confounding the artifacts and the signal. Biochemical and biomedical systems utilize various sensors and measurement technologies that are prone to artifacts. The effective detection and removal of artifacts from measured signals are important as they can lead to significant interpretive errors [2]. The problem of artifact removal is challenging because artifacts in the signals can arise from several unknown sources. Furthermore, it is not obvious whether fluctuations in the signal are due to underlying variations or artifacts. Models relating artifacts to the observed effects in the output measurements are also unavailable as the origins and the effects of the artifacts vary substantially.

Approaches for artifact removal have focused on passive and active filtering techniques. Passive techniques isolate the measurement devices from external factors, yet it may not be always feasible or practical based on the process design and operation setup. Active filtering relies on the use of a reference signal to adaptively deconstruct the measured signal into components associated either with the artifacts or the underlying signal. Adaptive filtering techniques have been effective in adaptive noise cancellation applications where the on-line estimation of the time-varying correlations between the reference signal and the artifacts is possible. Nevertheless, numerous applications do not have a highly informative and conveniently obtainable reference signal. Recent advances in signal processing and deep learning can further advance the tools available for artifact removal, particularly in highly susceptible and sensitive applications.

One application where artifacts are prevalent and pronounced is the measurement of heart rate using photoplethysmography (PPG), a non-invasive and low-cost optical technique used to measure the heart rate variation. The PPG technique measures the rate of heart beats by detecting changes in the backscattered light corresponding to volumetric changes in blood in peripheral circulation. The PPG measurements are highly susceptible to disruption from artifacts caused by motion and other noise sources such as ambient light interference and skin condition. Motion is a major challenge in obtaining accurate measurements as constant movements result in poor contact between the measurement skin surface and the photo sensor [3]. Motivated by the above considerations, a framework for artifact removal is developed with application to the heart rate estimation problem. The PPG signal is first divided into consecutive overlapped time-windows to facilitate batch-wise processing and a bandpass filter used to reject undesired variations. Wavelet approximation and orthogonal signal reconstruction are then applied to reject the artifacts and noise in the signal. A feedforward Neural Network has been utilized to estimate heart beats for a data set of 5 minutes experiment. The heart rate is estimated every two seconds. The comparison between estimated heart beats versus the actual values proves the fact that regardless of the type of physical activity, especially during running section which PPG signal is corrupted with intense motion artifact, the Deep Neural Network model is capable of the tracking the actual heart rate values.

References:

[1] Downs, J. J., and Vogel, E. F., "A Plant-wide Industrial Process Control Problem, "*Computers & Chemical Engineering*, vol.17, pp. 245–255 (1993).

[2] Roy, , Kumar, V. R., Kulkarni, B. D., Sanderson, J., Rhodes, M., and vander Stappen, M., "Simple denoising algorithm using wavelet transform," *AIChE Journal*, vol. 45, 1999, pp. 2461–2466.

[3] Zhang, , Pi, Z., and Liu, B., "TROIKA: A general framework for heart rate monitoring using wrist-type photoplethysmographic signals during intensive physical exercise," *IEEE Transactions on Biomedical Engineering*, vol. 62, pp. 522–531(2015).

561336 Nonlinear Dynamic Feature Extraction Based on Gaussian Process Dynamical Models for Jit-Based Adaptive Soft Sensors

Introduction

As industrial process plants, it is important to monitor product quality to assure both product quality and process safety. Soft sensors are models constructed between difficult-to-measure variables (y-variables) and easy-to-measure variables (X-variables). Values estimated with soft sensors are used to control plants rapidly. Partial least squares (PLS) [1] are the most popular linear regression methods. To handle nonlinear cases, support vector regression [2] and Gaussian process regression (GPR) [3] have been developed. Since the GPR method can generate a probabilistic model, an uncertainty estimation for the variable of interest can be obtained. However, the predictive ability of such techniques decreases due to process changes in chemical plants. To reduce the decrease of predictive ability, adaptive soft sensors have been developed [4]. We focus on just-in-time (JIT) soft sensors. The JIT model can be constructed online and can track process changes well. Locally weighted PLS (LWPLS) [5] is one of the nonlinear JIT methods. Since a set of hyperparameters in any JIT model has to be set beforehand and the hyperparameters of a JIT model will not always be the optimal ones for a query sample, the predictive ability of a JIT model decreases.

When constructing a soft sensor model for time series data, process dynamics are often considered to improve estimation performance [6]. In other words, not only the current time samples but also past samples are added to the X-variables to construct a model. On the other hand, if all these variables are used to construct the soft sensor model, it will increase the computation burden and make the model overfitting. Therefore, it is necessary to carry out dimensionality reduction and feature extraction before constructing the model. Principal component analysis (PCA) [7], as a traditional dimensionality reduction method, has been identified as an effective approach to transform observed variables to low-dimensional data. PCA is a linear method and developed in a deterministic manner, which lacks a probabilistic interpretation. As a matter of fact, process data are usually measured in a noisy environment and disturbances may happen. Also, there are nonlinearity between X-variables and temporal dependences between adjacent latent states and the data are autocorrelated. As an alternative, Gaussian process dynamical models (GPDM) [8] that is a nonlinear probabilistic dynamic latent variable model has been developed to model the process features. In GPDM, the latent states are assumed to be produced according to a first-order Markov chain, and each observation is generated from its current state by a nonlinear transformation. In this study, we propose to combine JIT and ensemble learning, and predict y-values with multiple JIT models, which are constructed using latent variables from GPDM and sets of hyperparameters are different. The proposed method is called ensemble just in time based on Gaussian process dynamical models (EJITGPDM). The weights of JIT models are determined based on Bayes' theorem, considering their predictive ability. We check that the proposed model has higher predictive accuracy than traditional models through two industrial data analyses.

Proposed method

In the proposed method EJITGPDM, y-values are estimated from X-values by weighting the y-values predicted by multiple JIT models constructed using different hyperparameters. When a query sample x_q comes, the Euclidean distance between each historical sample and the query sample is first calculated. Next, get the indexes of the N most relevant samples, which have a small Euclidean distance to the query sample. After extracting latent variables from GPDM, a GPR model is constructed between the latent variables of the sample with high similarity and y. In GPDM, it is necessary to determine the number of components, and in JIT model, the number of data for model construction. That is, when the number of GPDM components differs in m-values and the number of data for model construction differs in n-values, the number of JIT models is L (= m × n). The y-values for query sample are predicted with multiple JIT models, after that they are integrated with ensemble learning. Each weight of the predicted y-values are calculated with Bayesian theorem [9], taking into account the prediction ability of each JIT model. When S is the current (unobserved) state in a plant and M_i is the ith model, the probability of M_i given S, $P(M_i|S)$, is required to combine the prediction results of L JIT models. Using RMSE for the mid-points between the k-nearest-neighbor data points for the s data, which is called $RMSE_{midknn}$ [10], the predictive accuracy of nonlinear regression models can be evaluated. Therefore, using weight, that is, inverse of $RMSE_{midknn}$, y-values can be estimated while monitoring the predictive ability of each JIT model and weighting each model appropriately.

Results and Discussion

To verify the effectiveness of the proposed method, we analyzed two real industrial process datasets. The dynamic just in time based on Gaussian process regression model (DJIT-GPR) [11] and the dynamic LWPLS model (DLWPLS), the dynamic locally weighted principal component regression model (DLWPCR) [12] used for comparison.

Real industrial data in a debutanizer column [13]

We applied the proposed method to data obtained from the operation of a debutanizer column. The y-variable is the butane content in the bottom flow and the X-variables are top temperature, top pressure, reflux flow, flow to next process, sixth tray temperature and bottom temperature. We could use the dataset where the measurement intervals of the y-variable and the X-variables were 6 min and the measurement delay of the y-variables was 48 min. Data from 100 to 894 were used as training data and data from 895 to 1394 were used as validation data, and data from 1395 to 2394 were used as test data. Thus, we had 795 training data and 500 validation data, and 1394 test data.

The lowest r^2 values was 0.401 whose DLWPCR model produced. Using the DJIT-GPR model and the DLWPLS model, prediction accuracy improved and their values were 0.840 and 0.856. Between DJIT-GPR, DLWPLS, DLWPCR, and EJITGPDM, the highest r^2 was 0.898 whose EJITGPDM produced. The proposed model could predict y-values more accurately than the traditional models. The y-values predicted with the DLWPLS model have larger variation than those of the proposed EJITGPDM model, that is, the proposed EJITGPDM achieves stable and accurate prediction of y-values.

Real industrial data in a sulfur recovery unit (SRU) [14]

We applied the proposed method to data obtained from the operation of an SRU. The y-variable is the concentration of H_2S in the tail gas of Line 4 and the X-variables are gas flow MEA GAS, air flow AIR MEA, secondary air flow AIR MEA 2, gas flow in SWS zone, air flow in SWS zone. In industrial process plants, there are a case where the measurement interval of y-variables is larger than the measurement interval of X-variables. We verified the effectiveness of the proposed method in the case where the measurement interval of y-variables is one sixth of the measurement interval of X-variables. Data from 100 to 4101 were used as training data and data from 4102 to 6099 were used as validation data, and data from 6070 to 3982 were used as test data. Thus, we had 667 training data and 333 validation data, and 3982 test data.

As was the case in the debutanizer column data, EJITGPDM produced the r^2 value was the highest and its value was 0.693. It is more than 10% better than the value of DLWPLS. In time plots of actual y and estimated y for DLWPLS and EJITGPDM, The y-values, which were the unmeasured y-value part, have smaller variation than those of the comparison models.

Conclusions

We proposed the modeling method EJITGPDM, which takes into consideration the nonlinearity between X-variables and temporal dependences between adjacent latent states. Also, it is always possible to calculate predicted y-values of query samples using optimal hyperparameters. In order to verify the effectiveness of the proposed method, a case study was conducted using operation data measured by the debutanizer column, and operation data measured by the SRU. The simulation results show the effectiveness of the proposed method. Efficient and stable management and operation of chemical plants are expected by using the proposed method.

Reference

1. Joe Qin, Comput. Chem. Eng. 22, 503-514, 1998.
2. Yan, H. Shao, X. Wang, Comput. Chem. Eng., 28, 1489-1498, 2004.
3. B. Belhouari , A. Bermak, Comput. Stat. Data Anal., 47, 705–712, 2004.
4. Kaneko, K. Funatsu, AIChE J., 52, 1322-1334, 2013.
5. Kim, M. Kano, H. Nakagawa, S. Hasebe, Int J Pharm. 421, 269-274, 2011.
6. H. Kaspar, W. H. Ray, Chem. Eng. Sci, 48, 3447-3461, 1993.
7. Hotelling, Journal of educational psychology, 24, 417, 1933.
8. Wang, D. Fleet, A. Hertzmann, NIPS, 18, 1441-1448, 2005.
9. Khatibisepehr, B. Huang, S. Khare, J. Process. Control., 23, 1575–1596, 2013.
10. Kaneko, K. Funatsu, J. Chem. Inf. Model., 53, 2341–2348, 2013.
11. L. T. Chan, X. Wu, J. Chen, L. Xie, C. I. Chen, IEEE Trans. Semicond. Manuf., 31, 3, 2018
12. F. Yuan, Z. Q. Ge, Z. H. Song, Ind Eng Chem Res, 53, 13736-13749, 2014.
13. http://www.springer.com/us/book/9781846284793
14. http://www.springer.com/us/book/9781846284793

569019 Attention-Based Recurrent Neural Network for Multi-Step-Ahead Prediction

The advances in smart manufacturing, Industrial Internet of Things (IIOT), and data storage technologies have facilitated industries to collect, review, and utilize numerous types and huge amounts of data that could significantly improve plant operation and production of high quality products. The data can be used for different types of analytics including descriptive, diagnostic, and predictive analytics. However, process data are high dimensional, noisy, and dynamic with complex interaction among system variables, which makes data analytics very challenging. Recurrent neural networks (RNNs), a type of model for deep learning, can address some challenging issues effectively.

Motivated by [1, 2], we introduce an attention-based RNN for multi-step-ahead prediction that can have applications in model predictive control, fault diagnosis, etc. This model consists of an RNN that encodes a sequence of input time series data into a new representation (called context vector) and another RNN that decodes the representation into output target sequence. An attention model integrated to the encoder-decoder RNN model allows the network to focus on parts of the input sequence that are relevant to predicting the target sequence. The attention model is jointly trained with all other components of the model. By having a deep architecture, the model can learn a very complex dynamic system, and it is robust to noise. In order to show the effectiveness of the proposed approach, we perform a comparative study on the problem of catalyst activation prediction, by using Nonlinear AutoRegressive with eXogeneous (NARX) predictor, and conventional machine learning techniques such as Support Vector Regression (SVR).

[1] Cho, K., van Merrienboer, B., Gulcehre, C., Bougares, F., Schwenk, H., and Bengio, Y. Learning phrase representations using RNN encoder-decoder for statistical machine translation. In Proceedings of the Empiricial Methods in Natural Language Processing (EMNLP 2014).

[2] Dzmitry Bahdanau, Kyunghyun Cho, and Yoshua Bengio. 2015. Neural machine translation by jointly learning to align and translate. In Proceeding of the International Conference on Learning Representations (ICLR 2015).

---

* All correspondence should be addressed to Prof. Yinlun Huang (Phone: 313-577-3771; E-mail: yhuang@wayne.edu).

562148 A Deep Reinforcement Learning Approach for Production Scheduling

Scheduling in the chemical industry is a difficult task beset by uncertainties from maintenance issues, to demand uncertainty, to pricing changes, among others. For many supply chain scheduling problems, decisions need to be made in real-time when the situation in the plant changes, leading industrial operations to employ human schedulers in the decision making process who react to pressures in the organization, often rescheduling and creating suboptimal schedules. Shobrys and White [2002] estimate that "good" decisions for these scheduling problems can increase the profit margin by at least $10/ton of product. Given the thousands of tons produced by modern, industrial, chemical operations each day, there is a large financial incentive to improve scheduling processes and decision making under uncertainty.

Although there is a long history of optimization under uncertainty, many techniques are difficult to implement due to high computational costs, sources of uncertainty (endogenous vs exogenous), and their measurement [Grossmann et al., 2016]. The stochastic optimization approach deals with uncertainty in stages whereby a decision is made, then the uncertainty is revealed which enables a recourse decision to be made given the new information. For scheduling applications, Jung et al. [2004] develops a multi-stage stochastic optimization model to determine safety stock levels to maintain a given customer satisfaction level with stochastic demand. Sand and Engell [2004] developed a two-stage stochastic mixed-integer linear program to address the scheduling of a chemical batch process with a rolling horizon while accounting for the risk associated with their decisions. Englberger et al. [2016] develop and implement a two-stage stochastic optimization problem on a rolling horizon for integrated master production scheduling that reduced delays at the expense of higher safety stock.

In this paper we explore a new approach to scheduling using deep reinforcement learning (DRL) to train an agent to schedule a multi-product reactor under uncertainty and meet service level and profitability targets for the supply chain organization motivated by a real-world example. The deep reinforcement learning model recasts the scheduling problem as a Markov Decision Process (MDP) where the state is defined as the demand, forecast, inventory levels, current production, schedule, and time, and the decisions are related to products to be scheduled at given time intervals. This redefinition as a MDP enables a natural representation of uncertainty for the model. While training must take place off-line using a simulation, once trained, the model can then be deployed on-line in a production system to yield real-time schedules as new orders are entered into the enterprise resource planning (ERP) system and as the situation in the plant changes due to delays and unplanned events.

A drawback of DRL is the lack of theoretical guarantees of performance. To address this, we benchmark the trained DRL agent against stochastic mixed-integer linear programs (MILP) and deterministic models to understand how well the DRL solutions measure up. All models are validated using Monte Carlo simulations, which show that DRL performs quite well in comparison to MILP approaches. In addition, we explore integration of MILP's to leverage the optimality guarantees of MILP's with the efficient time-to-solution of DRL by training the agent to make optimal decisions in a supervised setting before allowing the agent to train in reinforced mode.

References

J. Englberger, F. Herrmann, and M. Manitz. Two-stage stochastic master production scheduling under demand uncertainty in a rolling planning environment. International Journal of Production Research, 54 (20):6192–6215, 2016. ISSN 1366588X. doi: 10.1080/00207543.2016.1162917.

ematical programming techniques for the optimization of process systems under uncertainty. Computers and Chemical Engineering, 91:3–14, 2016. ISSN 00981354. doi: 10.1016/j.compchemeng.2016.03.002. URL
http://dx.doi.org/10.1016/j.compchemeng.2016.03.002.

J. Y. Jung, G. Blau, J. F. Pekny, G. V. Reklaitis, and D. Eversdyk. A simulation based optimization approach to supply chain management under demand uncertainty. Computers and Chemical Engineering, 28(10):2087–2106, 2004. ISSN 00981354. doi: 10.1016/j.compchemeng.2004.06.006.

G. Sand and S. Engell. Modeling and solving real-time scheduling problems by stochastic integer programming. Computers and Chemical Engineering, 28(6-7):1087–1103, 2004. ISSN 00981354. doi: 10.1016/j.compchemeng.2003.09.009.

D. E. Shobrys and D. C. White. Planning, scheduling and control systems: Why cannot they work together. Computers and Chemical Engineering, 26(2):149–160, 2002. ISSN 00981354. doi: 10.1016/S0098-1354(01)
00737-2.

567760 Decision-Making of Online Rescheduling Procedures Using Neuroevolution of Augmenting Topologies

Online scheduling requires appropriate timing of rescheduling events, as well as the determination of relevant rescheduling horizon lengths. Optimal choices of these quantities are highly dependent on the uncertainty of the scheduling environment and may vary over time. We propose an approach where a neural network is trained to make online decisions on these quantities, as well as on the choice of the rescheduling method (mathematical programming or metaheuristics). In our approach, the neural network is trained using neuroevolution of augmenting topologies (NEAT) in a simulated environment.

NEAT, developed by Stanley and Mikkulainen [1], is a genetic algorithm that simultaneously evolves the topology and weight parameters of the neural network. Such et al. [2] report the performance of the evolutionary neuroevolution algorithms to compare well against the gradient-based backpropagation algorithms. A key feature of NEAT is that the evolution process is initiated from very simple neural networks, the topological complexity of which is then incrementally increased during the evolution. This feature reduces unnecessary complexity of the final neural network, and is not possible when using gradient-based algorithms. Recently, Hausknecht et al. [3] applied NEAT to train a neural network to play 61 different Atari 2600 games, where the controls for the player are to move a joystick and press a button. The controls for our neural network are also of a similar low level of complexity (see the first paragraph), but the critical aspect is their timely execution.

In many previous studies, neural networks have been used to learn dispatching rules from the historical data of scheduling decisions [4]. However, this approach can typically only mimic the historical decisions, and therefore the optimality of the scheduling solutions is highly dependent on the quality of the given historical data. Our neural network approach operates at a higher level; it exploits the strength of mathematical programming at finding the optimal, or near-optimal solution, and focuses on the allocation of the computational resources in the changing scheduling environment.

We benchmark the approach against periodically occurring and event-triggered rescheduling on a dynamic routing problem. The approach is also extensible to rescheduling of industrial batch processes.

References:

[1] K. O. Stanley, R. Miikkulainen, 2002. Evolving neural networks through augmenting topologies. Evolutionary computation 10 (2), 99–127.

[2] F. P. Such, V. Madhavan, E. Conti, J. Lehman, K. O. Stanley, J. Clune, 2017. Deep neuroevolution: genetic algorithms are a competitive alternative for training deep neural networks for reinforcement learning. arXiv preprint arXiv:1712.06567.

[3] M. Hausknecht, J. Lehman, R. Miikkulainen, P. Stone, 2014. A neuroevolution approach to general Atari game playing. IEEE Transactions on Computational Intelligence and AI in Games 6 (4), 355–366.

[4] P. Priore, A. Gómez, R. Pino, R. Rosillo, 2014. Dynamic scheduling of manufacturing systems using machine learning: An updated review. AI EDAM 28 (1), 83–97.

563451 Progan: Protein Solubility Generative Adversarial Nets for Data Augmentation in Dnn Framework

ProGAN: Protein Solubility Generative Adversarial Nets for Data Augmentation in DNN Framework

Generative Adversarial Networks (GANs), a state-of-the-art data augmentation algorithm in Artificial Intelligence (AI), provide an attractive and novel approach to generate mimic data which seems very realistic, such as genes and proteins in biological fields. GANs have achieved promising performance in the fields of image recognition and computer vision [1]. Those data augmentation algorithms may alleviate the problem of lacking data in some biological fields where producing data is time-consuming and labour-intensive. In literature, many machine learning models have been investigated to predict protein solubility from protein sequence, whereas the parameters of those models are underdetermined with insufficient data of protein solubility. However, the data augmentation algorithms have not been investigated to solve the problem of protein solubility and other problems in biological research. In this study, we developed a data augmentation algorithm, ProGAN to improve the prediction performance of protein solubility based on a deep neural network (DNN) regression model.

All the proteins used in the study were downloaded from eSol database [2], which is a unique database with continuous values of solubility, ranging from 0 to 1. The regression prediction model using DNN was developed and a data augmentation algorithm ProGAN was further applied to this dataset. Finally, we evaluated the model performance by using coefficient of determination (R^2) and other metrics calculated from predicted and actual solubility values.

At first, we propose DNN as a more accurate regression prediction model. A more accurate model is beneficial for guiding experimental validations and further optimization on the protein properties. In addition, to tackle the insufficient data problem, a novel data augmentation algorithm, ProGAN, was proposed for improving the prediction of protein solubility. By developing a data augmentation algorithm with latent label for protein solubility in the generator and an auxiliary regressor in the discriminator, tuning the hyperparameters of ProGAN and organizing the dataset, we achieved a R^2 value of 45.04%, which enhanced R^2 about 10% compared with the previous study using the same dataset. The mean squared error (MSE) was further reduced to 0.0563 on average and the analysis of variance (ANOVA) also demonstrated statistical significance. Data augmentation opens the door to applications of machine learning models on biological data, as machine learning models frequently fail to be well trained using small datasets. More importantly, this data augmentation algorithm provides an approach to generate mimic data *in silico* without hands-on experiments, which brings new insights into the fields where collecting massive data is challenging.

Reference:

[1] Goodfellow, Ian, Jean Pouget-Abadie, Mehdi Mirza, Bing Xu, David Warde-Farley, Sherjil Ozair, Aaron Courville, and Yoshua Bengio. "Generative adversarial nets." In *Advances in neural information*

processing systems, pp. 2672-2680. 2014.

[2] Niwa, Tatsuya, Bei-Wen Ying, Katsuyo Saito, WenZhen Jin, Shoji Takada, Takuya Ueda, and Hideki Taguchi. "Bimodal protein solubility distribution revealed by an aggregation analysis of the entire ensemble of Escherichia coli proteins." *Proceedings of the National Academy of Sciences* 106, no. 11 (2009): 4201-4206.

580967 Approaches to Knowledge Management and Data Transformations in Pharmaceutical Process Invention

Over the past decade, the increased molecular complexity of small molecule drugs has resulted in synthetic complexity. However, traditional unit operations (reactions, separations, crystallization, filtration, etc.) continue to be the basis for the manufacture of small molecule drug substances even as innovations in organic synthetic techniques such as new catalysts and coupling reagents are implemented to improve sustainability and efficiency. Chemical engineers are well-poised to support chemical route invention, process invention and process characterization activities to develop robust commercial manufacturing processes through utilization of limited data to quantify and metricize competing attributes of proposed synthesis approaches and process designs. Specifically, methods to understand and quantify process risks, process capabilities, and expected failure rates have the potential to speed development activities. Addressing these challenges requires innovations in tools supporting data generation and curation, and new approaches to metrics and modeling. This talk will cover some current work in this knowledge space and its potential to drive innovation in pharmaceutical development in the future.

581329 Leading the Way to Pharma 4.0

Manufacturing processes and plant & supply chain scheduling are becoming increasingly more complex. Pharma 4.0 is helping transform plants into smart factories. This is being done through improvements in sensor technology, real-time monitoring, advanced process control, continuous process verification, predictive maintenance, digital twin, and automated adaptive scheduling.

Applied Materials has assembled an integrated suite of advanced manufacturing software solutions which provide a real-time, data-driven plant environment that senses factory activity, predicts performance, and prescribes effective actions. SmartFactory Rx™ improves visibility, drives quality, safety, yield and throughput, reduces cycle times and costs, and increases profitability.

Applied Materials is the leader in factory automation solutions for high volume semiconductor manufacturing plants worldwide. SmartFactory Rx, for pharmaceutical manufacturers, is powered by technologies that are customer-proven and widely deployed.

Applied SmartFactory Rx enables companies to turn their data into action:

- ***Analytics & Control***: Increase yield, prevent deviations, and assure quality
- ***Operations Productivity***: Reclaim capacity, maximize throughput, providing real-time adaptive scheduling
- ***Predictive Maintenance***: Prevent costly unscheduled downtime
- **Knowledge Advisor**: Convert operator know-how & experience into process understanding & stability

Uses cases will be presented that illustrate the kinds of problems and solutions that can be realized in both small and large molecule manufacturing plants.

580759 Machine Learning in Pharmaceutical Process Development: Where Are We up to in Pharma 4.0?

Artificial Intelligence (AI) and Machine Learning (ML) have become ubiquitous terms in the past couple of years. This presentation inspects and compares the current AI, ML approaches with alternative techniques which are already well established within the process industries, and now also in many of the innovative Pharmaceutical companies.

Advanced Process Control (APC) and Multi-Variate Analysis are data driven techniques to build models, understand the product and process, monitor for abnormal operational events and more recently directly adjust the process to achieve closed loop control of product Critical Quality Attributes (CQAs). When we compare these "traditional" data-driven techniques with Machine Learning we see the same algorithms being applied and the common goal of improved decision making through data analysis, prediction and adjustment. There is a difference however, and we'll explore what it is within the talk.

Case studies from the Pharmaceutical industry are used to demonstrate the application of several forms of Machine Learning for process control and optimisation. By comparing these Machine Learning algorithms with other tools in the process optimisation "toolbox", we can examine the benefits and challenges to implementing this technology within the industry.

581055 Digital Twins, Microfactories and People for Future Pharmaceutical Manufacturing

The pharmaceutical industry is facing increasing demands to deliver a higher variety of products more quickly, economically and sustainably. In order to achieve this, the sector is making better use out of cyber-physical production systems with a specific focus on digital twins and how they can be used to hugely minimise the amount of material utilised in the design of a manufacturing process. However, these models which will be used to build a digital twin are not focused on a single piece of equipment, process or even physics, but are holistic models which aggregate data from multiple sources. This could include for example, crystal structure, solubility predictions, population balance, CFD, and DEM. Interfacing with all this data in a manageable manner is challenging, especially for a non-modeller. This presentation highlights CMAC's recent work in 1) creating digital twins that sit across multiple model types with an interface accessible for a non-modeller, 2) undertaking end to end manufacturing of an API mirrored by an end to end process model and 3) developing and delivering training to researchers to provide them with the skills needed to design, build and operate digital twins and microfactories across pharmaceutical manufacturing.

581330 The Art of the Possible When Digitising Your Pharmaceutical Operations

The Pharma and Life Sciences sector is in the midst of rapid, fundamental change. Manufacturers are embarking on digital transformation projects across the globe but there are few that have unlocked their full potential. From Industry 4.0, Digitisation to the opportunities that Continuous Manufacturing can bring, a transformation in the operations agenda is evident. For many manufacturers, a profound transformation is required, integrating business understanding and human insights with technology innovation. We will take a look at the manufacturers who have embarked on digital transformation and are leading the way, discuss what's holding some companies back, and discuss what can be done now to stay ahead of the global competition. With a lens on emerging Industry 4.0 technologies, how Continuous Manufacturing can be an enabler of the Digitisation agenda, we will consider some of the immediate challenges and the future opportunities that will deliver 'real' business benefit and ultimately patient value.

575192 Formulating Biologics in Polymeric Nanoparticles: New Methods to Overcome Manufacturing Barriers

Nanoscale polymeric vehicles can protect biologics such as peptides and proteins from clearance and produce a sustained release profile.1 The double emulsion process to form polymeric microparticles and nanoparticles is a commonly employed formulation technique but has well-documented trade-offs.2 This results in processes that are not economically viable because significant therapeutic quantities are lost as unencapsulated material. To achieve the same therapeutic dose with a low-loading formulation requires large excipient amounts and possibly high-volume injections.

We have developed a process that addresses these shortcomings by replacing the primary emulsion with an "inverse nanoparticle" containing the biologic. We employ inverse Flash NanoPrecipitation (iFNP) to produce these polymeric nanoparticles encapsulating a biologic at high loading (>50%).3,4 These inverse nanoparticles reduce process losses during production of the second emulsion due to the presence of a hydrophobic polymer shell around each biologic core. Alternatively, we have developed a process to introduce a dense PEG coating on the inverse nanoparticles to produce long-circulating water-dispersible particles structurally analogous to polymersomes or vesicles.

We detail the methods employed in forming these particulate delivery systems for sustained release applications. The loadings and encapsulation efficiencies (EE) achieved are in many cases an order of magnitude better than existing techniques. We will report approaches for tuning releases profiles and describe potential applications of the technology. The table below highlights process metrics for a range of biologics encapsulated in coated nanoparticles formed by iFNP.

Biologic	MW (g/mol)	EE (%)	Loading (%)
Lysozyme	14,400	97.7 ± 0.4	31%
Polymyxin B	1,302	80.4 ± 1.6	29%
HRP	44,000	63.7 ± 6.5	20%
MPEP-1	~3,410	29.8 ± 2.4	6%
Vancomycin	1,449	11.8 ± 3.9	2%

(1) Yu, M.; Wu, J.; Shi, J.; Farokhzad, O. C. Nanotechnology for Protein Delivery: Overview and Perspectives. J. Control. Release 2016, 240, 24–37.

(2) Pagels, R. F.; Prud'homme, R. K. Polymeric Nanoparticles and Microparticles for the Delivery of Peptides, Biologics, and Soluble Therapeutics. J. Control. Release 2015, 219, 519–535.

(3) Markwalter, C. E.; Prud'homme, R. K. Inverse Flash NanoPrecipitation for Biologics Encapsulation: Understanding Process Losses via an Extraction Protocol; 2017; pp 275–296.

(4) Pagels, R. F.; Prud'homme, R. K. Inverse Flash NanoPrecipitation for Biologics Encapsulation: Nanoparticle Formation and Ionic Stabilization in Organic Solvents; 2017; pp 249–274.

572141 Co-Immobilization of Enzyme Cascades Enables One-Pot Industrial Immobilized Biocatalysis

Enzymes are increasingly used to access novel chemical spaces and enable regio- and stereo-selective synthetic routes for a variety of drug substances. In many instances, biocatalytic transformations require concurrent operation of multiple enzymes to perform co-factor recycling or to enable cascaded transformations. While such transformations enable novel chemistry, they also increase the protein burden, which must be managed to both enable product isolation and mitigate potential off-target interactions from host-cell proteins which accompany unpurified recombinantly expressed proteins.

Immobilizing an enzyme to a porous resin solves many of these issues by creating a solid-supported catalyst with extremely high enzyme loads that can be prepared in advance, easily charged into a reactor, and rapidly isolated from product by filtration. We have developed a scalable enzyme-immobilization platform technology to selectively co-immobilize enzyme cascades onto a single resin directly from crude cell lysate and reject host-cell proteins. We have successfully utilized this technology to enable numerous multi-enzyme chemistries and applied it across multiple scales. In this talk we will discuss development and scale-up of multi-enzyme immobilized processes and the unique opportunities they afford for industrial immobilized biocatalysis.

573823 Application of Membrane Technology for Metal Removal from a Reaction Mixture

Transition metals are known for their good catalytic properties. Commensurate with the reputation, their use in various chemical transformations has gone up in recent years. However, their use in homogeneous catalysis in reactions often results in high metal content in the product. For many industries, especially for active pharmaceutical ingredients (APIs), such a high level of metal content cannot be tolerated, and the reaction mixture or product needs to be treated for metal removal.

Depending on the amount of metal contaminant, traditional, adsorption-based technologies may suffer from high adsorbent charge, long cycle times, and high costs. In this study we evaluate the application of membrane technology to remove metals from different reactions mixtures. In this presentation, we summarize the engineering methodology to develop a membrane process and touch upon scale-up considerations and challenges for successful technology implementation in a cGMP chemical pilot plant setting.

Mike W. Rasmussen, Manish S. Kelkar, John Hartung, Harsit B. Patel, Thaddeus S. Franczyk, Onkar Manjrekar, Geran Zhang, Elie Chaaya, and Nandkishor K. Nere are employees of AbbVie. The design, study conduct, and financial support for this research was provided by AbbVie. AbbVie participated in the interpretation of data, review, and approval of the publication.

574496 Applications of Soft Sensors in Pharmaceutical Manufacturing Processes

Drying is an essential processing step for residual solvent reduction in pharmaceutical processes. For the manufacture of Active Pharmaceutical Ingredient (API), drying is mainly used to produce crystals with the right form and/or particle size distribution that are suitable for down-stream processing. As an energy-intensive and time-consuming process, drying is usually a bottleneck of pharmaceutical manufacturing processes. Therefore, it is critical to well understand the drying process and improve the process efficiency. For the experimental characterization of drying progression, loss on drying (LOD) is a commonly used off-line measuring approach. LOD requires to take samples periodically and to weigh the loss in mass when the sample is continuously being dried until the weight becomes constant. However, LOD is a time-consuming and intrusive method that can disrupt the drying process and increase the cycle time. It is thus desired to develop on-line measurements to monitor the drying process.

Process Analytical Technology (PAT) has been developed as an in-line or on-line monitoring approach to provide real-time data about the process status and critical process attributes (CQA), which can facilitate the process control and optimization. Spectroscopy methods, such as near infrared (NIR) and Raman, have been widely adopted to enable the implementation of PAT in monitoring the removal of solvents in the drying process [1, 2]. However, such methods require the development of chemometric models, which typically requires large amounts of experimental data for calibration and validation. Additionally, mass spectrometry has been developed to predict the drying end point without the need of chemometric models [3]. The common feature of the mentioned PAT methods is that they usually require the installation of spectroscopic devices which can be prone to fouling and introduce additional capital costs [4].

Recently, soft sensors have emerged as a powerful tool to provide real-time process monitoring capabilities in many industry fields. Soft sensors are mathematical models that use easily accessible on-line data to estimate other variables of interest that can only be determined at low sampling rates or through off-line analysis [5]. For pharmaceutical processes, soft sensors have great potential to either replace the existing hardware sensor or work in parallel to provide redundancy and verify whether the hardware sensor is working properly [6]. However, applications of soft sensors in the pharmaceutical industry are still limited, partly due to the challenge of validation by regulatory agencies [7].

In this work, we proposed a soft sensor based on a hybrid modeling approach to estimate the product LOD in real-time during the drying process. The model combines first-principle relationships with empirical correlation models derived from historical data, which is more robust than data-driven models in handling variabilities in raw material and process conditions. The developed soft sensor has been successfully calibrated, validated with experimental data, and implemented in an API full scale commercial process that follows good manufacturing practices (GMP) procedures.

1. Hamilton, P., et al., *Studies of particle drying using non-invasive Raman spectrometry and particle size analysis.* 2011. **136**(10): p. 2168-2174.
2. Morris, K.R., et al., *Accelerated fluid bed drying using NIR monitoring and phenomenological modeling.* 2000. **26**(9): p. 985-988.
3. Dodda, A.G., et al., *Using online mass spectrometry to predict the end point during drying of pharmaceutical products.* 2014. **19**(1): p. 122-131.
4. Pla, D.L., et al., *Moisture soft sensor for batch fluid bed dryers: A practical approach.* 2018. **326**: p. 69-77.
5. Luttmann, R., et al., *Soft sensors in bioprocessing: a status report and recommendations.* 2012. **7**(8): p. 1040-1048.
6. Kourti, T.J.A. and b. chemistry, *The process analytical technology initiative and multivariate process analysis, monitoring and control.* 2006. **384**(5): p. 1043-1048.
7. Gernaey, K.V., et al., *A perspective on PSE in pharmaceutical process development and innovation.* 2012. **42**: p. 15-29.

571484 Process Performance of a Novel Continuous Process Line for Drying of an Active Pharmaceutical Ingredient

Lately, there has been increased interest in linking primary and secondary manufacturing. Especially, the increasing pressure towards cost effectiveness and cost reduction during phases II and III of drug product development is driving this demand.

After continuous API synthesis the steps filtration, washing and drying are still done in batch processes. This is mainly due to the lack of available equipment for pharmaceutical production. Additionally, drying processes are challenging. During crystallization particles are tailored with much effort and therefore, these properties must be maintained during subsequent process steps. However, during drying attrition and agglomeration compete, leading to a change of particle size distribution (PSD) of the primary particles.

This talk will show the design of a continuous process route for drying, considering also the upstream washing step, which strongly influences the agglomeration process during drying. This process line is an option to close the gap between primary and secondary continuous manufacturing.

The process starts with a continuous washing and filtration step. In the final filtration step the suspension is concentrated to the maximum solid content, which can be transported to the dryer. A novel contact convective drying technology was developed, which allows achieving a final moisture below 1% while not impacting the PSD. The process line is designed to dry cohesive temperature sensitive materials with a mass flow of 0.5 to 2 kg/h of dry substance. With this novel drying process, it is possible to gently dry particles with a particle size below 100 µm without changing the particle size. Also, this drying process guarantees that there is no agglomeration inside the dryer, ensuring a suitable residence time distribution. For this work Ibuprofen and other substances have been used to test PSD and residence time distribution, as well as drying efficiency. To prove that particle shape and surface did not change during the drying process, scanning electron microscopy was used.

In summary, it was shown that this novel contact convective drying technology is suitable to dry temperature sensitive cohesive particles at a mass flow range suitable for pharmaceutical production.

565447 On-Demand Drug Product Manufacturing in a Miniaturized System: Can Formulation be Simplified?

Oral direct compressible tablets are the most commonly used drug products. Tablets manufacturing requires design and development of formulations, which need to use several excipients. Choice of excipients depends on the concentration, manufacturability, stability, and bioavailability of the drug substance. Due to a large number of excipients typically used, formulation development becomes complex and tedious. At MIT, we developed a miniature platform for on-demand manufacturing of direct compressible tablets [1, 2]. On-demand manufacturing offers a great opportunity to simplify pharmaceutical formulations by minimizing the number of excipients required. In this study, direct compressible tablets of five pharmaceutical drugs, Diazepam, Diphenhydramine HCl, Doxycycline Monohydrate, Ibuprofen, and Ciprofloxacin HCl, with different drug loadings, were demonstrated. This miniature system is able to manufacture on-demand tablets from drug crystals by using the minimum number of excipients required for drug product performance. The critical quality attributes (CQA) of the tablets were assessed to meet the United States Pharmacopeia (USP) quality standards. All drug tablets met USP quality standards after manufacturing and after 2 weeks of accelerated stability test, except for slightly lower drug release for Ibuprofen. We foresee that advancing on-demand ready-to-use pharmaceuticals manufacturing with this type of flexible, miniature, emerging automated (integrated) system will simplify pharmaceutical formulation. This development route will provide significant economic benefits by reducing time-to-market and increasing product quality. In conclusion, this study confirmed that on-demand manufacturing can be used as a promising platform for simplifying the formulations design compared to traditional formulations.

References:

1. Azad M, Osorio Caicedo J, Brancazio D, Hammersmith G, Klee D, Rapp K, Myerson A 2018. A compact, portable, re-configurable, and automated system for on-demand pharmaceutical tablet manufacturing. Int. J. Pharm. 539 (1-2):157-164.
2. Myerson A, Azad M, Hammersmith G, Brancazio D, Osorio Caicedo J. Systems and methods for the fabrication of Ttablets, including pharmaceutical tablets. U.S. Provisional Patent Application US 2018/0271791 A1, Published on September 27, 2018.

568328 Slow Freezing Process Design for Human Induced Pluripotent Stem Cells By Modeling Intracontainer Variation

Human induced pluripotent stem (hiPS) cells are the emerging source of regenerative medicine products. Along with the successful clinical studies for treating diseases such as Parkinson's disease, or spinal cord injuries, the demand of hiPS cells is increasing. This trend indicates the necessity for establishing cryopreservation processes of hiPS cells for storage and transportation. Generally, there are two methods for freezing cells: slow freezing and vitrification. For hiPS cells, it is planed that the liquid containing hiPS cells are filled in to a container such as vial, cryopreserved, and then thawed at the point of use. Here, slow freezing, as compared to vitrification, can benefit from scalability, simplicity in operation, and no direct contact with freezing agent, such as liquid nitrogen. However, the process could suffer from the quality variation inside the vial (i.e., intracontainer), between vials (i.e., intercontainer), and between and among freezers (i.e., intra- and interfreezer).

We present a model-based design of slow freezing processes for hiPS cells considering the intracontainer quality variation. A single-cell model was developed by integrating models describing a radial and temporal temperature profile, cell volume change through transmembrane water transport, and intracellular ice formation during slow freezing. Given the cooling rate of the freezer, the vial diameter and material, and the cryoprotective agent type, the model can produce the maxima of cell volume change and the intracellular ice crystal volume as the cell quality indicators, and the required freezing time as the productivity indicator.

The developed model showed sufficient performance compared with a recently reported experimental study on hiPS cells. Upon this confirmation, the model was applied for three design cases. The choice of a cryoprotective agent, cooling rate, and vial diameter affected the quality evaluation more than the choice of the vial material. When considering the productivity as the design objective, the optimal vial diameter changed depending on the constraints of cell demand and the acceptable intracellular ice crystal volume. In the ongoing work, we are performing rigorous performance assessment of the model by extensive experiments. Advanced process design and optimization, e.g., dynamic temperature optimization, integrated design of freeze/thaw process, is another ongoing subject.

568904 Standing-Wave Design of Three-Zone, Open-Loop Non-Isocratic SMB for Purification

Chromatography with step changes in modulator properties such as pH, solvent strength, or ionic strength to facilitate desorption is widely used in the purification of proteins and other chemicals. Step changes can be incorporated into non-isocratic simulated moving beds; however, applications of such systems have been limited because one must select numerous operating parameters (zone velocities and port velocities). The operating parameters must be selected correctly to achieve high purity, yield, and productivity and depend on a large number of system parameters (feed, material, and equipment parameters). To address this challenge, the Standing-Wave Design method has been developed for three-zone, open-loop, non-isocratic, and non-ideal systems with both linear and non-linear isotherms. This method directly links the operating parameters to the system parameters. The operating parameters can be solved from a set of algebraic equations. This allows for operating parameters to be determined based on the new system parameters upon scale-up or scale down. Additionally the new productivity can be predicted without rate model simulations. In contrast, for non-ideal systems, previous literature design methods require extensive search using rate model simulations, which involve solving partial differential equations at each grid point. For each scale, a new search with rate model simulations is required to ensure the purity and yield is maintained. Two examples were tested for the effectiveness of the SWD method. In both examples, sorbent productivity was pressure limited. Higher pressure sorbents or equipment would lead to higher sorbent productivity. In the first example, a 3-zone open-loop simulated moving bed was designed and compared with an optimal batch step-wise elution system. Compared to batch step-wise elution systems, the 3-zone open-loop SMB can give an order of magnitude higher productivity in systems with weakly competing impurities and two orders of magnitude higher in systems with strongly adsorbing impurities. In the second example, an SMB designed using the Standing-Wave method achieved an order of magnitude higher productivity than a system designed using the Triangle Theory.

558662 Scaling Down to Scale-up: A Multi-Step Recrystallization for Particle Size Control

Particle size distribution control of active pharmaceutical ingredients is an important factor when developing a direct compression solid oral formulation. A variety of common methods exist for particle size control including controlled reactive, thermal, and anti-solvent crystallizations, as well as mechanical size reduction techniques such as milling.

This work demonstrates the utility of a multi-step recrystallization to control particle growth and final size distribution. A seeded, thermal, and anti-solvent recrystallization for particle size control will be presented, as well as the challenges associated with scaling up this complex process. These challenges can be understood through well-designed scaled-down experiments to map how process parameters impact material attributes.

558781 Solubility-Limited Impurity Purge in Crystallization – Theory, Applications and Development Strategies

Impurity control in pharmaceutical industry is imperative in ensuring the quality of the drug substance. Crystallization is a key unit operation in removing impurities and is relied upon across the synthetic scheme. Different mechanisms for rejecting impurities in the crystallization have been proposed, including surface adsorption, solid solutions, inclusions and solubility. Presented in this contribution is the mechanism of solubility-limited impurity purge in which the purge ability of the impurities is dependent on the respective solubilities in the crystallization solvent relative to the product. A theoretical framework is presented that is confirmed experimentally for three different types of solubility-limited impurity rejection mechanisms. A practical experimental test, the so-called *SLIP test*, is described that is able to identify the corresponding type of impurity purge mechanism. The novel approach to control purity in the crystallization is showcased using pharmaceutical compounds. Development strategies are presented that are based on the corresponding type of solubility-limited impurity purge mechanism.

572541 Understanding the Role of Cake Structure in the Filtration of Needle-like Crystals in the Pharmaceutical Industry

1. Motivation

The filtration time in the pharmaceutical industry can be critical for the feasibility of a process [1]. In this field, filtration is used to recover crystals of active pharmaceutical ingredients or intermediates resulting from purification processes [2]. The time required by this step is heavily influenced by the size and the shape of the crystals being filtered, which is a result of the nature of the compound and the way the particles are generated [3,4]. Crystals often have non-equant morphologies and, even though there is considerable knowledge on how to predict and control the final shape and size they will have, there is no direct way to forecast how they will filter. The size and shape of the particles are important for filterability as they influence the way particles pack, therefore the internal structure of the filter cake. This structure, in the end, determines the cake resistance, and hence the filtration time.

Being able to predict how populations of particles with different sizes and shapes will behave during filtration would enable us to find the optimal crystallization conditions that require the minimum combined processing time. It would improve and speed up process development allowing new drugs to get to the market sooner.

2. Methodology

Determining the way that particle size and shape affect the filter cake structure is fundamental in order to predict filterability. This prediction becomes increasingly difficult to be carried out for non-equant particles: here we present a work on needle-like crystals, which is a type of morphology that is often encountered in the pharmaceutical industry and strongly differs from ideal/model shapes such as spheres or cubes. We here propose two methods to tackle this understanding: the first is an experimental approach that takes advantage of X-ray tomography to investigate the internal structure of real filter cakes and the second one is a computational approach that simulates the packing of different size and shape distributions of particles. The two approaches together can be used to validate one another and to explore many scenarios without the need of further experiments.

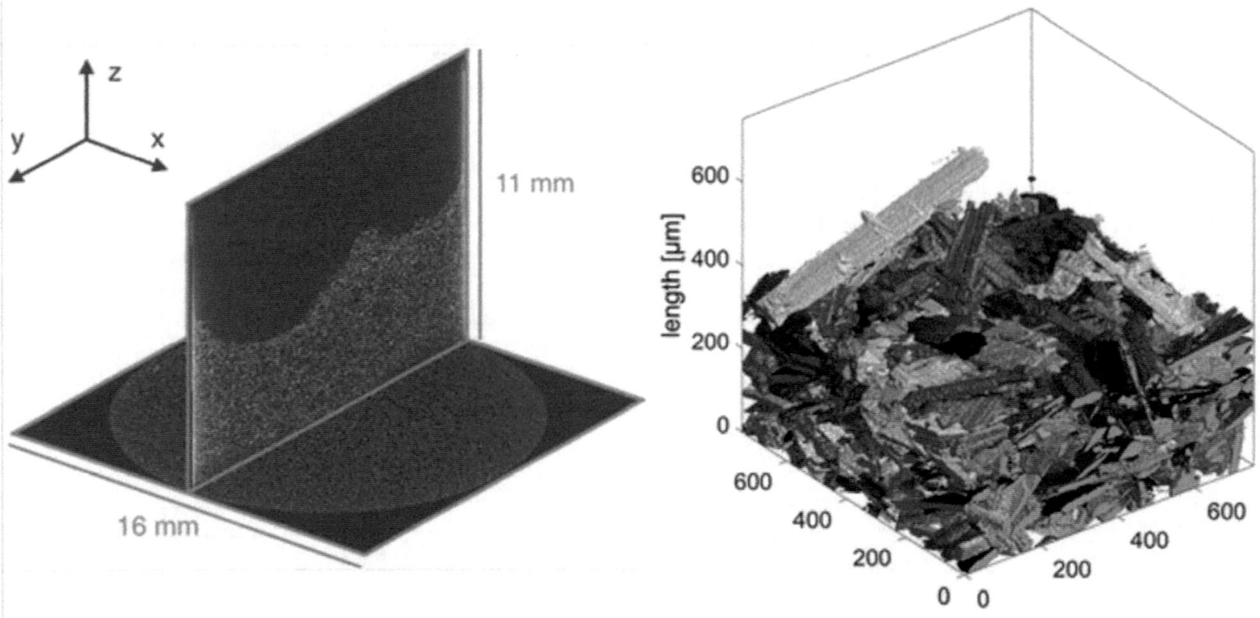

Fig. 1 – An example of the slices of a filter cake generated from X-ray tomography (left), plus the reconstructed 3-D composition for a small section of a cake.

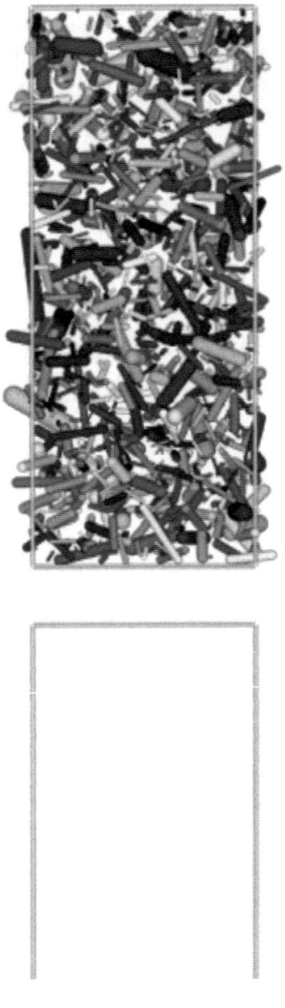

2.1. X-ray tomography

Micro X-ray tomography is a technology that enables us to recognize the composition of an object based on the density of its different components: elements with different densities will indeed cause different attenuation of the X-rays. In our work, this enables us to determine the position of particles and pores inside filter cakes, as we can detect objects down to the size of 1 micron. The output of these measurements is a series of slices that represent subsequent cross-sections of the cake. These slices together with a segmentation algorithm let us gather information on the single particles of the cake, as well as on the shape and connectivity of pores.

2.2. Random packing simulations

We aim to study systematically the effect of particle morphology (size, shape and polydispersity) on the cake structure. For this purpose, we use Monte Carlo simulations to model the final packing of particles in filter cakes. The resulting structures can be generated starting from a random configuration of dispersed particles to which a gravitational field (pressure gradient) is applied. For the simulation of needle-like systems, the spherocylinder (a cylinder topped by two spherical caps) was selected as model shape, as it is known to be an appropriate representation of elongated particles and to be computationally efficient [5].

3. Results

X-ray tomography has enabled us to notice the preferential orientation of needle-like particles inside filter cakes and how this impacts on the porosity of the system. Figure 1 shows a typical example of what can be extracted from such measurements. On the left two orthogonal slices of the cakes are shown,

Fig. 2 – An example of an initial population of sphero-cylinders before (top) and after packing (bottom).

highlighting the contrast between the particles (the lighter elements) and the pores (the darker parts). The image on the right shows the reconstructed 3-D structure of a small section of a cake, where different colors are used to distinguish the single segmented particles.

The results of Monte Carlo simulations also enable us to understand the way that particles tend to orient in the cake and how this changes by modifying the distribution of sizes and shapes, in particular its polydispersity. Figure 2 shows an example of how a system of particles resembling the experimental ones can be packed. After having checked the agreement between the experimental results and the computational ones we can use the knowledge gathered from the simulations to predict the cake structure of different populations of needle-like crystals.

References:

[1] MacLeod C. S. and Muller F. L. (2012) *Org. Process Res. Dev.*, 16, 3, 425–434.

[2] Murugesan S. (2010) *Chemical Engineering in the Pharmaceutical Industry*, 315-345.

[3] Perini G. et al. (2019) *Sep. Purif. Technol.*, 211, 768–781.

[4] Wakeman R. (2007) *Sep. Purif. Technol.*,58, 2, 234–241.

[5] Ferreiro-Córdova C. et al. (2014) *Chem. Eng. Data*, 59, 10, 3055–3060.

Acknowledgments:

The authors would like to thank EPSRC and AstraZeneca for funding the PhD project and the Henry Moseley X-ray Imaging Facility.

559709 The Effect of Particle Sedimentation on the Performance of Pressure Filters

The Effect of Particle Sedimentation on the Performance of Pressure Filters

Fragkopoulos, I.S.[1], Mitchell, N.A.[2], MacLeod, C.S.[3], Mathew, S.[4] and Muller, F.L.[1]

[1] *Chemical and Process Engineering, University of Leeds, Leeds LS2 9JT, UK*

[2] *Process Systems Enterprise (PSE) Ltd., London W6 7HA, UK*

[3] *Pharmaceutical Development, AstraZeneca, Macclesfield SK11 2NA, UK*

[4] *Global Research and Development, Pfizer, Sandwich CT13 9NJ, UK*

Pressure cake filtration is commonly used in the pharmaceutical industry for the separation of solids from the crystallisation slurry. The pressure applied has a huge impact on the particles within a cake and in turn affects the resistance of the liquid flowing through the cake [1]. This can cause the cake to compress and clog the filter medium, both of which slow down the filtration process. Although pressure can be increased to maintain a sufficient rate of filtration, this would cause the cake to further compress and/or lead to particle breakage. The filtration performance cannot currently be predicted accurately. Process scale-up is based on extensive experimentation, but scale-up surprises still occur between drug discovery and manufacturing. The main focus of this work was the development of a detailed filtration mechanistic model that takes experimental data and estimates filtration process parameters, so as to enable the prediction and control of the pressure filter performance at scale. The pressure filtration scale-up strategy that was followed in this study, is depicted in Figure 1.

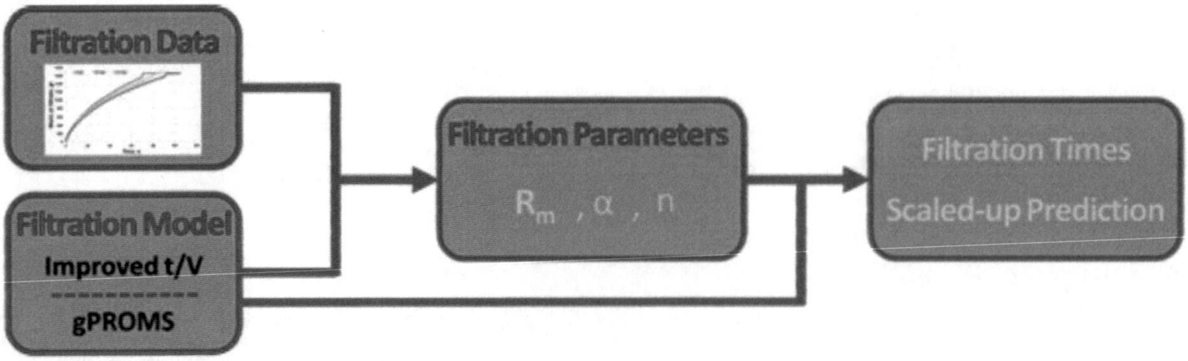

Figure 1: *Pressure filtration scale-up workflow.*

Lab-scale filtration experiments were performed for the investigation of the effect of size and shape of AZ and Pfizer materials on the filtration performance. An improved t/V filtration analysis model [2], which considers the fraction of solids that are already settled before filtering starts, was used in conjunction with the lab-scale filtration curves for the estimation of filtration parameters such as medium resistance, specific cake resistance and compressibility index of the cake.

Conventional Ruth-equation based filtration models are currently capable of describing only two extreme cases, where:

a) There is no sedimentation prior to filtration and cake builds-up while filtering, assuming a well-dispersed slurry (see Fig.2a).

b) Complete sedimentation is followed by filtration - permeation of liquid through a settled cake (see Fig.2b).

However, in reality the cake is commonly partly formed when filtration is initiated and also the slurry is not well-dispersed during the entire filtration duration (a layer of clear liquid at the top of the slurry is formed due to particle settling, as in Fig. 2c).

Figure 2: *Schematic representation of slurry in systems with (a) no crystal sedimentation prior to filtration, (b) entire crystal sedimentation prior to filtration and (c) partial crystal sedimentation prior to filtration.*

For this reason, a filtration-sedimentation lumped parameter model, currently being prototyped in PSE's gPROMS FormulatedProducts suite, was developed and used to show the effect of sedimentation during filtration on the process performance. Predicted filtration times at pilot-plant and manufacturing scales were found to increase by more than 10% and 15% respectively (for d50s of 10μm), when taking into account particle settling.

The support of the Advanced Manufacturing Supply Chain Initiative through the funding of the 'Advanced Digital Design of Pharmaceutical Therapeutics' (ADDoPT) project (Grant No. 14060) is gratefully acknowledged.

Keywords: particle settling, pressure filtration, parameter estimation, scale-up, digital design.

References

Hamilton, P., D. Littlejohn, A. Nordon, J. Sefcik, P. Slavin, J. Andrews and P. Dallin (2013). Investigation of factors affecting isolation of needle-shaped particles in a vacuum-agitated filter drier through non-invasive measurements by Raman spectrometry. Chemical Engineering Science, 101, 878–885.

Fragkopoulos, I.S., A. Shier, C.S. MacLeod and F.L. Muller (2019). Constant Pressure Filtration of Pharmaceuticals: An Integrated Experimental and Modelling Study. Chemical Engineering Science, to be submitted.

564014 Removal Dissolved Metal Impurities from API Stream Using Functionalized Silica Resins in a Packed Column

In recent years, the number of applications of transition metal-catalyzed transformations have increased due to their selectivity towards the desired product. Subsequently, the removal of metal residue from the reaction mixtures has become a challenge in the pharmaceutical industry. Purification of APIs from residual metal by traditional methods (chromatography, activated carbon, etc.) often leads to problems such as high costs, long cycle times, low efficiency, and reduced API yields. In the present work, we have demonstrated an application of a packed column with functionalized silica resin for removal of metal from an API stream. Present work shows initial screening at batch scale to identify ideal resin and then further optimization using a packed column to achieve desired metal removal from API stream. This study will also focus on the scale-up considerations for metal removal processes, optimization of process parameters (such as residence time and temperature), and modes of operation (re-circulation vs single-pass).

Disclosures:

All authors are employees of AbbVie and may own AbbVie stock. AbbVie sponsored and funded the study; contributed to the design; participated in the collection, analysis, and interpretation of data, and in writing, reviewing, and approval of the final publication.

569020 An Approach for Scaling up Jet Milling of Pharmaceutical APIs in Loop Mills of Similar Design

In the pharmaceutical industry a significant percentage of marketed drugs and new chemical entities are classified as BCS Class II therefore the need for particle size control of active ingredients in oral solid dosage forms continues to prevail. For such molecules bioavailability can be enhanced via particle size control, and milling remains a dominant approach to achieve this.

Several advances have been made with regards to available tools and methodologies to characterize the mechanical properties of APIs and predict particle breakage in milling processes. However some challenges that are still encountered include (1) Translating the characterization of mechanical properties, often conducted on single particles, to bulk milling behaviour; (2) Factors such as particle size distribution and particle shape can and do influence bulk flow and thus could impact bulk milling behaviour of pharmaceutical organic materials; (3) Some approaches take too much time or require too much material than is typically available at early stages of drug development; (4) The mode of stress operation during the mechanical properties assessment can be quite different to what actually happens during milling processes; (5) Some methodologies rank order materials with respect to milling propensity but don't necessarily quantify the amount of energy to apply in, for example, a jet milling process. Thus, some trial and error is still necessary for each material to be milled, most especially in cases where there are upper and lower particle size limits for the milled material.

This paper presents a modified approach for scaling up milling processes between lab and plant mills of similar design. Using small amounts of material the methodology involves characterizing the milling propensities of organic materials in a lab loop mill as a function of mechanistic energy input, which is then used to reproducibly scale up the milling process in a plant scale mill of similar design. An additional advantage of this approach is that since the milling propensity vs. mechanistic energy data is generated in a mill of similar design to the plant scale mill it can help with understanding, as early as possible, the risks associated with milling the material such as amorphous generation and operational issues. Furthermore, the material generated from the lab milling studies can be used for small scale development work in order to understand impacts on drug product formulation and API crystallization processes. Case studies will be presented where this approach of matching the milling energy across scales has been used to successfully achieve milled product with similar powder properties.

568834 API Micronization Strategy for De-Risking the Pathway to Commercialization

Fluid energy micronization of Active Pharmaceutical Ingredients (APIs) is a commonly utilized unit operation for enhancing active drug substance exposure related to Oral Solid Dosage (OSD) formulations of crystalline BCS Class II APIs. This exposure enhancement stems from an increase in API drug substance solubility, which results from an overall reduction in primary particle size of the API particles. In the pharmaceutical industry, the most dominant equipment used for this purpose is the jet mill micronizer, where two specific mill designs are directly applicable to these types of API particle size reduction processes. These are the spiral jet mill and the loop type jet mill. This presentation will detail the experience of scaling up the micronization of a specific investigational BCS Class II API that required a unique set of powder property specifications to prevent undesired processing problems during the downstream formulation activities. Due to this requirement for tighter control of the micronized API powder properties, and to de-risk the overall scale up to a commercial micronization process, a deeper understanding of how the selected jet mill design would influence process performance and robustness was required for maximizing the probability of a successful commercialization. In support, milling process equipment details from pilot and commercial scale activities will also be discussed along with associated data from pilot scale process development and commercial technology transfer activities. Finally, a summary of key lessons learned will be listed along with high-level guidance on how to leverage this information during process development to proactively screen for situations where equipment specific attributes related to the type of jet mill can become important towards ensuring a successful scale up of an API micronization process.

558678 Applications of Real-Time Machine Learning in Process Performance and Product Quality Prediction in Biopharmaceutical Manufacturing

In recent years, machine learning for predictive analyses has found a foothold in the biopharmaceutical industry. Various algorithms including linear regression, decision tree learning, gradient boosting, and neural networks are evaluated and optimized to deliver predictive capabilities to solve business problems. Models trained on historical data from various data sources (both discrete and continuous) are used to predict response variables of choice. At Amgen, these models are tied into an enterprise data lake in a way that enables predictions to be made in near-real time as data streams from the production line. In combination with domain knowledge, real time machine learning (RTML) has several advantages compared to more traditional prediction techniques and may be suitable for automated root cause analysis as well as driving open- or closed-loop control strategies in a GMP environment. The approach and benefits of RTML will be reviewed through case studies.

571346 A General Framework for Real-Time Risk Assessment in Pharmaceutical Processes: Application to the Automatic Rejection of Off-Spec Oral Drug Doses

A general framework for real-time risk assessment in pharmaceutical processes: application to the automatic rejection of off-spec oral drug doses

Francesco Rossi*, Sudarshan Ganesh, Linas Mockus, Gintaras Reklaitis

Purdue University, Forney Hall of Chemical Engineering, 480 Stadium Mall Drive, West Lafayette, IN 47907-2100, United States

frossi@purdue.edu

A risk assessment framework is a set of methods and tools for quantitative estimation of the risk associated with adverse events, which have potential to cause damage to people, property and the environment. Since risk is intrinsic to most human activities, risk assessment finds application in several different areas, e.g. finance (Iyer and Purkayastha, 2017), medicine (Leung et al., 2017) and geology (Cruden and Fell, 2017). In this contribution, we will focus our attention on a specific area, namely, the pharmaceutical industry.

In the context of pharmaceutical processes, risk can be defined as the probability of producing and distributing drug products that do not satisfy the quality requirements, imposed by the Food and Drug Administration (FDA). Therefore, it is evident that risk assessment must already play a very important role in the pharmaceutical sector. Indeed, it is currently used to assess whether critical quality attributes of the final drug products, produced by pharmaceutical plants, meet the FDA regulations (Figure 1). We refer to this type of risk assessment as offline risk assessment, as product quality is monitored only at the very end of all of the manufacturing stages.

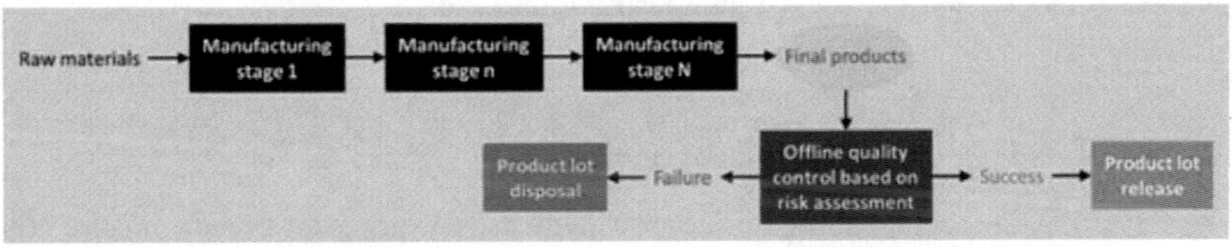

Figure 1: Conventional use of risk assessment in the pharmaceutical industry.

The use of offline risk assessment enables us to detect the presence of off-spec drug products but only after an entire batch of material has been produced, thus may lead to unnecessary economic losses and delays in meeting demand. To mitigate this problem, we propose to use risk assessment in real time, so as to detect and automatically reject off-spec material before it contaminates on-spec production (Figure 2). We refer to this type of risk assessment as online risk assessment, as product quality is continuously monitored over time, during all manufacturing stages.

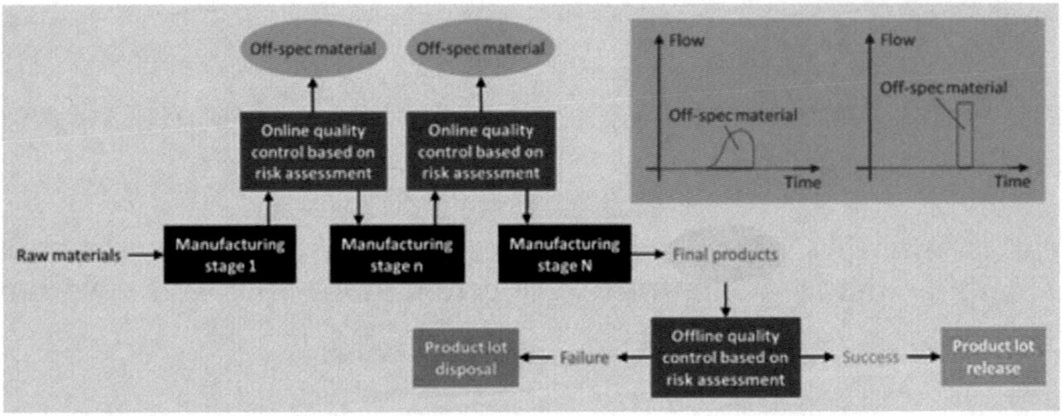

Figure 2: Innovative use of risk assessment in pharmaceutical processes.

The systematic use of online risk assessment may benefit both continuous and batch drug manufacturing plants by decreasing their environmental impact, lowering drug production costs and reducing the probability of delays in production (disposal of off-spec drug products often requires incineration, which is expensive and has non-negligible carbon footprint). On the other hand, frameworks for real-time risk assessment require efficient computational methods for estimation of probability distributions (PDFs) and for uncertainty propagation through systems of nonlinear equations, which may not be readily available. In this contribution, we overcome this difficulty by utilizing the new, efficient PDF estimation strategies, proposed by Rossi, et al. (2018).

The potential of real-time risk assessment is demonstrated on a pilot-scale drug product manufacturing plant for production of oral drug doses, equipped with an OSIsoft PI system that enables us to retrieve data from equipment sensors, in-line and at-line sensors. The results of our simulations show that the systematic use of online risk assessment can improve process economics and reduce the amount of waste, produced by the aforementioned pilot-scale plant. We also demonstrate that real-time risk assessment is superior to other conventional techniques for online detection and rejection of off-spec products, which rely on dynamic data reconciliation and/or residence time distribution.

References

Cruden, D.M., Fell, R. (2017). Landslide risk assessment. Taylor & Francis, New York (NY).

Iyer, K.C., Purkayastha, D. (2017). Credit risk assessment in infrastructure project finance: relevance of credit ratings. Journal of Structured Finance, 22, 17-25.

Leung, A.A., Daskalopoulou, S.S., Dasgupta, K., McBrien, K., Butalia, S., Zarnke, K.B., Gelfer, M. (2017). Hypertension Canada's 2017 guidelines for diagnosis, risk assessment, prevention, and treatment of hypertension in adults. Canadian Journal of Cardiology, 33, 557-576.

Rossi, F., Mockus, L., Manenti, F., Reklaitis, G. (2018). Assessment of accuracy and computational efficiency of different strategies for estimation of probability distributions applied to ODE/DAE systems. Computer Aided Chemical Engineering, 44, 1543-1548.

564354 Towards a Comprehensive Ontology to Describe a Pharmaceutical Product

The design of pharmaceutical chemistry manufacturing & controls (CMC) ontologies to fully describe a pharmaceutical product hold significant business value, given the multiple applications in efficient resource and activity planning, accelerating submissions, and building a core digital foundation for a multitude of analytics applications. The flexibility afforded by the inherent data structure of an ontology affords the adoption of an 'agile development' paradigm, wherein targeted data sub-sections are captured, while others are captured later without compromising the comprehensiveness of the solution. This talk will initially showcase a top view ontology based on product requirements, customers, decisions, activities and resources. A specific example of such an ontology instanced for multiple small and medium molecule assets will be shared. The middle layer of the ontology is based on risk-based development, and the integration of a comprehensive array of risk grids to the top ontology layer. The integration of these risks into the ontology allows for a dynamic digital connection of CMC risks directly to product requirements (that come from patients, payers, providers, manufacturing, clinic and the business), and enables powering decision making (activity selection, resource allocation) to optimally reduce CMC risks subject to constraints. Lastly, specific risk grid attributes within the risk grids are linked to instrument data (powered by instrument-specific ontologies such as those from Allotropė) to build the connection to the specific data elements. A case study showcasing the power of setting up product requirements, that lead to decisions to commission activities which consume resources, but also reduce the CMC risks by generating specific data, will be shared. Longer term plans to design and install these ontologies is a pre-competitive collaboration topic across pharma and will be discussed

571944 PRIME: Process Ranking of Inputs from Manufacturing and Environment

This work presents a newly developed knowledge-based assessment tool intended to evaluate and classify API synthesis processes, this way guiding drug development in providing reliable and efficient processes at a rapid pace. This tool incorporates commonly known green chemistry metrics, including atom economy, E-factor, volume-time-output, and the semi-quantitative EcoScale tool.[1], [2] Additionally the recent concept developed by the IQ consortium of the innovation green aspiration level (iGAL) is also being calculated in the tool.[3]

The concept of green chemistry proposed to change the way chemistry and chemical engineering was done, by enabling a "greener" and sustainable process, which involved waste prevention, lower energy consumption, synthetic efficiency, and reduced hazardous components (used and generated). To reduce the subjectivity inherent in this type of evaluation, scientists have developed several quantification metrics over the years, which would allow for a clear and fast way to obtain information on the greenness of any type of chemical process, resorting to simple mass and energy calculations, health, safety and environmental (HSE) considerations, and lifecycle impact approaches. These metrics would also enable to predict how a certain modification, such as replacement of a solvent or elimination of a unit operation, would influence the environmental impact and efficiency of a process.

However, these metrics often represented individual approaches of evaluating one single aspect of the API process, this way not providing enough information to fully evaluate its efficiency, for it depends on numerous factors. Therefore, only a holistic approach of assembling several relevant green metrics can provide an adequate assessment of chemical processes.

By gathering all these inputs and assembling them in a structured framework, chemical processes can be evaluated in terms of synthesis strategy, waste generation, productivity, quality, process conditions, raw materials classification, and health, safety and environmental considerations, achieving a final classification based on every single one of these key aspects to truly determine process efficiency. The developed assessment tool was successfully implemented on various drug development projects, providing cross-project comparison and the creation of a centralized database for the company's process knowledge. Additionally, a critical aspects analysis allowed for a rapid detection of what criteria should be improved on a given process, and a case study evaluation of a project with multiple process revisions over time allowed for its improvement-evolution assessment.

References:

[1] R. Dach, J. J. Song, F. Roschangar, W. Samstag, and C. H. Senanayake, “The Eight Criteria Defining a Good Chemical Manufacturing Process,” *Org. Process Res. Dev.*, vol. 16, no. 11, pp. 1697–1706, Nov. 2012.

[2] D. J. C. Constable, A. D. Curzons, and V. L. Cunningham, “Metrics to ‘green’ chemistry - Which are the best?,” *Green Chem.*, vol. 4, no. 6, pp. 521–527, 2002.

[3] F. Roschangar *et al.*, “Inspiring process innovation via an improved green manufacturing metric: iGAL,” *Green Chem.*, vol. 20, no. 10, pp. 2206–2211, 2018.

561716 Natural Language Processing - What Can Chemistry Learn from Finance?

If a machine were to understand human language at scale, there are so many tasks that can be automated; there are so many insights that can surface easily. While the discipline of Natural Language Processing (NLP) has been around for decades, it has become especially meaningful with recent developments in Artificial Intelligence. This presentation will the compare latest approaches for NLP across sectors with their pros and cons, and suggest some best practices on how to start with AI and NLP with your team.

560367 Economic Evaluation Tool for Holistic Design of Solid Drug Product Manufacturing Processes Considering Continuous Technology

Continuous technology is now becoming a realistic option in addition to the conventional batch technology, which makes process design more complicated. In the clinical development of drugs, designers have to decide a process from the various alternatives considering product characteristics and uncertainty. In the literature of continuous technology, modeling of unit operations, e.g., modeling of a milling process using discrete element method [1], is advancing. However, generalized models and tools are required to support holistic design that includes the technology selection.

This work presents an economic evaluation tool for the design of solid drug product manufacturing processes considering continuous technology. The tool integrates our previous studies [2, 3] and consists of three steps: (i) comprehensive alternative generation, (ii) input parameters settings, and (iii) stochastic economic evaluation. In the first step, 4,477 alternatives of processes including the choice of continuous technology are generated by representing possible combinations of unit operations using a superstructure. Estimated values of input parameters are set considering characteristics of target products, and probability density distribution of all parameters are defined in the second step. Finally, in the third step, stochastic economic evaluation is performed for all alternatives by Monte Carlo simulation. Net present value was applied to an evaluation indicator, which considers clinical development cost as well as annual cash flow in the production stage. Estimated ranges of net present value for all alternatives are output at the end.

We performed several case studies, which differ in product characteristics such as demand and raw material price, to demonstrate the effectiveness of the developed tool. Appropriate technology was found to depend on defined probability density distribution of input parameters. For some cases, batch technology was better in terms of expected values while continuous technology was better in terms of probability. Sensitivity analysis was performed for the situations which are difficult to interpret, and then, high effective parameters were analyzed. In these ways, we confirmed that the developed tool works for the rational design including the technology selection and the investigation of high effective parameters for decision making.

[1] N. Metta, M. Ierapetritou, R. Ramachandran, Chem Eng Sci, 178, 211–221 (2018)

[2] K. Matsunami, T. Nagato, K. Hasegawa, H. Sugiyama, Int J Pharm, 559, 210–219 (2019)

[3] K. Matsunami, T. Miyano, H. Arai, H. Nakagawa, M. Hirao, H. Sugiyama, Ind Eng Chem Res, 57, 9798–9809 (2018)

559263 Impact of Material Properties and Process Parameters on Residence Time Distributions in Pilot Scale Continuous Powder Blending

The U.S. Food and Drug Administration (FDA) has taken multiple steps to support the implementation of the emerging technology continuous manufacturing (CM), including the release of a recent draft guidance regarding quality considerations for CM [1, 2]. Continuous processes are dynamic systems and knowledge of process dynamics is essential for identification and mitigation of risks to product quality. In this regard, residence time distributions (RTDs) offer powerful insight into flow processes. RTDs can be used to trace the flow and dispersion of material through a continuous process. Development of traceability algorithms for RTDs enables prediction of how disturbances will propagate through a process, allowing for isolation of out-of-specification material. For these reasons, interest in implementing and expanding the application of RTD models for powder flow processes in pharmaceutical industry is growing. Given the high impact of using RTD models for feed-forward process control, understanding how material properties and process parameters involved in the establishment of such models impacts their performance is critical to ensure product quality in continuous powder flow processes.

This work investigates the effects of material properties and process parameters on the RTD of active in continuous powder blending at the pilot scale (10–20 kg/hr). Indomethacin is used as the model active pharmaceutical ingredient (API), while the bulk material consists of a combination of two commonly used tablet fillers (lactose and microcrystalline cellulose). Material properties in the powder blend are manipulated via flow rate adjustment of the individual components fed into the blender. Separately, blending process parameters investigated include blender speed, screw element configuration, and total throughput. The effects of changing to another API with similar flow properties and switching to a different batch of the same API are also evaluated. In all cases, the composition of the exiting blend is monitored in-line via near infrared spectroscopy and confirmed via off-line measurement. These measurements are used to assess the RTD of API associated with each set of conditions, which are then compared to elucidate the major factors influencing the flow rate and degree of dispersion of API in continuous powder blending at pilot scale.

Disclaimer: This work reflects the views of the authors and should not be construed to represent FDA's views or policies.

1. Lee, S.L., et al., *Modernizing Pharmaceutical Manufacturing: from Batch to Continuous Production.* Journal of Pharmaceutical Innovation, 2015. **10**(3): p. 191-199.
2. FDA, *Quality Considerations for Continuous Manufacturing - Guidance for Industry*. 2019.

571674 Advanced Quality Control Method for a Continuous Wet Granulation Tableting Line

Recent success in academia in combination with the guidance and support of regulatory bodies (FDA, EMA, etc.) have increased the attractivity of continuous manufacturing for pharmaceutical producers. The acceptance and approval of continuous production methods is especially high in secondary production of solid oral dosage forms, which can be realized through a continuous wet granulation tableting line (GEA ConsiGma 25). The line consists of feeders for pre-mix and granulation liquid, a twin screw granulator, a fluid bed dryer, a mill, further feeders for dosing of an external phase, a blender and a tablet press. The material is either transported continuously (feeding, granulation and tableting) or in "mini-batches" (drying, milling and blending) through the single unit operations. The size of the "mini-batches" or so-called product keys (PKs) is defined by the line throughput and the filling time of the dryer cells. Those PKs are then kept separated from each other, i.e., intermixing between the PKs is avoided until the PKs enter the tablet press. From a material tracking point of view, this approach offers simplicity, since no residence time distribution (RTD) of the material has to be considered for the mentioned unit operations. However, the twin screw granulator and the tablet press (including hopper and feed frame) show a RTD. Knowledge of the complete RTD, and therefore, the behavior on material propagation through the line can be used to support quality assurance. Under application of split feeding or when API is dosed via granulation liquid, mass flow disturbances of the feeders or deviations in pre-mix concentration can lead to out of spec granules. By means of RTD modeling the PK and tablet concentration can be predicted after the granulation and the tableting, respectively.

In our study, we have experimentally captured the residence time distribution of the granulator and the tablet press. Mathematical models for single unit operations were parameterized and combined to a RTD model of the entire line, which enables material tracking as well as prediction of PK and tablet concentration. The model was validated by an experimental characterization of the complete line (from outlet of the first feeder until the tablets exiting the press). The presentation highlights the utilization of this model for quality assurance of the final dosage form.

571800 Investigating Effects of Dynamic Process Variability in Continuous Direct Compression on Tablet Quality Attributes through a Science-Based Digital Twin

Continuous direct compression (CDC) is emerging as a common platform for continuous tablet manufacturing. In a CDC line, several sources of process variability are introduced through individual unit operations, including feeder flow rates (both due to refill events and higher frequency noise), micromixing of the blend, and tablet press flow and die filling behavior. These dynamic disturbances can influence the final tablet composition, porosity, strength, and mass.

In this work, the combined effects of dynamic sources of process variability on tablet quality attributes will be investigated through a science-based digital twin encompassing loss-in-weight feeders, refill units, continuous blenders, a surge hopper, a feed frame, and a tablet press, shown as a flowsheet model in Figure 1. Model validation requirements and workflows for the loss-in-weight feeder and continuous blender will be presented, and the impact of feeder control modes and settings will be explored. The significance of the height of the material in the surge hopper on disturbance propagation will be demonstrated. Dynamic variability in the tablet press will be introduced through variations in blend composition and bulk density in the die. The tablet-to-tablet variability in quality attributes will be quantified as distributions in quality attributes as predicted by the model for the tablet tester. The implications for process risk assessment and mitigation and process control strategies will be discussed.

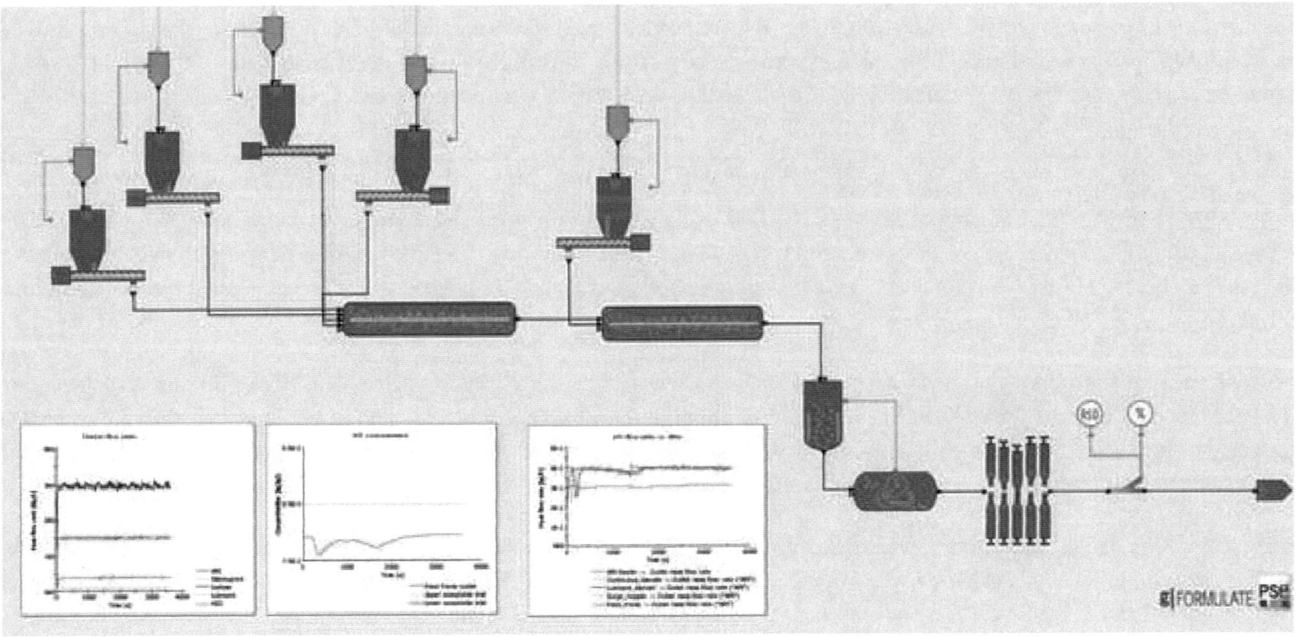

Figure 1: Flowsheet model of continuous direct compression line, including five feeders, two blenders, a surge hopper, feed frame, and tablet press and associated sensors. Results figures show fluctuations in feeder flow rates and blend composition throughout the system in response to disturbances.

572418 Use of Blue Dye Powder Tracer to Evaluate the Residence Time Distribution of Continuous Manufacturing Direct Compression Equipment

Use Of Blue Dye Powder Tracer To Evaluate The Residence Time Distribution Of Continuous Manufacturing Direct Compression Equipment

Authors Bill Engisch, Mike Kennedy, Steve La

Affiliation/Company Amgen, Inc.

Purpose

Continuous manufacturing (CM) of solid oral dosage drug products is a rapidly emerging technology being implemented by the pharmaceutical industry to reduce development and commercial costs, improve product quality and operational efficiency, and provide a flexible manufacturing network. The residence time distribution (RTD) of a CM process is a powerful characterization tool that can be used to develop robust process control through predictive modelling. There would be great advantages to being able to understand the RTD through a visual API surrogate that can be qualitatively observed, quick to implement, and independent of traditional NIR methods that are instrument dependent. This has a huge advantage early in development when API may be limited, analytical methods such as NIR have not yet been developed, or equipment access is limited. Therefore, the purpose was to develop and verify a visual API surrogate method for evaluating RTD.

Methods

The work presented here is from a case study for testing the feasibility of manufacturing a solid oral dosage formulation via continuous direct compression (DC) on a GEA CDC-50 (commercially available integrated CM DC line). During the study, a method was developed to assess residence time through tracer-pulse addition experiments using a small quantity of FD&C Blue No. 2 Aluminum Lake dye as a tracer and collection of time-stamped samples from the output of the tablet press.

Results

Prior to an API trial on the CM equipment, a placebo trial using a similar formulation composition with a surrogate for the API was executed. During these placebo runs, the residence time as a function of CM process conditions was evaluated with pulse additions of blue dye. Collected tablets were compared to a pre-made set of color standards with known concentration for quick qualitative analysis.

Post-analysis involved quantitative measurement of the blue dye tablets via a photo spectrometer. Color difference (delta E*ab, calculated based on baseline white tablets via the CIE76 formula) is plotted against sampling time in the figure below (red diamonds). The shade of blue from the pulsed-in dye correlates with concentration. The resulting concentration-time profile of tablets from the CDC-50 experiments (shown as blue diamonds) was translated into a RTD via a fit to the axial dispersion model.

During the API trials, samples from a transition experiment (change between API loads) were collected and afterwards assayed by HPLC, verifying the RTD prediction derived from the quantitative analysis of the blue dye containing placebo tablets. See the figure below; the model and assayed tablets are overlayed.

Conclusion

The blue dye method provided information needed to predict system response without consuming a limited API supply. RTD is a tool that provides valuable insight into any continuous process, and in this case, it was used to predict the transition time needed to reach a steady state following process or formulation changes.

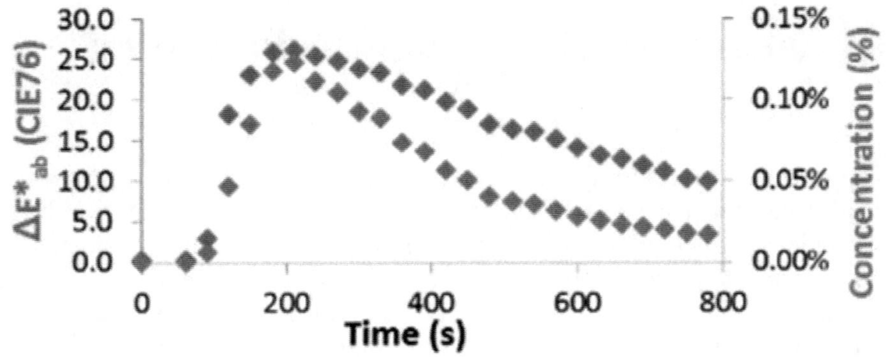

30.0
25.0
20.0
15.0
10.0
5.0
0.0
ΔE*ab (CIE76)
0.15%
0.10%
0.05%
0.00%
Concentration (%)
0
200
400
600
800
Time (s)

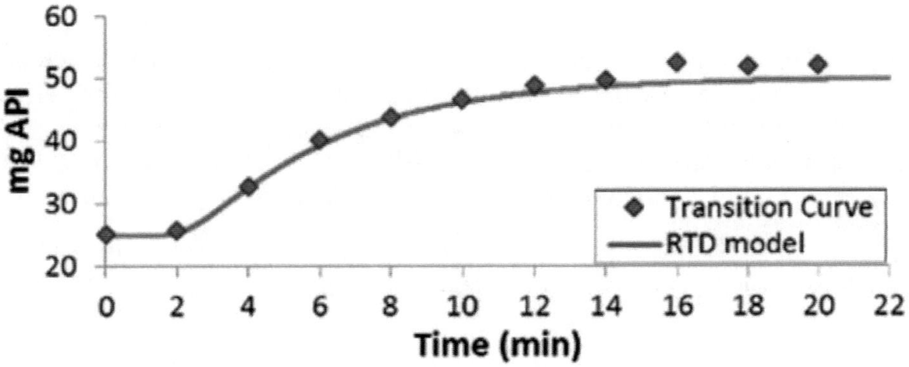

60
50
40
30
20
mg API
Transition Curve
RTD model
0
2
4
6
8
10
12
14
16
18
20
22
Time (min)

574212 Transient Mean Age Analysis in Continuous Processing Systems

Transient mean age analysis is an approach for predicting spatial variations in local age and local mean residence time across an open system. Like tracer particle-based approaches, transient mean age can be used to predict the total tank mean residence time and quantify non-idealities in the total tank residence time distribution (RTD). Beyond tracer particle-based approaches, however, transient mean age illuminates the specific regions and pathways inside the tank driving the exit RTD. In this way, transient mean age analysis provides quantified and actionable insights into the internal flow field driving the total tank properties measured at the exit.

This work, we show how transient mean age can be combined with large eddy simulations to perform real-time, transient mean-age analyses. After a brief review of the transient mean age theory, we perform self-consistent simulations to predict a total tank residence time using both mean-age and tracer particles approaches. We compare results and discuss the strengths/convergence behavior of each methodology. We then validate these predictions to experimental data, obtained for system with the same geometry and flow configurations used in the models.

577524 Product Robustness within a Dry Granulation Continuous Manufacturing Line

Successful manufacturing of pharmaceutical dosage forms require that the concentrations of formulation components are on target and consistent throughout product manufacturing. For continuous manufacturing of oral solid dosage forms the concentration of each formulation component is controlled by the dosing of loss-in-weight (LIW) feeders as well as mixing that occurs downstream of the feeders. Several authors have demonstrated that the residence time distribution of the system can be used to predict product concentrations given upstream disturbances for a direct compression line[1] or a continuous blender[2].

The current work looks to employ a similar methodology to predict the effect of concentration disturbances at the feeders within a dry granulation line which contains more unit operations that can impart more mixing. A case study showing the residence time distribution within the dry granulation line and the robust range of disturbances that can be accommodated will be presented. The system dynamics along with the response rate of the LIW feeders indicate that short-term fluctuations are low risk to the product. Strategies to ensure long term variation does not occur will be discussed.

References

1. García-Muñoz, S. , Butterbaugh, A. , Leavesley, I. , Manley, L. F., Slade, D. and Bermingham, S. (2018), A flowsheet model for the development of a continuous process for pharmaceutical tablets: An industrial perspective. AIChE J., 64: 511-525. doi:1002/aic.15967
2. Sebastian Escotet-Espinoza, Sara Moghtadernejad, Sarang Oka, Zilong Wang, Yifan Wang, Andres Roman-Ospino, Elisabeth Schäfer, Philippe Cappuyns, Ivo Van Assche, Mauricio Futran, Fernando Muzzio, Marianthi Ierapetritou,Effect of material properties on the residence time distribution (RTD) characterization of powder blending unit operations. Part II of II: Application of models,Powder Technology,Volume 344,2019,Pages 525-544,ISSN 0032-5910,https://doi.org/10.1016/j.powtec.2018.12.051.

571653 Development of a Segregated Cell Growth Model Structured By Dynamic Flux Balance Analysis

Development of a Segregated Cell Growth Model Structured By Dynamic Flux Balance Analysis

Barbosa R[1], Skeene K[2], Diaz-Fernandez P[2], Finka G [2], Kontoravdi C[1]

[1]Department of Chemical Engineering, Imperial College London

[2]GlaxoSmithKline, Stevenage Gunnels Wood Road SG1 2NY

Fed-batch Chinese Hamster Ovary (CHO) cell systems are the most efficient processes for large-scale manufacturing of Monoclonal Antibodies (mAbs), introduced as cancer biotherapeutics. CHO cells undergo changes in their internal controls in response to intracellular and extracellular conditions affecting their growth and the quality of mAbs they secrete. Predictive mathematical modelling can be applied as a basis for optimisation and control of often expensive, time consuming and hard-to-analyse biological and process parameters that affect cell proliferation and product titer. The cellular transition through different metabolic states displays random and deterministic features, which can be integrated into a more refined modelling approach. Despite recent advances in mathematical bioprocess engineering, most cell growth model structures lack the intrinsic balance among the series of dynamic behaviours under different metabolic states. To address this issue, we have designed a novel segregated kinetic model describing by the transitions in metabolic behaviour, which are linked to cell cycling activity. We also integrate in the model formulation exercise the results of Flux Balance analysis (FBA) as a means to determine model structure and introduce reasonable constraints for metabolic parameter values and therefore expedite the parameter estimation exercise. Simulation results show a better agreement with the observed data than previous unsegregated models as the co-existing cycling and resting cell population dynamics are incorporated and validated. This modelling approach further aids systematic experimental design through quantitative and qualitative closed loop analysis, with the aim of predicting, controlling and optimizing the manufacturing of mAbs."

571282 A Holistic Modelling Framework Describing CHO Cell Metabolism and Both Antibody and Host Cell Protein Glycosylation

The quality of pharmaceutical proteins, including monoclonal antibodies (mAbs), is greatly influenced by N-linked glycosylation. Glycosylation initiates in the Endoplasmic Reticulum (ER) followed by extensive modifications in the Golgi apparatus of cells. While several experimental strategies have been implemented to control recombinant protein glycosylation through feeding strategies and gene editing techniques, mathematical models have been also widely used to describe, predict and, ultimately, influence the produced glycoform *in silico*. Among mathematical models, kinetic models, although often detailed and mechanistic, are usually limited to specific operating conditions because of the inability to predict *a priori* regulation over enzyme expression in response to changes in the cellular environment.

In the present work, a holistic model is utilized to describe changes in mAb and host cell protein (HCP) glycoform distribution in fed-batch cultures of Chinese hamster ovary (CHO) cells under different strategies of galactose and uridine feeding, accounting for enzyme regulation. Specifically, a previously developed software (GLYMMER) is used to construct the network for HCP glycosylation and subsequently reduce network's complexity to accurately describe the experimentally observed glycoprofile of HCPs in CHO cells. Stoichiometric tables extracted from GLYMMER are then fed to a kinetic modelling framework describing (i) CHO cell metabolism, (ii) Nucleotide Sugar Donor (NSD) synthesis and (iii) mAb N-linked glycosylation, resulting in a model that is able to simultaneously describe both HCPs and mAb N-linked glycosylation, based on cellular metabolism. The framework includes a refined glycosylation model that describes the reactions taking place in both the ER and Golgi apparatus, introducing this way the variability of initial, high-mannose structures as an input to the processing steps in the Golgi apparatus. The model uses five continuous stirred tank reactors to describe and segregate the reactions in the ER and Golgi apparatus, including four ER-localized and thirteen Golgi-localized enzymes. Substrate-driven enzyme regulation is also introduced in the model to account for the overexpression of glycosylation enzymes in response to culture supplementation with NSD precursors.

The modelling framework successfully describes the experimentally measured glycoform distribution of the secreted mAb and the intracellular HCPs of CHO cells in a fed-batch culture experiment, within a maximum 3% error range, while at the same time closely describing CHO cell metabolism and antibody production ($R^2 > 0.9$). The proposed holistic modelling framework sets the basis for the *in silico* prediction and control of the product-of-interest and HCP glycoprofile, providing insights on the usage of NSDs towards mAb and HCP glycosylation. In addition, the glycosylation model can be easily adapted to broader reaction networks yielding highly complex glycan structures, using stoichiometric tables obtained by the GLYMMER software and automatically generating the reaction network while minimizing the simulation time by approximately 90% compared to the initial modelling framework.

568248 Dynamic Simulation and Process Assessment of Cultivation Modes in Monoclonal Antibody Production Using Pilot-Scale Experimental Data

The market size of biopharmaceuticals has been expanding, especially, for monoclonal antibodies (mAbs). In the cultivation step of the production, host cells such as Chinese Hamster Ovary (CHO) produce mAb. There are mainly three established cultivation modes, Batch, Fed-batch, and Perfusion. Recent advancements in cell design include the development of host cells which can tolerate high cell density cultivation. Development of continuous operation of the downstream processes following cell cultivation is ongoing to improve the overall process productivity. Such developments enable the connection of the output of cultivation units run in perfusion mode, where continuous feeding and removal of mAb are conducted, can be connected to the continuously run downstream processes. The number of process options is expanding in lieu with developments in the cell performance, however process assessment efforts have been using fixed expected cell performance results, e.g. output titers without considering recent developments or fully exploring the potential range of design options.

The purpose of this work is to integrate recent experimental cell developments with the developed cultivation models to map out preferential production modes through dynamic simulation and process assessment with respect to productivity and economic aspects. The work also explores the impact of potential future developments in the cell design on the process performance. The approach consists of three steps: (I), Based on the previous work [2][3], the mathematical cultivation model for each mode was developed. Model parameters were fitted using experimental data from a pilot-scale research facility; (II), dynamic simulation was conducted using the developed models to explore the impact of process parameters, e.g. production scale and size and the produced mAb titer on the process performance and preferential production mode. (III) As the final step, a map was created for the mode minimizing the cultivation time and operating cost.

The results showed that the perfusion-favorable operating range was larger than batch and fed-batch especially at larger production scales, however the fed-batch favorable operating area widened for processes employing cells with higher doubling rate. Results were validated with experimental data from the pilot-scale facility. Similar analysis is conducted for further cell parameters.

[1] Bunnak, P., Allmendinger, R., Ramasamy, S. V., Lettieri, P., & Titchener-Hooker, N. J. (2016). Life-cycle and cost of goods assessment of fed-batch and perfusion-based manufacturing processes for mAbs. Biotechnology Progress, 32(5), 1324–1335. https://doi.org/10.1002/btpr.2323

[2] Xing, Z., Bishop, N., Leister, K., & Li, Z. J. (2010). Modeling kinetics of a large-scale fed-batch CHO cell culture by markov chain monte carlo method. Biotechnology Progress, 26(1), 208–219. https://doi.org/10.1002/btpr.284

[3] Kornecki, M., & Strube, J. (2018). Process Analytical Technology for Advanced Process Control in Biologics Manufacturing with the Aid of Macroscopic Kinetic Modeling. Bioengineering, 5(1), 25. https://doi.org/10.3390/bioengineering5010025

560868 Mathematical Modeling of the Freezing of Pharmaceutical Solutions

AIChE 2019 Abstract

Mathematical Modeling of the Freezing of Pharmaceutical Solutions

A convenient formulation for (bio)pharmaceutical drugs is as a freeze-dried product. The product is formed by lowering the temperature of an aqueous solution to below freezing to generate ice, and then a vacuum is applied to sublimate the ice and remove water in the form of vapor. This process is carried out at low temperatures and is suitable for thermally sensitive drugs, and a freeze-dried product is less costly to store and transport than use of freezing alone. Freeze-dried products can be reconstituted quickly and easily.

A typical freeze-drying cycle consists of three phases, namely freezing, primary drying, and secondary drying. The product is firstly poured into single-dose containers, which are then loaded onto the shelves of the chamber of a freeze drier. A technological fluid flows inside the shelves and removes the heat from the product, thus inducing the crystallization of the (free) water; part of the water remains bound to the product molecules without crystallizing. Then, in the primary and secondary drying phases, respectively, the frozen and bounded water is removed. In the primary drying, the pressure is reduced well below the triple point of water, and heat is provided to sublimate the ice crystals. In secondary drying, the bounded water absorbed into the solid structure is desorbed by increasing the temperature until a desired value of the relative humidity is reached.

The freezing step plays a key role in the overall process, as the nucleation and crystal growth kinetics determine the number and size distribution of crystals formed. During the sublimation stage, the water vapor flows from the frozen product to the drying chamber through the porous structure. A highly porous structure enables a higher sublimation flux, due to the lower transport resistance of the dried product, while having an opposite effect on the rate of the secondary drying, as the surface area to volume ratio is lower. As the porous structure directly influences the drying rates, and product temperature results from a thermal balance between the heat transferred to the product and the heat removed through sublimation, the porous structure also affects the spatial temperature distribution during the drying stages. The temperature is a critical variable that must be carefully monitored to avoid jeopardizing product quality.

A full understanding of the freezing step would enable a real-time prediction of the axial distribution of the pores inside the product, which would advance the monitoring, control, and optimization of the freeze-drying process. The nucleation of ice crystals is intrinsically stochastic, which can induce high variability of the porous microstructure of the material inside the vials. The differences in the nucleation temperatures experienced by the different vials are one of the main sources of in-batch variability of a freeze-drying process.

Empirical approaches have been proposed for estimation of the porous structure from the variables that are directly accessible. A series of empirical formulas have been proposed to link the pore diameters to the velocity of the freezing front and the thermal gradients inside the frozen layer. The empirical dependency of the parameters in these models on the application, product, and process conditions – together with the lack of any physical foundation in this approach – makes the results obtained neither reliable nor accurate. Arsiccio et al. (2018) first proposed a mechanistic model, based on an energy balance applied to a slice of the product, for the prediction of the pore average diameters once the information on the thermal evolution of the process is available. This approach, although much more physically grounded than those previously discussed, neglects the nucleation rate and only focuses on the crystal growth kinetics. Furthermore, the model does not provide a full characterization of the pore distribution.

This presentation develops and validates a new mathematical model of the freezing step of a (bio)pharmaceutical solution. Both nucleation and crystal growth kinetics are modeled and included in a one-dimensional population balance that describes, given a product temperature measurement, the evolution of the pore size distribution. The kinetic parameters are estimated from least-square fitting on the experimentally measured induction times obtained on a set of seven batches of ten vials each. The nucleation rate is modelled using a stochastic model in the form of a Master equation, following the approach of Goh et al. (2010), which assumes that the nucleation of crystals is a nonstationary Poisson process.

The experiments were performed inside laboratory-scale equipment (Telstar, Spain) and, in each batch ten vials, 10R (ISO 8362-1) are filled with 5 ml of a 5% b.w. solution of sucrose. Each test was 2.5 h long and the cooling policy was to simply cool to –50°C as fast as possible. The vials were placed inside the chamber and monitored using the sensor presented by Lietta et al. (2019) based on infrared imaging.

The crystal growth rate was experimentally observed to be linearly dependent on the chemical potential difference between the ice and the liquid solution. The chemical potential of the solid phase was modelled according to Corti et al. (2010), while the parameters for the UNIQUAC model of the sucrose solution were taken from Cattè et al. (1994). Experimental evidence indicated fast nucleation, with the material changing its optical properties (turbidity) within less than one second. To model this behavior, nucleation was implemented as a boundary condition on the population balance. An analytical solution of the model was derived using the method of characteristics, making the model suitable for real-time application.

A more general mathematical model, including the aforementioned population balance, of half a vial filled with a solution of sucrose was developed and used to validate the approach. A modified porous media approach was used to model the heat and mass transfer occurring in the system. The freezing rate (that is, the mass rate of conversion of water into ice), which couples the solution depletion in the mass balance and the heat generation, was calculated from the third moment of the number density function obtained from the population balance. The nonlinear set of differential equations was solved via a finite volume numerical scheme. Both the simulated temperature profile and the forecast pore structure were validated. The spatiotemporal evolution of the temperature of the product in different axial position across the vial wall was extracted from real-time infrared video images, while the experimental pore distribution was obtained from the analysis of Scanning Electron Microscope (SEM) images of two vials after being completely freeze-dried. In all cases, a good agreement was observed between model and experiments. When the equilibrium was reached between the technological fluid at –50°C inside the shelves and the product, the model predicts a 18% b.w. water bounded to the solid structure. Experimental measurements ranging between 16% to 20% are reported in literature. The end time of the primary drying stage, and the maximum temperature inside the material, simulated through a simplified model of the process and the pore distribution forecast, were in good agreement with the experimental values.

The purpose of the developed model was for coupling with the real-time measurements obtained from the infrared video camera. The resulting Process Analytical Technology (PAT) has the potential of accelerating the development and optimization of a freeze-drying cycle and the implementation of a physically grounded Quality-by-Design approach for the manufacturing of pharmaceuticals.

Future developments of this work should validate the performance of this model on different case studies, namely active pharmaceutical ingredients, cooling strategies, and controlled nucleation techniques such as ice fog and vacuum- and pressure-induced nucleation. Also of interest is the development of a deeper understanding and modeling of the dynamics of ice crystal nucleation – including its unstable propagation in a short period of time – and of asymmetries observed in the pore distribution of the drug product.

Keywords: Master equation; mathematical modeling; pore distribution.

References:

Arsiccio, A., Barresi, A.A., Pisano, R. (2017) Prediction of ice crystal size distribution after freezing of pharmaceutical solutions. Crystal Growth & Design 17, 4573-4581.

Catté, M., Dussap, C.G., Achard, C., Gros, J.B. (1994) Excess properties and solid-liquid equilibria for aqueous solution of sugar using a UNIQUAC model. Fluid Phase Equilibria 96, 33-50.

Corti, H.R., Angell, C.A., Auffret, T., Levine, H., Buera, M.P., Reid, D.S., Roos, Y.H., Slade, L. (2010) Empirical and theoretical models of equilibrium and non-equilibrium transition temperatures of supplemented phase diagrams in aqueous systems. Pure and Applied Chemistry 80, 1065-1097.

Goh, L., Chen, K., Bhamidi, V., He, G., Kee, N.S., Kenis, P.J.A., Zukoski, C.F., Braatz, R.D. (2010) A Stochastic model for nucleation kinetics determination in droplet-based microfluidic system. Crystal Growth & Design 10, 2515-2521.

Lietta E., Colucci D., Distefano G., Fissore, D. (2019) On the use of IR thermography for monitoring a vial freeze-drying process. Journal of Pharmaceutical Sciences,108, 391-398.

576055 Combining Numerical Modelling (CFD), Statistical Modelling (MVDA) and Laboratory Experiments for the Optimization of Large Scale UF/DF Operations

Plant scale experiments are both costly and time consuming to run. Typically, in pharmaceutical manufacturing, these experiments involve large quantities of high value product. APC combined multiple models; numerical, statistical and laboratory-scale, to optimise large scale UF/DF process runs. Computational Fluid Dynamics (CFD) used direct numerical simulations to model the developed flow and associated velocity and shear profiles. A representative laboratory scale unit operation was built, used to test the outcomes from CFD and develop a predictive MVDA model in a time and cost-effective manner.

CFD modelling was deployed to gain an understanding of two UF/DF manufacturing unit operations at two manufacturing sites. One site was exceeding the specified protein aggregation level while the other met the criteria. CFD determined the hydrodynamic environment generated within each site and potential factors (pipe network, elbow joints, control valves, etc.) contributing to the protein aggregation levels. CFD showed that shear stress within the systems was related to flowrate, pipe diameters, elbow joints in the pipe network and the actuator valve.

A numerical lab-scale model that could be constructed in a laboratory and replicated the shear effects present in the failing manufacturing UF/DF process was developed. The lab-scale unit operation accurately represented the process by including steel tubing, elbow joints and a T-valve (to mimic the disruption to flow caused by an actuator valve). A 0.29 % increase in aggregation, similar to levels observed at the failing site, was achieved during runs of the CFD developed lab-scale model. MVDA models were built to correlate the Raman spectral data to off-line SE-HPLC protein aggregation data, enabling the prediction of protein aggregation.

This integrated approach between mathematical models and experiments allowed for the rapid assessment of potential practical solutions on a laboratory scale unit operation before large scale pharmaceutical manufacturing runs. While the MVDA model allowed for real time assessment of critical quality attributes.

574797 Engineering-Driven Application of Quality By Design to the Gmma Platform: Enabling the Production of Affordable and Effective Vaccines for Low and Middle Income Countries

Gram-negative bacteria naturally shed outer membrane vesicles (OMV) during growth. GMMA (Generalized Modules for Membrane Antigens) derive from bacteria engineered to enhance native OMV formation and to reduce GMMA reactogenicity. The GMMA surface is a representation of the outside of the bacteria, with multiple membrane antigens presented to the immune system in their native conformation and correct orientation (1,2). Large body of preclinical data shows that GMMA are highly immunogenic and elicit substantially higher antibody responses against key vaccine candidate antigens, whether these are polysaccharide or protein moieties, compared with corresponding purified antigens delivered as glycoconjugate vaccines or recombinant formulations (3).

GMMA is produced by bacterial fermentation with simple filtration-based downstream processing. Such process has been already scaled-up for producing GMP-quality GMMA at large scale for *Shigella sonnei* and nontyphoidal *Salmonella*, pathogens of particular relevance for LMIC. The most advanced GMMA-based vaccine is *S. sonnei*, already tested in clinical trials showing to be safe and immunogenic in adults (4,5).

In this work we present a techno-economic analysis of the GMMA production process, using SuperPro Designer. We investigate different operating scenarios and scales and monitor their impact on operational and capital costs of the facility. Current efforts are focused on a systematic and quantitative identification of quality attribute-process parameter relationships based on first-principles and data-driven modelling. We will present an illustrative example for such an implementation of Quality by Design principles to the separation unit of tangential flow filtration. The latter is based on independent experimental runs under varying conditions that indicate critical process parameters and allow for a thorough risk assessment. Lastly, we demonstrate a science-driven framework to assist with: (a) identification of the critical quality attributes & critical process parameters and (b) quantification of the impact of the critical processes parameters on the monitored attributes.

Deep process understanding combined to a robust analytical control strategy will support sustainable manufacturing of safe and efficacious GMMA based vaccines, focusing on both product quality and process performance.

1. Gerke C, et al. PLoS One. 2015 Aug 6;10(8):e0134478.
2. De Benedetto G, et al. Vaccine. 2017 Jan 11;35(3):419-426.
3. Micoli F, et al. Proc Natl Acad Sci U S A. 2018 Oct 9;115(41):10428-10433.
4. Launay O, et al. EBioMedicine. 2017 Aug;22:164-172.
5. Obiero CW, et al. Front Immunol. 2017 Dec 22;8:1884.

574832 IQ Consortia and Pre-Competitive Information Exchange

Case studies in successful collaboration for driving innovation and establishing best practices for the pharmaceutical industry.

The International Consortium for Innovation and Quality in Pharmaceutical Development (IQ) is a not-for-profit organization of pharmaceutical and biotechnology companies with a mission of advancing scientifically-driven standards and regulations for pharmaceutical and biotechnology products worldwide. The Active Pharmaceutical Leadership Group (API-LG) and affiliate groups are focused on advancing science-based phase appropriate strategies in the industry to create efficient and sustainable processes to deliver high quality API in alignment with Quality-by-Design principles and in compliance with regulatory guidelines and good manufacturing principles.

The API leadership group consists of members from approximately forty pharmaceutical and biotechnology companies, and they coordinate and oversees the activities of the topically focused working groups. This group is responsible for ensuring representatives from member companies represent the activity at the working group to their individual organizations.

The working groups typically consists of interested scientist from the member companies, and regularly meets to drive innovation and quality through the sharing of non-competitive best practices. Information is shared with the industry by publication, presentation or collaboration to peers, agencies or other key stakeholders. These communications typically contain case studies, survey standard approaches or practices within the industry, or provide interpretation of various regulatory guidelines.

573907 Lhasa Limited-Achieving Innovation through Collaborative Progress

For many years, Lhasa Limited has been instrumental in setting up and chairing various pre-competitive consortia mainly within the pharmaceutical industry. These consortia allow various companies who usually compete against each other to come together and find solutions for problems which they commonly share. Such collaborations save time and costs, and the shared knowledge will contribute to advancements in the drug development process in general which ultimately will benefit patients worldwide.

Lhasa Limited was founded on the idea of sharing proprietary chemical data and knowledge in order to deliver mutual benefit to the scientific community. Founded as a not-for-profit organisation and educational charity, we pioneered precompetitive chemical data sharing for mutual benefit within chemistry and the life sciences. The data and knowledge shared with us was used to generate synthetic rules for the LHASA system that did not compromise the competitive advantage of the collaborating organisations whilst delivering mutual benefit for all. Since 1983 we have delivered solutions in the fields of toxicology, metabolism, degradation, and purge chemistry, through our world-leading software tools.

573786 Niimbl -the National Institute for Innovation in Manufacturing Biopharmaceuticals

NIIMBL is one of 14 Manufacturing USA institutes. NIIMBL's mission is to accelerate biopharmaceutical manufacturing innovation, support the development of standards that enable more efficient and rapid manufacturing capabilities, and educate and train a world-leading biopharmaceutical manufacturing workforce. This talk will focus on some of the lessons learned in managing pre-competitive collaboration in the first two years of the National Institute for Innovation in Manufacturing Biopharmaceuticals.

573852 Sage Bio Networks-Searching for Mutual Incentives: The Sweet Spot for Collaboration

This talk will explore case studies that look at different models of collaboration and issues to consider when working in a decentralized landscape. Sage Bionetworks is a nonprofit biomedical research and technology development organization that was founded in Seattle in 2009. Our focus is to develop and apply open practices to data-driven research for the advancement of human health. Data-driven research has become an important component of biomedicine, but it's not always easy to understand how to apply computational approaches appropriately or how to interpret their results.Our interdisciplinary team of scientists and engineers work together to provide researchers access to technology tools and scientific approaches to share data, benchmark methods, and explore collective insights, all backed by Sage's gold-standard governance protocols and commitment to user-centered design.

568313 Enabling Technologies Consortium: A New Paradigm for Pre-Competitive Collaboration on Technology

The Enabling Technologies Consortium (www.etconsortium.org), as a forum for pharmaceutical companies to discuss ideas, share information, and collaborate on the development of new enabling technologies used in pharmaceutical drug substance and product development, has now been operating since 2017. ETC gives member companies the ability to partner with external collaborators to further the mission of developing pre-competitive enabling technologies for pharmaceutical development and manufacturing. Pre-competitive projects, defined here as collaborations between multiple pharmaceutical companies and third parties, are in progress in the areas of solubility, crystallization, modeling, analytical instrumentation, and high throughput automation. The goal is to produce an efficiency-enhancing advancement or refinement that will ultimately be made broadly available to the public, either through publication, commercialization of a new product, or other means. ETC has been able to streamline these partnerships by eliminating the need for parallel agreements between each member company and the external party, providing a governance framework for the advancement of collaborative projects, and allowing for companies interested in enabling technologies to invest equally in their development. ETC members are also able to provide feedback on newly developed technologies upon request from vendors. This talk will provide an update on the progress of the ETC and some of the lessons learned.

568359 FDA's Emerging Technology Program: Innovation and Collaboration to Advance Product Quality

Over the past decade, there has been substantial progress towards modernizing pharmaceutical manufacturing. However, at the same time, pharmaceutical manufacturing continues to confront a number of challenges, which result in unacceptably high occurrence of product quality issues. Working with industry to encourage the modernization of pharmaceutical manufacturing technology is a mutually beneficial approach to increase manufacturing efficiency, improve overall drug quality, avoid shortages, and prepare for future regulatory challenges.

To encourage the adoption of innovative approaches in pharmaceutical manufacturing and to prepare for reviews and inspections involving technology for which there is little experience, the Agency established the Emerging Technology Program. This program features a cross-functional Emerging Technology Team (ETT) which works directly with industry to identify and resolve potential scientific and policy issues that may impact technologies new to the pharmaceutical industry. In a relatively short period of time, the Emerging Technology Program has facilitated the approvals of the first 3D-printed drug product, the first drug product made via continuous manufacturing, and the first switch to a continuous manufacturing process for an approved drug product.

The other key components of the Emerging Technology program are stakeholder engagement and advancing regulatory science. The program components aim to complement each other. Early stakeholder engagement and communication can be used to identify innovative technologies that have significant potential to advance product quality. Collaborating with public-private partnerships or other pre-competitive groups can potentially accelerate the implementation of emerging technologies by fostering multi-disciplinary collaborations among researchers, manufacturers, and government organizations to enable the type of feedback interactions critical to the innovation process. As part of the partnership or through other avenues FDA can support regulatory science activities to address any knowledge gaps that could impact the commercial implementation and assessment of the novel technology. Then as industry begins to develop the technology for commercial use, ETT provides a mechanism for early one-on-one engagement to promote shared learning and address specific product and process issues.

This talk will describe the current state and the continuing maturation of the Emerging Technology program towards a scalable, forward looking, and globally influencing program.

572858 Designing a Continuous Flow System to Safely Nitrate Alkylamines to Nena's

Nitroethylnitramines (NENA-compounds) are useful in propellants and explosives as energetic plasticizers. The NENA compounds of interest include methyl, ethyl, propyl, butyl NENA's. These Alkyl-NENA's contain a nitrate ester (-ONO_2) as well as a nitramino (-NH-NO_2) functional groups and are known to effectively plasticize cellulosic polymers such as nitrocellulose. Manufacture of NENA compounds presents significant process safety challenges due to the potential energy of the materials, thermal stability, and the highly exothermic reactions. Previous investigators have provided stern warnings of fire and "spark spheres" forming in the reactors at laboratory scale. Highly exothermic reactions pose significant scaleup challenges that are compounded with unstable, high energy products. These challenges can be addressed with chemical engineering solutions that scale such as continuous processing. The various phases of the two-step nitration have been studied for years with a focus on safety. Heat flow calorimetry, flow NMR, off-line analytics (e.g. HPLC, GC), and in situ spectroscopy contributed to establishing reaction kinetics and overall process understanding to safely produce this family of compounds. In addition, critical safety data were generated using thermal analysis (e.g. DSC, TSu, ARC) of the various processing streams. The combined results and design of a safe continuous manufacturing process for NENA's will be discussed.

563926 Continuous Enzymatic Production of Enantiopure Amines in a Packed Bed Recycle Reactor

Continuous enzymatic production of enantiopure amines in a packed bed recycle reactor

Robert D. Franklin‡, Adam A. Caparco‡, Joshua A. Whitley‡, Bettina R. Bommarius‡, Julie A. Champion‡, Andreas S. Bommarius‡#

‡School of Chemical and Biomolecular Engineering, Georgia Institute of Technology,

Parker H. Petit Institute for Bioengineering and Bioscience,

Engineered Biosystems Building, 950 Atlantic Drive N.W., GA 30322, USA

#School of Chemistry and Biochemistry, Georgia Institute of Technology,

950 Atlantic Drive, Atlanta, GA 30332-2000, USA

Scalable manufacturing of chiral amines is of major importance in the pharmaceutical industry. Biocatalytic production of these compounds offers key advantages including high enantioselectivity, mild operating conditions, and sustainable catalyst production and disposal. However, enzyme stability and retention pose challenges which must be addressed prior to scale-up, a task frequently accomplished by enzyme immobilization. The authors report the first use of a recently reported leucine zipper system[1] to noncovalently immobilize an amine dehydrogenase[2] (AmDH) and formate dehydrogenase (FDH) onto methacrylate copolymer beads, which were subsequently employed in a packed bed reactor with recycle. AmDHs convert prochiral ketones into chiral amines with the addition of ammonia and oxidation of NADH. FDH consumes formate to regenerate NADH from NAD^+, and generates carbon dioxide in the process. Key reactor parameters, including residence time, reaction temperature, and recycle ratio, were explored to determine their effects on space time yield and total turnover number. Interesting flow patterns were observed and analyzed using high-speed micro flow sensors. Finally, process control techniques allowed for steady flow rates over long operation times, despite degradation of pump tubing. As the biocatalytic production of pharmaceutical ingredients matures, more work must be focused on bridging the gap between protein science and process development. The reported work represents an important contribution to that effort.

1. Caparco, A. A.; Bommarius, A. S.; Champion, J. A., Effect of peptide linker length and composition on immobilization and catalysis of leucine zipper-enzyme fusion proteins. *AIChE Journal* **2018,** *64* (8), 2934-2946.

2. Abrahamson, M. J.; Vazquez-Figueroa, E.; Woodall, N. B.; Moore, J. C.; Bommarius, A. S., Development of an amine dehydrogenase for synthesis of chiral amines. *Angew. Chem.-Int. Edit.* **2012,** *51* (16), 3969-72.

572989 3D Printed Fixed Bed Catalytic Monolith for Continuous Chemistry

Various support structures (metal, ceramic, polymer) are used to anchor or enhance heterogeneous catalysts. Metal and ceramic monolith support structures are typically extruded. Extruded monoliths contain regular, parallel channels enabling high throughput because of the low pressure drop accompanying high flow rate. However, extruded channels have low surface-area-to-volume ratios resulting in low contact between the reagent and catalyst. Three-dimensional printing (3dP) can be used to overcome disadvantages of extruded monoliths while also preventing channeling due to a fixed bed. 3dP enables precise control over all design parameters of the monolith. The polymer, reactor design, and bed geometry can be easily tuned to match different reaction systems based on reaction kinetics and desired throughput. Additionally, 3dP allows for reactor design based on system constraints (e.g. design a less dense bed geometry to minimize backpressure for a peristaltic pump). We have implemented a rational design approach to 3dP monoliths (made from commercially available filament) that continuously catalyze cross-coupling reactions. This system is reproducible, versatile, and scalable.

561680 Surface Plasmon Resonance Spectroscopy As a Tool for Distinguishing Homogeneous Versus Heterogeneous Catalytic Pathways for Metal Nano-Catalyzed C-C Coupling Reactions

Carbon-Carbon (C-C) coupling reactions are the most widely used reactions used for making variety of molecules, including polymers and pharmaceuticals. These reactions have been traditionally carried out using homogeneous, expensive and toxic palladium (Pd) catalysts. Nano-metal catalyzed cross-coupling reactions will have several advantages as they can be used as a platform to develop energy efficient and environmentally friendly flow processes. Recently, promising results have been reported in the literature on the use of less toxic metal nanocatalysts such as gold (Au) and copper (Cu) for cross-coupling reactions. However, identifying the catalytic pathway (i.e., heterogeneous versus homogeneous pathway) and monitoring the stability of functioning nanocatalysts remain highly challenging tasks.

In this contribution, we show that localized surface plasmon resonance (LSPR) spectroscopy can be used as platform to distinguish homogeneous and heterogeneous pathways, and monitor stability of functioning nanocatalysts in nano-metal catalyzed cross-coupling reactions. Using finite-difference time-domain (FDTD) optical simulations, we show the use of UV-Vis extinction spectra to distinguish catalytic pathways and monitor nanoparticle stability. We demonstrate that for a variety of nanocatalysts including monometallic catalysts (e.g., Au, Pd and Cu), and bimetallic core/shell nanocatalysts (e.g., Au/Pd and Ag/Pd), there is a measurable SPR peak shift, change in SPR intensity and full-width at half-maximum (FWHM) in the surface plasmon extinction spectrum. This novel and simple UV-Vis surface plasmon spectroscopic technique will help to (a) fundamentally understand the catalytic pathway, and (b) develop environmental friendly metal nanocatalysts for C-C, C-N, C-S and C-O coupling reactions.

Reference

1. Mohammadparast et.al "C-C Coupling Reactions Catalyzed by Gold Nanoparticles: Evidence for Substrate-Mediated Leaching of Surface Atoms Using Localized Surface Plasmon Resonance Spectroscopy", J. Phys. Chem. C, 2019, In press. https://pubs.acs.org/doi/abs/10.1021/acs.jpcc.8b12453

566232 Integrated Thermodynamic and Mechanistic Model of Homogeneous Catalytic *N*-Oxidation Processes

Integrated thermodynamic and mechanistic model of homogeneous catalytic *N*-oxidation processes

Jingyao Wang[a,b], M. Sam Mannan[a,b], Benjamin A. Wilhite[a,b]

a Artie McFerrin Department of Chemical Engineering, Texas A&M University, 3122 TAMU, College Station, Texas 77843, United States

b Mary Kay O¡¯Connor Process Safety Center, Texas A&M University, 3122 TAMU, College Station, Texas 77843, United States

The homogeneous phosphotungstic acid catalyzed *N*-oxidation of alkylpyridines by hydrogen peroxide has important applications in pharmaceutical and fine chemical industries. Current industry practice is to employ a semi-batch reactor with gradually dosing of hydrogen peroxide into an alkylpyridine/catalyst solution under isothermal conditions. However, due to lack of understanding of reaction mechanism and thermodynamic behavior, this system is subject to significant risk of reactor overpressure due to hydrogen peroxide decomposition. In this study, semibatch *N*-oxidation process was executed in an isothermal reaction calorimeter (RC1) over a wide range of temperature, catalyst amount and oxidizer dosing rates. Reactor pressure, reaction heat generation rate and *in-situ* FTIR spectra of liquid phase species were recorded in real-time during experiments, and final product was quantified using HPLC and GC-MS analytical tools. An integrated thermodynamic and mechanistic model of homogeneous *N*-oxidation reaction was proposed based on experimental results. More specifically, Wilson excess Gibbs model was employed to estimate activity coefficients of highly non-ideal liquid mixture. Ideal gas law was found satisfactory in calculating incondensable oxygen pressure. First principle reaction mechanism and kinetics parameters of (i) catalytic *N*-oxidation reaction; (ii) catalytic hydrogen peroxide decomposition reaction; (iii) non-catalytic *N*-oxidation reaction; (iv) non-catalytic hydrogen peroxide decomposition reaction was derived based on experimental findings of this study and past literature. The proposed integrated thermodynamic model and mechanistic model successfully predicted highly nonlinear reactor pressure, species mole and reaction duty profile of homogenous catalytic *N*-oxidation reaction. The optimal reactions conditions with maximum *N*-oxide product yield and minimum reactor pressure and catalyst usage was theoretically identified and further verified by experiments. The obtained model can be used for inherently safer reactor design and application to other homogeneous tungstic acid catalytic hydrogen peroxide oxidation processes.

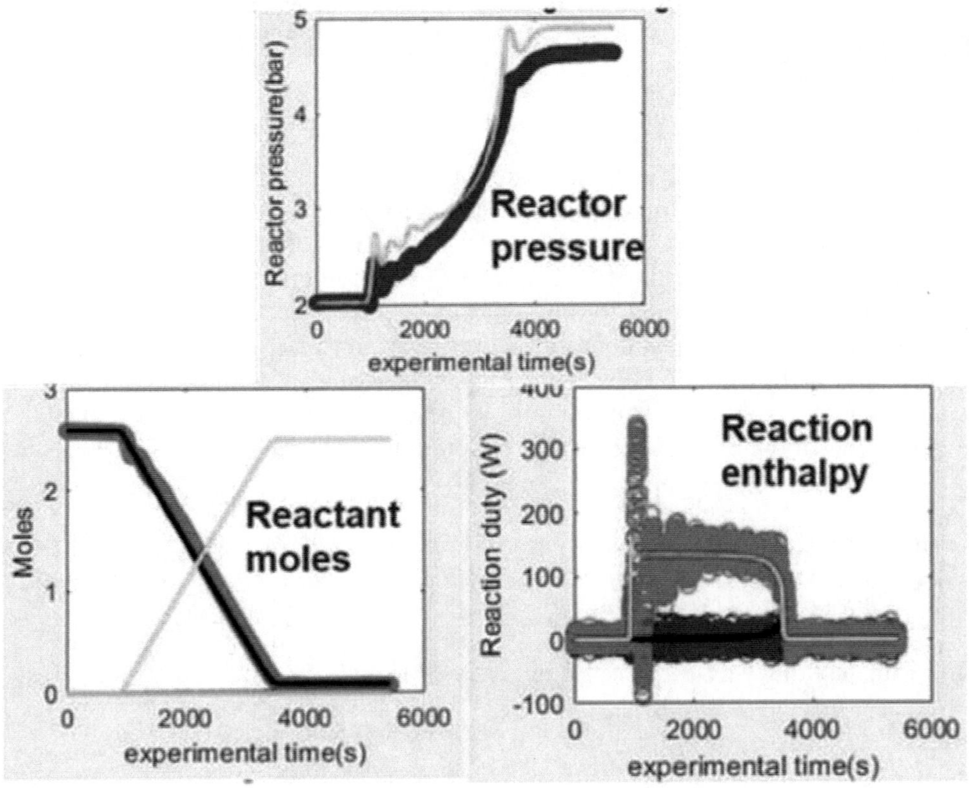

Reactor pressure
Reactor pressure(bar)
experimental time(s)
Reactant moles
Moles
experimental time(s)
Reaction enthalpy
Reaction duty (W)
experimental time(s)

579446 Impinging Jet for a Flow Synthesis of Mixing Sensitive Pharmaceutical Molecule

For a mixing sensitive complex reaction network, impinging jet mixer have shown to be an efficient reactor design due to the generation of high turbulent energy dissipation (>1 kW/kg) to provide intensive mixing. As a result, impinging jet provides reproducible mixing conditions and enables sensitive control of fast reactions. In this study, an impinging jet flow process was developed under GMP conditions to address the formation of a problematic dimer impurity in a mixing sensitive reaction network. Efficient process development of the robustness of the impinging jet required an intricate understanding of intrinsic kinetics, transport phenomena, and analysis of flow patterns within these jets. A computational fluid dynamics (CFD) model was developed to design the impinging jet mixer and mixing time was calculated and verified experimentally. Rational exploration of these factors enabled us to successfully scale-up this process to a multi-kilogram amount of API.

All authors are employees of AbbVie and may own AbbVie stock. AbbVie sponsored and funded the study; contributed to the design; participated in the collection, analysis, and interpretation of data, and in writing, reviewing, and approval of the final publication.

575950 Challenges and Successes in Continuous Chemical Processing

Process development for chemicals involves various stages that challenge chemical engineers to become innovative and creative with respect to the design of various unit operations. Batch processing simplifies unit operations but is not the most efficient or safest method to produce chemicals. In addition, there are processes that simply cannot be done using batch processing. These challenges frequently present themselves at the kilo lab scale. Nalas scales up dozens of chemical processes each year at their kilo lab that frequently require continuous process unit operations. These challenges have afforded excellent opportunities to apply chemical engineering principles across a multitude of chemical reactions, extractions, distillations, crystallizations, and isolations. No two processes are the same, each with their own unique challenge in solution.

Traditional reaction challenges include poor heat and mass transfer as reactor volume increases. Batch reactors are limited with respect to cooling capacity due to the reduced surface area to volume ratio of larger vessels compared to smaller vessels. Reaction temperature that was easily controlled in the laboratory becomes increasingly difficult or impossible to control at larger scales. This poses a scaleup safety and sustainability challenge. The use of smaller, continuous reactors has historically obviated these challenges. Recent advances in reactor design have afforded the opportunity for process intensification, reducing waste streams and frequently improving yield. These advances also result in higher throughput and smaller square footage reactor design that result in lower capital and operational costs. A continuous reaction often requires additional unit operations for purification and isolation of the desired material. Continuous extraction also reduces waste streams through process intensification. For example, the use of counter current centrifugal extractors has been demonstrated to dramatically reduce the required extraction solvent volume. The reduction of extraction solvent results in reduced demand on the subsequent continuous distillation. Again, this results in increased throughput with the benefit of lower capital and operational costs.

Several vignettes have been prepared for discussion based on our experience with continuous unit operations.

564925 Chemical Processes Optimization By Hydrodynamic Simulations – Implementation in the Solid Phase Peptide Synthesis

In this presentation, we will describe the concept of Scale Down – Scale Up (SDSU) for Batch and Semi Batch operations in stirred reactor systems. The concept is based on hydrodynamic studies of the mixing procedure obtained from experiments in small scale mixed rectors and process simulation by VisiMix software. We describe how the hydrodynamic simulations on small and large scale reactions can aid in process optimization and scale up. The use of the SDSU concept and methods described in this article, will offer the user the possibility to achieve the best results during production stage, saving time and currency, and at the same time increasing the knowledge of the performed process. Several examples in the article demonstrate the benefits of the proposed method.

The brittle nature of the polymer beads used in SPPS dictates mild mixing techniques with low mass transfer. We demonstrate that using vigorous overhead mechanical stirring with superior mass transfer properties kept the beads intact and significantly accelerates reaction kinetics and efficiency.

567533 Modelling, Design and Operation of a Continuous Cryogenic Reaction

In the past decade, Pharmaceutical companies have focused on continuous manufacture of small molecules due to quality and control advantages over traditional batch processes. In addition, flow chemistry can enable a diversity of reactions, facilitating the synthetic route to a molecule that may in turn produce a simpler and more effective manufacturing process. Finally, telescoping multiple unit operations together in a continuous supply chain could potentially reduce inventories and lower labour and overhead overall cost.

Until recently, the development and the control of processes in the pharmaceutical industry have been dominated by experimentation. The definition of the optimum process and the control strategy has been supported by Design of Experiments (DoE) feeding into statistical models. Such approaches are extremely cost and time ineffective due to high consumption of material, numerous experiments and high labour requirement.

On the other hand, the mechanistic modelling of drug substance unit operations, such as reactions and separations can yield a high degree of understanding of the chemical and physical phenomena occurring in the process. Furthermore, process optimization can lead to maximizing yield and quality, reducing manufacturing cost and supporting the control strategy through understanding and tracking of critical quality attributes (CQAs) across the process.

In this work the development of a cryogenic reaction process will be discussed. A detailed system model of the process has been developed, which was then used for process design, process optimization and support the QbD strategy. Finally, the developed model was used for online monitoring of the process in GSK's manufacturing facilities during the clinical batches.

563576 Packing Prediction of Pharmaceutical Powders Via Bond Number Estimation

The packing of pharmaceutical powders plays an important role in various unit operations such as die filling, compaction, and feeder performance. In the case of pharmaceutical tablets, the final tablet porosity can dictate critical tablet properties such as tensile strength and dissolution kinetics, and is dependent on blend porosity. Therefore, pharmaceutical powder or blend porosity is an important property that requires better understanding of how it is impacted by both mixing process parameters and constituent material properties. Recent work has introduced mechanistic models to predict the porosity of pharmaceutical powders, which employs a semi-empirical approximation using commonly measured particulate properties such as particle size, particle density, and surface energy. Here, this initial work has been extended to include a wider range of active pharmaceutical ingredients (APIs) and high functional pharmaceutical excipients, both before and after dry coating with nano-silica. Results show that the recently introduced porosity prediction model can predict most pharmaceutical powder porosities fairly accurately, but such is not the case for certain types of materials including functional excipients such as Avicel grade microcrystalline cellulose (MCC) excipients. These functional excipients have a distinctly rougher particle surface, compared to smooth particle surfaces for most other pharmaceutical powders. This distinction could be the key to improving the proposed porosity prediction model, and such a topic has been discussed in previous literature. In this work, we also examine how a prevalent approximation of natural surface asperity size of 200 nm diameter requires re-examination. The specific surface area, internal void volume and particle shape descriptors of the pharmaceutical powders were experimentally measured, and the dependence of powder packing on these characteristics was investigated. Additionally, binary blends of APIs and commonly used excipients were tested for their bulk density (packing). Results show that API dry coating significantly alters the interparticle cohesiveness of these binary blends, and their packing behavior is more similar to non-cohesive glass beads. Such processing significantly improves the design space for micronized, cohesive APIs for use in processes such as high speed direct compression. However, extensions of the current models are also necessary for accurate blend porosity prediction if dry coating processing was used for one or more blend constituents. The experimental work presented here illustrates the need to improve the current semi-empirical, mechanistic porosity prediction model and may help assess if our initial attempts are adequate.

571265 Theoretical and Experimental Investigation into the Impact of Tribo-Charging on the Performance of Dry Powder Blends

Introduction

Powder charging is known to greatly affect the industrial processing and performance of particulate and granular materials. The underlying process of contact charging is referred to as tribo-electrification[1]. Pharmaceutical materials are usually insulators and often accumulate electrostatic charges that can detrimentally affect their processability[1]. Thus, tribo-electric charging can be expected to provide an important contribution to particle-particle and particle-wall interactions during handling and transport of powders of dielectric materials. It would therefore be desirable to systematically determine the factors that contribute to powder tribo-charging. Generally, the extent of tribo-charging is expected to be related to the difference between the work function (WF) values of the two contacting surfaces[2]. Other material properties relevant to tribo-charging include surface energy, hygroscopicity/wettability, surface roughness, and particle shape[1]. Therefore, it is anticipated that different materials as well as particulate properties will be crucial in the tribo-charging of different pharmaceutical blends. Finally, also, the effect of unit operations involved in the manufacturing of pharmaceutical products must be taken into account. Thus, it was the aim of our work to systematically understand the factors critically affecting this phenomenon, and its respective impact on formulation performance.

Design of the studies

Model pharmaceutical materials were selected and their material and particle properties characterized. Tribo-charging acquired during the flow over a stainless steel tube was investigated in order to obtain relevant information of their behavior during unit operations such as feeding, transport, blending and capsule filling etc. The results of relative charging in the materials were interpreted considering their distinct chemical and physical properties. More concretely, evaluation of the net charge of the different materials was carried out using Granucharge™ (Granutools). Measurements were performed under defined environmental conditions. First, initial charge to mass ratio (q0) was recorded by dispensing the powder directly into a Faraday pail. In a consecutive step, the powder was transferred to a vibrating stainless-steel V-tube and the charge to mass ratio (q1) was recorded again after the powder was transported through the tube. The acquired charge during flow, thus charging tendency, was determined as $\Delta q = q1 - q0$.

Tribo-charging trends predicted from first principles

The tribo-charging of five different pharmaceutical materials acetylic salicylic acid (Merck KGaA), Ibuprofen (Merck KGaA), a-lactose monohydrate (LH100, Meggle), trehalose (Cargill Deutschland GmbH), and D-Mannitol (P160C, Roquette) were experimentally investigated. In order to concentrate on the effect of electronic properties of the materials, other characteristics and experimental parameters, in particular the particle size (distribution) of the used powders, were kept identical for the five investigated materials. The measured tribo-charging trends were compared with the energies from electronic structure calculations.

It was found that varying the level of theory (from semi-empirical/PM6, over DFT/GGA, to DFT/meta-GGA) affects the correlation between tribo-charges and calculated energies only weakly. However, of the calculated values (highest occupied molecular orbital (HOMO), energy levels, vertical and adiabatic ionization potentials (IP), and work functions (WF)) only adiabatic IP values showed a useful correlation with tribo-chage, with a pearson correlation around $r = 0.9$. The calculated WF values however, the measure that is traditionally used for prediction of contact-electrification, showed a poor correlations with our experimental data. This unexpected result is presumably due to a combination of two facts: 1) experimental IP and WF values usually show a good correlation, and 2) the lowest unoccupied molecular orbital (LUMO) energy, which is used here to estimate electron affinity and for the calculation of the WF, is notoriously difficult to calculate accurately, this carrying over to the calculated WF values. We conclude that since the IP can be calculated more accurately, and it correlates with the WF, the IP can be used as a replacement for the WF, in particular if one is primarily interested in relative values. Although limited in number, our data suggests that electronic structure calculations can be used to estimate relative tribo-charging trends for pharmaceutical materials, provided that other material properties are comparable.

Tribo-charging trends of different particulate materials

Although prediction of tribo-charging trends using first principles is desirable to determine *a priori* the molecular properties affecting the contact of different materials, its applicability is still limited as it does not discriminate between different solid and particle states of materials (as discussed in the previous section). Therefore, the tribo-charging of different salbutamol sulphate (SBS) (Selectchemie) particles engineered using jet-milling (Spiral air Jet Mill 50 AS, Hosokawa Alpine AG) and spray-drying (Nano spray-dryer B-90, Büchi) was experimentally investigated and compared[4].

It was demonstrated that different surface chemistries arising from distinct solid-states of the same material, different contact areas due to particle properties like shape, size and morphology, as well as variable water contents can critically impact the polarity and magnitude of tribo-charging. Amorphous, spherical shaped SBS particles with a high water content (generated via spray-drying) attained positive charge of a higher magnitude. In contrast, the milled fine particles with irregular shape and lower water content tended to charge negatively and to a lower extent[4].

Tribo-charging trend of different material blends

To understand how different particulate characteristics can affect the tribo-charging of pharmaceutical powder blends, the distinctly engineered fine particles of SBS (1-5 μm) were mixed with coarse D-mannitol (about 100 μm) (MAN) at 2 drug loadings (2 wt% and 5 wt% of SBS). Likewise, the tribo-charging of the obtained powder blends was evaluated[4].

Comparison of the two crystalline powders (jet-milled SBS and MAN) revealed that theoretically calculated WFs can provide a rank order of the sign of charge when the materials were in contact with metal, supporting our previous conclusions that electronic structure calculations can be used to estimate relative tribo-charging trend. Concerning the charging tendency of the powders blends, an increased charge mitigation effect was observed when a higher concentration of fine SBS particles were blended with the coarse MAN. Likewise, it was found that the charging tendency of the coarse particles is less relevant when blended with fine materials. Consequently, it was possible to conclude that evolution of charge in the blend seemed to be dominated by the nature of fines and their concentrations contributing to inter-particle contacts[4].

Tribo-charging trends arising from distinct process parameters

In a final step, powder blends have to be processed into a final dosage form to be administered to the patient. Capsules are one of the most common solid oral and inhalation dosage forms. Therefore, the capsule filling process was selected to investigate the impact that processability can have on the tribo-charging. For that, size 3 gelatin (Quali-G™, Qualicaps) as well as thermally (Vcaps® Plus, Capsugel) and chemically gelled HPMC capsules (Quali-V®-I, Qualicaps) were filled (Labby, MG2) targeting distinct fill weights (25 and 75 mg) and the effect on tribo-charging was evaluated and compared[5].

Comparing the charge tendency of the different capsules, it was seen that gelatin (Gel-Cap) and the thermally gelled HPMC (TG-HPMC-Cap) capsules showed a higher tendency to acquire charge during contact with stainless steel than chemically gelled HPMC (CG-HPMC-Cap). When different fill weights were used, it was observed that the Gel-Cap and TG-HPMC-Cap had a decreased tendency to acquire charge as the powder fill level increased. In contrast, the powder fill level had negligible impact on the charging properties of CG-HPMC-Cap capsules. This revealed the similar charging tendencies of capsules manufactured with different materials and the dissimilarity between capsules of the similar material[5].

Conclusion

Tribo-charging is a complex and very relevant phenomenon that is still largely unpredictable. In this work, we propose a bottom-up approach to the understanding of tribo-charging trends of pharmaceutical materials. From molecular properties to particulate and powder bulk characteristics to the influence of process conditions on tribo-charging, providing a systematic understanding the factors critically affecting this phenomenon, and its respective impact on dry powders performance.

References

[1] Wong J, Kwok PCL, Chan HK. Electrostatics in pharmaceutical solids. *Chem Eng Sci*. 2015; 125:225-237.

[2] Helmholtz H. Studien über electrische Grenzschichten. *Ann der Phys und Chemie*. 1879; 243(7):337-382.

[3] Brunsteiner M, Zellnitz S, Pinto JT, Karrer J, Paudel A. Can we predict trends in tribo-charging of pharmaceutical materials from first principles?. *Powder Tech.* 2019; Submitted.

[4] Zellnitz S, Pinto JT, Brunsteiner M, Schroettner H, Khinast J, Paudel A. Tribo-charging behaviour of inhalable mannitol blends with salbutamol sulphate. *Pharm Res*. 2019; Accepted.

[5] Wutscher T, Zellnitz S, Kobler M, Buttini F, Andrade L, Daza V, Mercandelli A, Biserni S, Probst SE, Khinast J, Paudel A. Influence of Storage Humidity and Powder Fill Level on Charging Behaviour of Different Capsules Types. RDD Europe 2019; Lisbon, Portugal.

571333 Influence of an External Lubricant on the Tribo-Charging Behaviour of Capsules

Introduction

During the manufacture of empty two-piece hard capsules, an external lubricant is applied to the outer surface of the capsules to prevent sticking to each other after packaging and to improve transfer of the empty capsule during filling operations. This lubricant also helps the capsule to minimize friction in a capsule filling machine and therefore is needed for a smooth processing [1, 2]. For the external lubricant, the same substances are used for capsules comprised of either gelatin or hydroxypropyl methylcellulose (HPMC). These lubricants, for example, are magnesium stearate (MgSt), sodium lauryl sulphate (SLS) and carnauba wax (CW). The structural differences among lubricant can lead to a different work function of the material (the minimum energy needed to withdraw an electron of the surface of a solid), which places them at different positions of the tribo-electric series. When materials with different work functions brought into contact and are separated, electrons are transferred and tribo-electric charges are created. Tribo-charges are often responsible for many manufacturing issues and can influence the quality, safety and efficiency of pharmaceutical products [3] or for example the aerodynamic performance of a dry powder inhaler (DPI). Tribo-charges can also cause problems in terms of powder mixing homogeneity and dosing accuracy [4], and can lead to particle adherence on surfaces and agglomeration of particles during manufacturing or powder release from the capsule [5].

This study examines the impact of the charging behaviour of capsules that are externally lubricated with different materials, before and after transport over stainless steel and PVC. The transport over steel was intended represent the transfer from the bulk container to a filling machine hopper-feeder, whereas the transport over PVC could give an idea of how the capsules behave/or will be charged within a packaging material.

Material and Methods

Different types of size 3 gelatin capsules (Quali-G®) and chemically gelled HPMC capsules (Quali-V®-I) were received from Qualicaps (Qualicaps Europe, S.A.U., Spain) with various variations of colorant and external lubricant. Gelatin capsules were supplied with the external standard lubricant SLS, unlubricated, MgSt, and CW. Batches with 1g and 2g of MgSt on 10kg of capsules were received, to study the influence of the amount of external lubricant. Gelatin capsules with MgSt or CW as external lubricant were further divided into two groups, in which different colorants were used. HPMC capsules were also used in an unlubricated state, as well as with CW, which is the standard lubricant for HPMC capsules, SLS, and MgSt.

Charge measurements were performed in triplicate using 2.0g ±0.15g capsules on the GranuCharge™ (Granutools, Belgium) instrument. The room was conditioned to a relative humidity (RH) 50% ±3% at 23.5°C ± 0.6°C, whereas the conditions inside the device remained constant at 50% ±2% at 23.5°C ± 0.5°C. Charge measurements started with the measurement of the initial charge to mass ratio (q0) by pouring the capsules directly out of the bag into the inbuilt Faraday cup of the device. Afterwards, the capsules were transferred to the grounded vibrational feeder of the GranuCharge and fed into a V-shaped steel or PVC tube to drop into the Faraday cup again after transport. Then, the charge to mass ratio was calculated for a second time (q1). By subtracting the initial charge to mass ratio from that obtained after sliding through the V-tube yielded the charge to mass ratio (Δq = q1-q0) that describes the charge gained.

To differentiate the surface energy (SE) of the capsules with different external lubricants, contact angle measurements were made using the pendant drop method (Easydrop system, Krüss GmbH, Germany) at 22°C ± 0.3°C. For this purpose, the contact angle of water and diiodomethan was measured on the outer surfaces of the capsules (as received). Surface energy was calculated via Owens-Wendt-Rabel-Kaelble method (OWRK) [6].

Results and Discussion

Overall, the charge to mass ratio (Δq) of capsules after the transport over steel and PVC was positive, although a higher charge to mass ratio was observed when capsules were in contact with PVC tube walls. Gelatin capsules on steel charged in a range of 0.63 (± 0.05) to 2.87 (± 0.07) nC/g whereas HPMC charged from 0.43 (± 0.06) to 1.45 (± 0.06) nC/g. Charges of gelatin and HPMC capsules after passing over PVC surface were from 0.72 (± 0.27) to 2.62 (± 0.71) nC/g and 1.61 (± 0.05) to 3.05 (± 0.13) nC/g, respectively. Capsules with MgSt showed an exception because they were charged to the same amount on both materials. As Staniforth and Rees [7] stated, MgSt is an electropositive and therefore able to lower the charging effect of the capsule material and wall interactions. Within one type (gelatin or HPMC) of capsules, the un-lubricated capsules showed the highest charging and those lubricated with SLS showed the lowest charging. It can be considered, that the electron transfer of the HPMC and gelatin is hindered due to the work function of the contacting surface of the capsule material or the external lubricant. The batches with two different amounts of MgSt showed that a lower amount of lubricant gave higher charging. This seems to fit to the theory that lubricants sometimes work as antistatic agents as it is also reported for coated particles [8]. A part of the lower charging tendency of HPMC capsules compared to gelatin capsules can be explained by the antistatic properties of HPMC itself [9] although other ingredients of the capsule formulation like the gelling agent or different colorants also play a role . The effect of a lower charging tendency due to lubricants is supported by the fact that a lower amount of external lubricant lead to an increase in charge. In order to assess the batch to batch variability, two batches of gelatin capsules without lubricant were measured. The two batches gave the same results in charging, although the surface energy was slightly different. Overall, SE showed results between 3.18 to 8.08 mN/m for HPMC capsules and 26.6 to 38.59 mN/m for gelatin capsules, respectively. The HPMC capsules with the highest SE showed the lowest charging tendency within the group of HPMC capsules, whereas in gelatin the highest SE also gave the highest charging tendency. It can be assumed that the lubricants cover the main part of the capsule surface but do not fill all cavities of the underlying material and therefore the differences between HPMC and gelatin occur.

Conclusion

In summary both, the external lubricant and the capsule materials themselves have an influence on the tribo-charging behaviour of the capsules. The low amount of lubricant on a single capsule (only 2% of the average capsule weight), seems to smoothen the surface of the capsule. Since it is not known how uniform the distribution of the external lubricant is, it is possible that some regions of the capsule surface are not covered enough to protect the capsule material from directly interacting with the environment. Hence, the results for charging can be a mixture of the interactions of the capsule material and the external lubricant. The work functions of HPMC and gelatin on one hand and the work functions of the lubricants on the other hand do not explain the differences on their own. Calculated work function values depend on well-defined molecular structures, therefore real work function values are hard to define for HPMC and gelatin as impurities in the material and the orientation of functional groups on the surface also have an impact. Overall, tribo-charging effects of capsules seem to be an interplay of capsule materials, parameters like surface roughness of the capsule and the interacting machine, water content and lubricant attributes as well as the lubrication level. It could be shown, that lubrication of capsules plays a role in tribo-electrification.

579032 Investigation into the Effects of Roller Compaction on Granule and API Properties

Solid oral dosage forms are the most popular drug delivery system. The pharmaceutical industry is a very highly regulated one and as such the processes involved in the manufacture should be carefully controlled. One key aspect to control is the particle size and shape of the Active Pharmaceutical Ingredient (API), as this has been shown to have significant effects on prcoessibility (e.g. punch sticking) and the performance of the final product (e.g. dissolution rate) (Leane et al. 2015).

However, this can prove challenging as the API is subjected to harsh environments in various unit operations, such as roller compaction, which can cause attrition of the particles (Gamble et al. 2014). For control purposes it is important to understand what is occurring to the API within these processing steps, however, tracking changes in the API's properties once it has been incorporated into a blend is challenging. Consequently the performance of the end product are often related to the input properties, which are likely to have changed during processing.

A modern approach combining static image analysis (SIA) and Raman Spectroscopy can be utilised to measure the particle size and shape of a single component within a multi-component system (Gamble et al. 2015) thus enabling changes in API particle size during processing to be measured. This approach has been applied to investigate the effects of roller compaction process parameters on API properties. The results of this study provide an insight into the dry granulation processing route and the effects of processing conditions on the properties of API particles and the subsequent granule. The presentation will also address topics including the nature of the API within granulate, both in terms of size and location, and the possible impact on to downstream processing such as tabletting.

References:

Gamble, J. F., Hoffmann, M., Hughes, H., Hutchins, P. and Tobyn, M. (2014). "Monitoring process induced attrition of drug substance particles within formulated blends." International Journal of Pharmaceutics **470**(1): 77-87.

Gamble, J. F., Tobyn, M. and Hamey, R. (2015). "Application of Image-Based Particle Size and Shape Characterization Systems in the Development of Small Molecule Pharmaceuticals." Journal of Pharmaceutical Sciences **104**(5): 1563-1574.

Leane, M., Pitt, K. and Reynolds, G. (2015). "A proposal for a drug product Manufacturing Classification System (MCS) for oral solid dosage forms." Pharmaceutical Development & Technology **20**(1): 12-21.

568348 In-Depth Experimental Analysis of Pharmaceutical Continuous Twin-Screw Wet Granulation in View of the Development of a Generic Compartmental PBM

INTRODUCTION

Recently, a mechanistic Population Balanced Model (PBM model) has been developed and calibrated for twin-screw wet granulation for a few specific formulations [1]. In-depth data collection and analysis was performed to investigate the influence of process parameters on granule formation mechanisms along the twin-screw granulator barrel and resulting granule Critical Quality Attributes (CQAs) [2]. A logic next step is to make this model applicable for any formulation by developing a generic PBM for continuous twin-screw wet granulation, predictive for granule size distribution (GSD) in function of both process parameters and raw material properties. Given the potential importance of granule porosity on granule formation and final drug product, information with regard to this property was collected per granule size fraction along the length of the granulator barrel. This information will serve for the development of a generic 2D-PBM model.

MATERIALS AND METHODS

Different commercially available Active Pharmaceutical Ingredients (APIs) with major differences in physicochemical properties, primarily solubility and wettability, were selected. Milled and non-milled APIs were utilized to investigate the effect of API particle size on granule formation and final granule CQAs.

Blends were prepared using a fixed formulation base consisting of microcrystalline cellulose, HPMC, lactose, and 5% or 50% API for the low and high dose formulations, respectively.

APIs, excipients and blends were thoroughly characterized for particle size distribution, bulk and tapped density, dissolution rate, water holding capacity, solubility, residual moisture content, contact angle, ring shear test, true density, angle of repose, dispersion charge, and specific surface area.

Granules were produced using the high shear twin-screw wet

granulation module of the ConSigma™-25 system (GEA Pharma systems, Collette, Wommelgem, Belgium).

A fixed screw configuration was used with 2 kneading compartments, each consisting of 6 kneading elements arranged at 60° stagger angle. Both kneading zones were separated by a conveying element and at the end of the granulator barrel, an additional conveying element was added together with three size-control elements.

In order to characterize the granule size distribution and granule porosity along the length of the twin-screw granulator barrel, samples were collected in four well-defined compartments: (i) the wetting zone, (ii) after the first kneading zone, (iii) after the second kneading zone and (iv) at the granulator outlet (after the size control elements) using the method described by Verstraeten et al. [2].

A 5-level Rotatable Central Composite Circumscribed Design of Experiments (CCC) (MODDE 12, MKS Umetrics, Umeå, Sweden) was performed to evaluate the effect of process parameters liquid-to-solid ratio (L/S), material throughput and screw speed on local and final granule size distribution and porosity. The granulator barrel temperature was kept constant at 25 °C. Granules collected at the different locations were dried at ambient temperature and relative humidity (25 °C, 40% RH) for at least 24 hours and granule size distribution (Partan^{3D}, Microtrac Inc, Montgomeryville PA, USA) and granule porosity by mercury intrusion porosimetry (Porotec, Thermo Finnigan, California, US) were measured. The flowability of the granules at the end of the granulator was evaluated by the Hausner ratio. Also, friability was measured at the end of the barrel to evaluate the effect of granulation process parameters and raw material characteristics on granule strength.

RESULTS AND DISCUSSION

The experimental results not only showed prominent similarities but also differences with what has been observed before due to the different starting material properties [2]. The effect of TSG process parameters material throughput, screw speed and L/S ratio on granule size distribution along the length of the barrel was highly dependent on formulation properties. Furthermore, L/S ratio remained the most important factor dictating the size distribution of granules collected at the outlet of the twin-screw granulator.

However, barrel fill level, dictated by screw speed and material throughput, also had a relevant effect on granule size distribution. While the impact of the narrow chopper section on the granule particle size has been shown to be insignificant at low L/S ratios [2], in this study, the size control elements at the end of the granulator barrel appears to have a major effect on granule size, both at low and high L/S ratios.

CONCLUSION

This experimental study provided valuable information on the granulation mechanisms in function of process parameters and raw material attributes. Interestingly, the results indicate that the size-control elements and raw material properties have an important effect on granule formation and final granule CQAs. This extensive data collection and analysis is a first crucial step in the development of a generic 1D as well as a 2D PBM model for twin-screw wet granulation.

References

[1] Van Hauwermeiren, D., Verstraeten, M., Doshi, P., am Ende, M. T., Turnbull, N., Lee, K., ... Nopens, I. (2018). On the modelling of granule size distributions in twin-screw wet granulation: Calibration of a novel compartmental population balance model. *Powder Technology*, *341*, 116–125. https://doi.org/10.1016/j.powtec.2018.05.025

[2] Verstraeten, M., Van Hauwermeiren, D., Lee, K., Turnbull, N., Wilsdon, D., am Ende, M., De Beer, T. . (2017). In-depth experimental analysis of pharmaceutical twin-screw wet granulation in view of detailed process understanding. *International Journal of Pharmaceutics*, *529*(1–2), 678–693. https://doi.org/10.1016/j.ijpharm.2017.07.045

567489 The Use of T-PLS to Investigate the IMPACT of Raw Material Properties on Granule Quality Attributes Obtained after Continuous Twin-SCREW WET Granulation

INTRODUCTION

Since 15 years, the interest for continuous manufacturing has steadily increased in the pharmaceutical industry. The advantages offered in terms of process efficiency, costs reduction, flexibility and environmental footprint reduction have been explained in multiple articles and reviews. Among the different continuous drug product manufacturing processes currently available, twin-screw granulation (TSG) is the most evaluated continuous wet granulation process. However, a predictive platform linking the influence of raw material properties, the ratio's in which the raw materials are mixed, and the applied process settings with the granule quality attributes is not yet available.

PURPOSE

In this study, T-shaped partial least squares (T-PLS) was used to study the combined influence of raw materials properties, formulation composition and process settings on granule quality attributes.

MATERIALS AND METHODS

A set of 25 powders consisting of 13 active product ingredients (APIs), 9 fillers and 3 binders for immediate release was characterized using 15 different techniques covering dry powder characteristics (particle size, compressibility, flow properties, ...) and wet powder characteristics (solubility, contact angle, dissolution rate, ...). Principal Component Analysis (PCA) was then performed on the API and filler raw material characteristics subsets to elucidate correlations between the powders and the raw material properties. Based on this analysis, formulation blends were selected, ensuring that a maximal area of the material variability space was covered. Each formulation consisted of an API, a filler and a binder. Herewith, three different blend ratios were used for each formulation, corresponding with an API content of 10, 40 or 70%. As the binders were intended for immediate release, their content was varied from 2 to 5%. Each formulation blend was granulated using the twin-screw wet granulation unit of the ConsiGma™ 25 system (GEA Pharma systems, Wommelgem, Belgium). A 2-level full factorial Design of Experiments (DoE) was performed on each formulation blend to investigate the influence of the granulation process parameters mass flow rate (MFR) and liquid to solid ratio (L/S ratio) on the resulting granule quality attributes. The MFR was varied from 10 to 20 kg/h, while the applicable L/S ratio ranges were experimentally determined for each formulation blend, because this depended on the properties of the selected raw materials. Granule size and shape (PartAn^{3D}, Microtrac, Montgomeryville, United States), friability (Friabilator, PTFE Pharma Test, Hainburg, Germany) and flow properties (GranuDrum, GranuTools, Awans, Belgium) were measured for the obtained granules after each granulation experiment.

RESULTS AND DISCUSSION

The T-PLS model was able to indicate which raw material properties affected the most the granulation process. It was seen that dissolution rate, solubility, porosity, specific surface area and water binding capacity had the highest impact on the variability in granule quality attributes. Furthermore, it was also possible to highlight the most influencing raw materials for each granule quality attribute. For example, the granule size fraction that remained ungranulated (i.e. granule size fraction <150 μm) was correlated with water binding capacity, while it was anti-correlated with solubility, dissolution rate and compressibility. The T-PLS model indicated as well that L/S ratio affected more the granulation process than MFR. But, in general, the variation in granule quality was more influenced by raw material properties than by process parameters.

CONCLUSION

Besides an enhanced process understanding of the impact of raw material properties on granule quality attributes, the developed T-PLS model has also the potential to predict a reasonable starting point for formulation and process parameters for new API's. This approach can reduce the amount of API and the time needed to find optimal process parameters during product development.

567670 Visualization of Granule Temperature Along the Length of the Barrel Using Thermal Imaging to Improve Process Understanding of Pharmaceutical Twin-Screw Melt Granulation

INTRODUCTION

The pharmaceutical industry has been showing an increasing interest in continuous manufacturing for the past 15 years. In order to improve process efficiency, reduce costs and decrease the environmental footprint, innovative continuous manufacturing processes are gaining more interest. Twin-screw melt granulation (TSMG) is an interesting technique to perform agglomeration due to its multiple advantages. For instance, the use of a solid meltable binder to achieve agglomeration instead of using a solvent like in wet granulation enables the use of moisture-sensitive drugs and the exclusion of a drying step. Additionally, the high-dosed formulations with up to 95% of API and the ability to modulate release kinetics can be added to the extensive list of advantages.

In batch melt processes, the heat transfer on powder particles is not uniformly distributed and is uncontrolled. In combination with a long residence time, batch melt processes are not desirable to be used in pharmaceutical industry to obtain uniform quality attributes of granules. On the contrary, in continuous TSMG heat transfer is uniformly distributed and therefore also granule growth can be controlled because every powder particle is exposed to an equal amount of heat and mechanical stress. Furthermore, the residence time in continuous melt granulation is much lower than in batch melt granulation.

The aim of this study is to increase process understanding of the granulation mechanism in twin-screw melt granulation by evaluating the influence of different screw configurations on granule formation, granule temperature and the relation between this.

MATERIALS AND METHODS

Soluplus was used as a thermoplastic binder in the 2 formulations selected for this study. Metoprolol tartrate (MPT) and Caffeine anhydrous (CAF) were chosen as model drugs due to their different miscibility with Soluplus [1]. The former is miscible, while the latter is immiscible with Soluplus. Physical mixtures (85% API/binder w/w) were processed using a twin-screw extruder (Prism Eurolab 16, Thermo Fisher Scientific) with a screw configuration containing 2 kneading zones of each five kneading elements and a screw mixing element at the end of the screws. A D-optimal screening design was performed with screw speed (150-500 rpm), material throughput (0.4kg/h-0.8kg/h), barrel temperature (60°-80°C for MPT and 90°-110°C for CAF), stagger angle (30°, 60° and 90°) and direction of the kneading zone (forward and reversed) as factors.

Granule and process properties were evaluated as responses like friability, granule size, shape, density, porosity and torque. These properties were evaluated for samples collected at 4 different locations along the length of the granulation barrel to visualize and understand the granule formation along the length of the barrel [2]. Samples were collected after only conveying elements, after the first kneading zone, after the second kneading zone and after the screw mixing element at the end of the barrel.

Furthermore, the granule temperature and barrel temperature were monitored at these 4 locations. Therefore, a FLIR A655sc Infrared Camera (FLIR systems, Frankfurt am Main, Germany) was used. Heat can be transferred onto the granules from the heated barrel wall, but heat can also be generated due to the shear created by the friction forces of the kneading elements. The temperature of the barrel wall was monitored with the data logging system of the granulator, but this does not represent the actual temperature of the granules. The maximum and the mean granule temperature was calculated and the difference between the programmed barrel temperature and the measured granule temperature was calculated for the different runs of the DoE to evaluate the influence of the different factors on the granule temperature. Granule formation, granule temperature and resulting granule properties are closely related. Therefore, the link between granule and process properties and granule temperature was also investigated in this study.

RESULTS AND DISCUSSION

The most influencing factors of the DoE on the granule temperature increase is the reversed configuration with a small stagger angle and as to be expected the programmed barrel temperature. The higher granule temperature for this screw configuration can be linked with an increased torque value and therefore a higher amount of shear. Because of this, the binder can be melted sufficiently to attain granule formation. Granules that are obtained with this kneading zone configuration also have the best quality attributes, e.g. the granules produced are the largest and strongest and have a monomodal granule size distribution.

It can be observed that granule formation occurs in the zones along the barrel that create higher shear and subsequently higher temperature which ensures enough melting of the binder. Therefore, no granule formation is occurring in the conveying zones or in the kneading zones with forward configuration. The torque value for the forward configuration is not increasing sufficiently and hence not enough shear is created to form granules with good quality attributes. This can be countered by changing the screw configuration or by elevating the barrel fill level. For these two options there will be an increase of the torque value to ensure enough shear at the kneading zones to obtain granules with good quality attributes.

CONCLUSION

Heat transfer in continuous melt granulation occurs in a more controllable and uniform way compared to batch melt granulation, resulting in granules with more uniform quality attributes. By applying different amounts of shear and barrel temperature, different granule properties can be obtained. The smallest stagger angle in reversed configuration creates the highest shear and the largest granule temperature increase and therefore, more melting of the binder and densification occurs and better quality attributes of granules are obtained. Because the solid state of various active pharmaceutical ingredients are not resistant for all levels of shear and temperature, the screw configuration can be optimized for every formulation so that optimal quality attributes can be obtained.

REFERENCES

[1] T. Monteyne *et al.*, "The use of rheology to elucidate the granulation mechanisms of a miscible and immiscible system during continuous twin-screw melt granulation," *Int. J. Pharm.*, vol. 510, no. 1, pp. 271–284, 2016.

[2] M. Verstraeten *et al.*, "In-depth experimental analysis of pharmaceutical twin-screw wet granulation in view of detailed process understanding," *Int. J. Pharm.*, vol. 529, no. 1–2, pp. 678–693, 2017.

562508 3D Printing: A New Era of Precise Manufacturing of Individually Developed Pharmaceuticals

The growing demand for customized medicines increases the impact of 3D printing as one of the emerging technologies for precise manufacturing of individually developed dosage forms. In all 3D printing technologies, large number of materials, geometrical and process factors affect the quality of these computationally designed dosage forms and safety of their use. In addition, for printing an identical dosage form, the quality may vary when built using different 3D printers, even when the same model of the printer, printing parameters, process steps, and raw materials are used. Therefore, knowledge of how the variability of each input parameter and processing step affects the final finished product is critical to ensuring its quality. There is no specific guidance concerning the design, manufacturing process and quality testing considerations for 3D printed solid dosage forms. This presentation will then highlight some of specific methodologies and control strategies considerations over the material, geometrical and process parameters for 3D printing of solid dosage forms. A case study of employing 3D pneumatic assisted microextrusion based 3D printing to manufacture modified release tablets and polypills will be presented. considerations for process validation will be also discussed so that quality for all printlets is maintained when built in a single cycle, between cycles, and by different printers. The issue of identifying and linking the critical factors to the critical quality attributes of the printlets will be also highlighted with an emphasis on the regulatory considerations involved.

561294 Smart Design and Virtual Product Analysis for 3D Printing Pharmaceutical Tablets for Patient Centric Product Development

The Additive Manufacturing is an innovative and promising method that is suitable for the fabrication of personalized dosage forms of tablets. The formulation selection and geometry design of the tablets are key elements for product design. Tuning the formulations and geometry and design of tablet would be based on personalized dose and release profile, and tablets can be smartly designed to provide adequate physical strength and desired drug release profile.

In this work, two mathematical models were developed and validated by experimental data:

The first model is a solid mechanics model by Finite Element Method technique. The model studies the various internal geometry and design of the tablet on tablet axial and radial strength. The sensitivity analysis demonstrates how changing the printing parameters and internal structure of the tablet will impact the tablet hardness. Experimental data for tablet axial and radial strength are in confirmation with the model's results.

The second model is a dissolution and drug release profile simulation. Effects of formulation variabilities (soluble to insoluble ingredients ratio) and tablet geometry on drug release profile have been studied. The model is developed based on unsteady diffusion of water to the solid tablet and swelling (moving boundary PDEs) of the tablet, and unsteady diffusion of drug from the tablet to the dissolution medium.

The models demonstrate how the smart design and virtual product analysis for the 3D printed tablets could be used for patient centric product development.

Disclaimer: This scientific publication reflects the views of the authors and should not be construed to represent FDA's views or policies.

573787 Rapid on-Site Manufacturing Approach with 3DP Capsules: Delayed Burst Release for Regional Absorption

A critical gap existing in our first-to-human strategy is the ability to evaluate complex dosage forms under the on-site formulation (OSF) strategy. We propose to use 3D printing (3DP) to fill this gap by enabling OSF with custom dosage forms using this flexible, high quality, highly reproducible, rapid, and material sparing approach. The 3DP platform has the potential to enable project teams to gain a better understanding of the complex in vitro-in vivo relationships across different drug release mechanisms through accelerated pharmacokinetic screening clinical trials during or immediately following Phase I, and potentially diminish a portion of the decade-long timing from molecule patent to commercialization. Specifically, human regional absorption and permeation is difficult to determine in early clinical phases. The models we use today for calculating regional permeation in the GI tract are largely based on datasets collected in the last few decades using one of two approaches: 1) a double lumen or triple lumen catheter system, or 2) either a single or double tube-balloon system, where the tube would be fed through the intestinal tract and the balloons would be inflated to isolate a portion of the intestines during the study. Both methods are invasive and not feasible as a routine study in humans. An oral delivery system has been developed by Medimetrics called IntelliCap, which is capable of investigating regional absorption; however, this technology is proprietary and therefore may not be cost effective to use for routine studies.

A Hyrel 3D System 30M printer along with various modular heads were purchased from Hyrel 3D, whose heads were modified or custom designed to be able to print new polymer filaments that are semi-brittle, soft, or flexible, or a combination thereof, or to be capable of injecting fluids with viscosities between 20-3000 cP for a single-click print-and-fill dosage form. In this presentation, the development of a set of capsule formulations for delayed burst release with controlled timing is shown to facilitate regional absorption studies on new active pharmaceutical ingredients. By manipulating the thickness of the capsule walls, the onset of release can be controlled. Due to the swelling behavior of PVA, it was discovered that the mechanical properties were critical for robust formulations due to delamination failure that would occur if the layer-layer weld during the printing process was not sufficiently strong. It was also found that fillings containing a large fraction of swellable materials, such as water, affected the physical stability of the formulations. As a first step to enable robust formulations, finite element analysis was used to understand the effect of water-containing drug fillings on the dissolvable capsule shell, and the shell geometry was optimized to reduce swelling stresses that may cause delamination in- vitro or vivo. Second, software techniques were developed, called zoning, to print capsules with no functional defects. Additionally, due to the nature of the 3DP process, in vitro analysis revealed a unique anisotropic dissolution behavior. Custom USP2 hanging baskets were designed and manufactured via 3DP for orientation control during dissolution. Lastly, an in vivo (beagle dogs) study was run using two 3DP capsule formulations and an immediate release compacted tablet as a control, and found a good in vitro – in vivo relationship.

557916 Production of Small-Batches of 3D Printed Multi-Compartment Capsular Devices for Oral Drug Delivery Using High-Throughput Robotic Manufacturing Technologies

In this talk, we present an automated production process enabling the manufacturing of multi-compartment capsular devices for oral drug delivery. This process includes *i)* the production of capsule parts and *ii)* filling the capsules with formulations containing active pharmaceutical ingredients. In this respect, our technology allows fast and efficient fabrication of fixed-dose combination products in small, personalized batches.

The capsular devices manufactured with this process have a structure that is different from the traditional capsules that are already on the market – generally made either of gelatin or hydroxypropyl methylcellulose. Our capsular devices have a modular structure that consists of multiple inner compartments, which are independent of one another. Each compartment can be filled with different active ingredients and with a range of doses, as well as formulations, of the same drug.

Moreover, these capsular devices transform the polymeric barriers of traditionally coated drug delivery systems into self-operating shells. Therefore, they represent a step forward in controlled release – separating the container (i.e. the capsule shell) from the content (i.e. the drug formulations contained in each compartment). The release performance of a compartment is determined by the composition and design features (i.e. morphology and thickness) of its walls. By varying the composition and thickness of each compartment, its release profile can be modified without affecting the properties of the other compartments. In other words, these multi-compartment capsules products yield multiple release kinetics in a single unit.

Multi-compartment capsular devices are especially advantageous for therapies involving multiple drugs that are mutually incompatible or interact in the gastrointestinal tract. Using multiple compartments simplifies the dosing schedule and has a positive impact on overall patient compliance. Moreover, these capsular devices enable the independent development of the drug formulation and the capsule shell, reducing time-to-market and development cost for new drug delivery systems.

The talk will present an example of this new class of capsular devices. The system consists of a two-compartment capsule shell composed of three parts: two hollow halves and a middle part, acting both as a joint and a partition. The joint allows the hollow parts to be assembled into a closed capsular device while dividing the internal cavity into two separate compartments. The hollow parts can have different shapes and thicknesses, resulting in compartments with different release properties and internal volume.

As a proof of concept, we will discuss the fabrication of capsule shells composed by 400 μm-thick and 800 μm-thick compartments using hydroxyl propyl cellulose (Klucel™ LF). We selected 3D printing by fused deposition modeling (FDM) as a suitable manufacturing technique for small-scale batches. This required the development of an industrial-grade FDM system able to realize pharmaceutical products. The talk will discuss in detail the development steps required to manufacture both the filament (through hot melt extrusion) and the capsule shells (through FDM 3D printing).

The FDM process requires as input material thermoplastic polymers supplied in the form of filaments – with suitable diameter and mechanical characteristics. We designed custom extrusion and 3D printing systems, and validated them to fulfill strict pharmaceutical quality requirements. In both cases, the study of the process was aimed at the identification of the critical operating parameters that affect the quality attributes of the final product. We assessed the stability of hydroxyl propyl cellulose after hot-processing and 3D printing, ruling out contamination and the formation of any hazardous degradation product. This led to the formation of an internal protocol for evaluation of by-products. These results demonstrate that the filaments and prototype capsule parts meet the target specifications – in terms of composition, absence of microbiological contamination, and absence of degradation by-products.

Once filled and assembled, the two-compartment capsular devices showed the expected two-pulse release behavior, in agreement with the swellable/erodible nature of the polymer. Furthermore, we compared the dimensional and release characteristics of multi-compartment capsules prepared by fused deposition modeling with those of analogous devices obtained by injection molding (IM). The real-time prototyping capabilities of 3D printing can significantly speed up the R&D process for IM. Both represent viable manufacturing techniques. 3D printing enables the rapid and inexpensive realization of small on-demand batches. Then, once the prototype capsules have been tested and refined, IM allows the large-scale production of capsule shells with the desired behavior.

After the capsule shells are produced, a proprietary robotic system fills each compartment with specific drug formulations. This part of the process relies on a set of automated filling modules working simultaneously, in parallel. The flexibility of this configuration is particularly suitable for the efficient production of small batches of fixed-dose combination products. The automated deposition stage allows the rapid and accurate adjustment of the combination and dosage strength of the active ingredients. Cross-contamination is avoided by enclosing each filling station in a dedicated clean room. A robotic arm (designed to safely collaborate with human operators) moves the capsular devices form one filling station to the next. The capsule filling systems were validated with preliminarily filling trials, which used powders with different particle size distribution and bulk density. We considered 25 batches of 50 capsule each for each powder formulation. In all cases, the amount of active ingredient deposited by the robots had an average coefficient of weight variation (i.e. the ratio between the standard deviation and the average value of the weight of the deposited powder) below 3%. These results demonstrate the accuracy and the flexibility of the powder deposition systems.

This robotic capsule filling system enables the realization of batches that are much smaller than the ones typical of traditional large-scale production processes. This level of personalization requires innovative in-process, non-destructive controls in order to guarantee the quality of the products in a cost-effective manner. We developed quality control machines that are able to check the thickness and weight of the capsule parts. Additionally, we use spectroscopic techniques to check the weight uniformity and identity of the pharmaceutical formulations.

566475 Pharmaceutical Application of Inkjet Based 3D Printing Process

Pharmaceutical Application of Inkjet Based 3D Printing Process

Koyel Sen [1], Sameera Sansare [1], Si Wan Li[2,3], Bodhi Chaudhuri [1,2,3], Anson Ma [2,3]

[1]Department of Pharmaceutical Sciences, University of Connecticut [2]Department of Chemical and Biomolecular Engineering, University of Connecticut [3]Institute of Material Sciences, University of Connecticut

Correspondence: koyel.sen@uconn.edu

Background:

There has been a growing interest in using 3D printing (3DP) for drug tablet manufacturing. This is motivated mainly by 3DP's potential in achieving different dosage geometries, attaining complex drug release kinetics, and combining multiple drugs efficiently for personalized medicine applications. The aims of this study are to: (i) experimentally evaluate an inkjet-based 3D printing method for manufacturing of tablets, (ii) understand the underlying physics, and (iii) to compare against tablets produced by the conventional tableting process.

Methods:

In an inkjet-based 3DP process, a liquid binder is printed onto a layer of excipient powder to bind the powder in a pre-determined pattern. Then, a fresh layer of powder is then spread evenly on top, followed by another round of binder printing. The process is repeated until the construction of the 3D object, or drug tablet in this case, is finished. The surface tension, viscosity, and density of the binder solutions were measured to predict their jettability. This was achieved using dimensionless groups that compare the relative importance of inertia, surface forces, and viscous forces. Two different types of powders and three different types of binder solutions were studied. Successfully printed tablets were further characterized for their weight variation, hardness, disintegration, true density, crystallinity, and physical stability.

Results and Conclusions:

Inkjet-based 3DP was performed using the selected pharmaceutical excipients and binder solutions. The final attributes of the printed tablets were clearly linked to the physical properties of the powder and binder solutions. Such physical understanding was used to create a "design space" in terms of material selection for successful 3D printing of drug tablets. This research lays the foundation for improving the precision and accuracy of the 3DP process for personalized medicine and controllable drug release applications, which will in turn improve patient compliance, lower drug price, and fulfill public health mission.

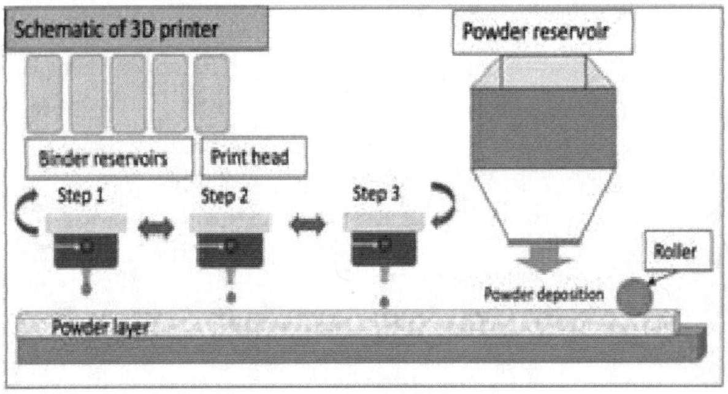

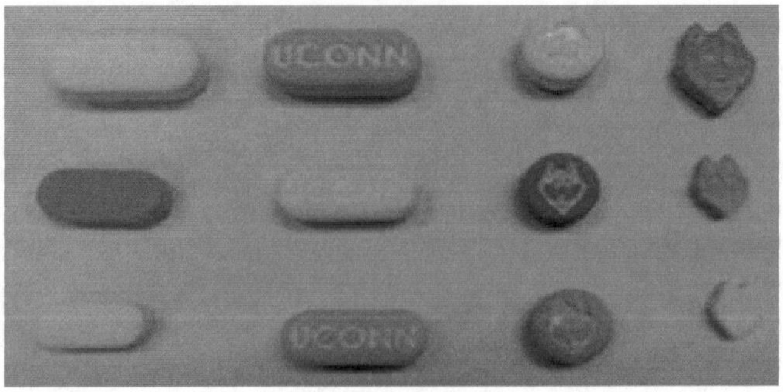

Figure 1- a) Schematic diagram of the inkjet based 3D printer used for the study (left) b) Non-pharmaceutical grade 3D printed tablets using inkjet based printing method

570318 3D Printed Tablets Using Binder Jetting and Pharmaceutical Excipients

Binder jetting is an additive manufacturing (AM) technology used to create three-dimensional model constructs. The materials used in binder jetting are relatively limited generally due to insufficient knowledge surrounding powder technology. The ability to self-formulate powder materials for use in binder jetting machines is not common practice today as not enough rigorous knowledge exists for the requirements of a 3D printable powder.

In this study we characterised several commercial 3D printable powders used in a 3DS Projet460 binder jetting printer, using techniques central to powder flow and powder wetting. The goal was to determine the key powder parameters that are likely to result in a successful binder jetting process, and then to use these parameters to screen standard pharmaceutical excipients for their likelihood to be suitable for 3D printing in a standard commercial printer.

Five different commercial binder jetting powders of various generations from different manufacturers were used as the benchmarks for a binder-jettable powder. Particle size and shape was measured using a Camsizer X2 and the powder flow (bulk and dynamic) was characterised using a Freeman FT4 powder rheometer. All five commercial powders had similar d_{10} and d_{50} values and similar shape distribution profiles. The FT4 showed that all powders fell within a narrow range for the parameters comprising basic flowable energy, specific energy, flow rate index and compressibility.

Drop penetration tests were used to determine the likely construct quality prior to the powder being 3D printed. The dried nuclei from these tests were analysed for spread, depth and hardness, which correlates to the dimensional accuracy of the powder after 3D printing. Drop penetration time seemed to have a wide range of acceptable values, but nuclei construct strength was important. A preliminary framework highlighting the quantitative parameters for binder jetting powders was created.

Finally, to test the framework approach, a simple formulation of 20% VA64 Kollidon and 80% Microcrystalline cellulose was used to successfully 3D print a range of basic tablet shapes in a Projet460 binder jet printer. The use of these commercial powders and characterisation techniques ultimately provide a useful benchmark for determining the suitability of standard pharmaceutical powder APIs and excipients to achieve good quality printed constructs, and sets out the considerations for specially engineered pharmaceutical particles and formulations that would need to be developed in future.

575111 Potential of 3D Printing in Precision Combination Therapy: A Case Study of Cardiovascular Polypills

3D printing is one of the emerging manufacturing technologies that is well suited to produce precision (personalized) medicine. The unique versatility of 3D printing to engineer desired designs with specific geometry allows for achieving targeted release profiles of printed drugs. It has a high potential for individualization of drug therapy combination through the so called "Polypill printlets" by which all the drugs required for the therapy can be printed into a single dosage unit. In this regard, this presentation will highlight a case study of engineering and manufacturing a cardiovascular polypill of Metoprolol (MT) and Atorvastatin (AT) using 3D pneumatic assisted micro-extrusion (PAM) technique. The polypills were designed to provide extended release properties of MT and immediate release properties of AT. The effect of printing, geometrical and formulation parameters on breaking strength and disintegration of the polypills and release properties of the drugs will be discussed. Various designs of the polypills will be demonstrated to modulate the release rate of the drugs from the polypills. The design spaces for formulating these polypills will be discussed in terms of identifying the critical material attributes and key geometrical parameters that achieve independent release properties of both drugs from the polypills. In addition, the regulatory considerations and failure modes of the polypill printlets will be also shared.

DISCLAIMER

This abstract reflects the views of the presenter and should not be construed to represent FDA's views or policies.

574482 *In Vitro* biomimetic Tumor Model for High Throughput Drug Screening

Although much scientific effort has been expended for the treatment of breast cancer, this disease is still a serious threaten to woman life. Due to the limitations and shortcomings of current therapies, more effective drugs or therapeutic strategies are strongly required in the market. An efficient and accurate drug screening assay contributes to the acceleration the drug development. Thus, we have developed an assay for drugs that are targeted on the promotors specifically overexpressed in tumor tissues. Those promotors were designed to initiate the transcription of enhanced green fluorescent protein(EGFP) in the breast cancer cells (MCF-7). With the treatment of promotor inhibitors, we found that the EGFP signal was closely correlated to the promoter gene expression level in MCF-7 cells, indicating the EGFP signal can be used to identify the drug efficacy. This approach is easy and high throughput for the screening the biomarker inhibition drugs. On the other hand, considering part of drug failures in clinical phase are caused by the poor predictability of early drug assay using two dimensional cell culture system, our research then focused on the construction of three dimensional cancer cell models with the simulation of tumor microenvironment. We have co-cultured MCF-7 cells and mouse fibroblasts (NIH-3T3) in the 3D scaffold, and conducted the co-culture in both static and dynamic culture modes. The results demonstrated that the drug resistance of MCF-7 was significantly increased either by the 3D structure or by the interaction with NIH-3T3 cells. This novel 3D co-culture assay recapitulated the tumor microenvironment effects on drug efficacy and had improved predictability for in vitro drug screening. Meanwhile, the MCF-7 and NIH-3T3 cells were transfected with EGFP and Ds-red, respectively. The fluorescence of each protein was verified to be sensitive enough to monitor individual cell proliferation. Combing this dual fluorescence system, we are able to produce a simple, real-time and high throughput 3D co-culture assay for a variety of anti-cancer drug screening.

570460 Automated Nanomole-Scale Platform for Rapid Screening and Optimization of Electroorganic Synthesis

Electroorganic synthesis has recently emerged as a powerful tool for synthetic chemists to access unprecedented transformations under mild, and environmentally friendly conditions. However, the vast parameter space and prolonged reaction time of electroorganic synthesis create challenges to discover, optimize, and understand newly developed electrochemistry.

Here, we demonstrate a fully automated droplet-based screening platform to enable rapid examination of the electrochemical reactions with nanomole amount of reagents, which can potentially facilitate the mechanistic understanding of the electrochemical systems. This platform achieved 10-fold increase in screening throughput compared to traditional batch screening approach. Electrochemical radical cross-coupling reactions are examined on this system. Additionally, we demonstrate the optimal electrochemical reaction condition obtained from the screening platform can be directly applied to the large-scale electrosynthesis to quickly access the desired substrate.

577918 Virtual Rodents: Design and Translation of Glucose-Responsive Insulins Aided By Pharmacokinetic Modeling

Despite considerable progress, the development of glucose-responsive insulins (GRIs) still largely depends on empirical knowledge and repeated experimentation. In this work, we established a mathematical model for GRIs in rodents so as to assist the therapeutic's rational design and clinical translation. We combined the chemical kinetics of GRIs with a physiological model describing the full-body glucoregulation of rats and mice, adapted from our prior human model. The resultant *Virtual Rodent* is used to explore the GRI parameter space and thereby identify the optimal designs. Blood glucose profiles predicted by the parameterized Virtual Rodent match the experiments. The simulated GRI action profiles and set of optimal designs are consistent with literature. We further explored the translatability between animal and clinical studies via a comparative analysis on the rat, mouse, and human results. It was revealed that the optimal GRI spaces of rodents and humans overlap to a limited extent, mirroring the concern that animal models may not accurately emulate human physiological processes in the case of diabetes therapy. We showed that Virtual Rodent predicts the clinical viability of a candidate found effective in preclinical trials, while also watches out for the *false-negatives* wrongly discontinued from further testing. We condense such information onto a *translatability grid*, a straightforward visual guide for research planning and decision-making. We find Virtual Rodent a promising tool for the rational design of GRIs and their translation to the clinic.

562602 Combined Synthesis Planning for Synthetic Drugs on the WHO Essential Medicines List

The WHO Model List of Essential Medicines (EML) was created in 1977. It contains medicines that are needed to satisfy the priority healthcare demand and should be in sufficient amount in any time[1]. However, up to today, access to essential medicines is still problematic for about one-third of the world's population. Prices for some medicines make them unaffordable in less developed areas[2], and drug shortages have been long and widespread challenges[3]. While this is a complicated problem and no single solution can suffice, it is desirable to develop strategies to lower the manufacturing costs of the medicines and simplifying the supply chain for the essential medicines for quicker response to drug shortages.

In this work, we take advantage of the recent development in artificial intelligence applied in synthetic planning to explore solutions to the above challenge. Using a state-of-the-art synthesis planning tool, we try to minimize the chemical inventory needed to access the synthetic drug compounds in the WHO EML. The problem is solved in two stages. In the first stage, retrosynthesis analysis is performed for all the target molecules. Machine learning models are used to help find synthetic pathways[4], propose reaction conditions[5], and evaluate the feasibility of the suggested reactions[6]. In the second stage, a multi-objective mixed-integer programming problem is solved[7] to minimize the chemicals used in all the syntheses while maximizing the likelihood of success for all the pathways. Results show that by planning the syntheses in a combined manner we can reduce the number of chemicals by 30% compared to planning the syntheses separately, without significantly sacrificing the likelihood of success of the syntheses. In the meantime, this work demonstrates the potential benefit of simultaneously planning the syntheses for multiple targets, which can also be applied to other multitarget lists, e.g., a compound library during drug discovery.

Reference

1. The World Heath Organization. WHO Model Lists of Essential Medicines Available online: https://www.who.int/medicines/publications/essentialmedicines/en/.
2. Hill, A.M.; Barber, M.J.; Gotham, D. Estimated costs of production and potential prices for the WHO Essential Medicines List. *BMJ Glob. Heal.* **2018**, *3*, e000571,
3. Gray, A.; Manasse, H.R. World Health Assembly». *Bull. World Health Organ.* **2012**, *90*, 158–158A.
4. Segler, M.H.S.; Waller, M.P. Neural-Symbolic Machine Learning for Retrosynthesis and Reaction Prediction. *Chem. - A Eur. J.* **2017**, *23*, 5966–5971
5. Gao, H.; Struble, T.J..; Coley, C.W..; Wang, Y.; Green, W.H.; Jensen, K.F. Using Machine Learning to Predict Suitable Conditions for Organic Reactions. *ACS Cent. Sci.* **2018**, *4*, 1465–1476
6. Coley, C.W.; Jin, W.; Rogers, L.; Jamison, T.F.; Jaakkola, T.S.; Green, W.H.; Barzilay, R.; Jensen, K.F. A graph-convolutional neural network model for the prediction of chemical reactivity. *Chem. Sci.* **2019**, *10*, 370–377.
7. IBM Co. IBM ILOG CPLEX. **2018**

570471 Epitope-Targeted Peptide Immunostimulants to Combat Antibiotic-Resistant Bacteria

Antibiotic resistance in bacterial pathogens poses a rapidly emerging health threat. Today, many bacteria have acquired resistance to all available antibiotics, while the discovery of new antimicrobials has slowed. Without a reliable and perhaps general method to generate antibiotics against newly emerging resistant bacterial strains, antibiotic resistance will grow as a health threat. We envision a synthetically versatile molecular platform, called epitope-targeted peptide immunostimulants (EPIs), as a sustainable therapeutic countermeasure against drug-resistant pathogens. EPI molecules comprise of two chemical moieties: (1) a polypeptide ligand that binds a protein at a bacterial surface and (2) an immunogenic group that recruits endogenous antibodies. EPIs adsorb to the surfaces of the targeted bacteria, promote opsonization, and ultimately enhance phagocytic killing by innate immune cells. To test this, we used a powerful combinatorial chemistry technique, the protein-catalyzed capture agent (PCC) approach, to identify macrocyclic peptide ligands that bind epitopes at the surface of carbapenem-resistant *Klebsiella pneumoniae*, which are of high clinical significance. Guided by multi-omic and bioinformatics analyses to select target epitopes, the PCC approach was employed to develop EPI ligands against four 14-residue epitopes on the highly-abundant fimbiral subunit protein, MrkA, on *K. pneumoniae*. These screens yielded several EPI ligands, and the top-performing ligand exhibits high-affinity (EC_{50}~30 nM) to the full-length MrkA protein and selectively binds the surfaces of MrkA-expressing *K. pneumoniae*, versus other pathogenic bacterial species. EPIs that are conjugated with an immunogenic 2,4-dinitrophenyl (DNP) moiety enabled the recruitment of anti-DNP antibodies to *K. pneumoniae* surfaces and, further, that *K. pneumoniae* opsonized in this manner were more susceptible to phagocytosis by macrophages than non-treated cells. While *K. pneumoniae* was targeted in this study, the PCC approach enables EPIs to be developed against virtually any epitope and so EPIs could easily be generated for other resistant pathogens. Our work demonstrates EPIs as a promising all-synthetic solution to antibiotic resistance.

564305 Mathematical Modeling and Optimization of the Upstream Monoclonal Antibody Production

Mathematical Modeling and Optimization of the Upstream Monoclonal Antibody Production

Ou Yang, Marianthi Ierapetritou

Department of Chemical and Biochemical Engineering, Rutgers—The State University of New Jersey, 98 Brett Road, Piscataway, New Jersey 08854-8058, United States

Abstract

Mathematical modeling has been widely used in upstream biomanufacturing to simulate bioreactor operations and optimize and control the cell culture systems[1-2]. A critical component of cell culture performance for monoclonal antibody (mAbs) production is glycosylation[3-4]. The work done in the literature for cell culture simulation can be distinguished into kinetic-based model and stoichiometric method[1]. Kinetic model such as cisternal maturation model is mainly used in Golgi and glycosylation simulation[5-8] which is a powerful way to understand dynamic change of glycosylation process and to predict product components under different operating conditions. However, the kinetic model involves a large number of parameters and requires extensive experimental data. In addition, many components concentration such as nucleotides and nucleotide sugars, glycoprotein distributions inside of CHO cells that are important inputs for model building, are hard to be accurately measured. Flux based method and Markov chain models on the other hand, apply stoichiometric method which do not require intricate parameter estimation but use flux to represent the possibility of each reaction happens in side of the cell[4, 9]. This method is good for process understanding and product prediction with mutated cell lines but it assumes the quasi-steady state for intracellular metabolites and hard to predict product as operating conditions change.

In this study, a simplified single cell model is developed to dynamically simulate the CHO cell culture process in both fed-batch and perfusion bioreactors. The model contains unstructured cell culture model which is coupled to structured glycosylation model with reduced number of parameters. Cell culture model simulates cell growth and death, nutrients consumption and metabolites' concentrations. The glycosylation model predicts major glycoprotein concentrations varying with time including intermediates glycosylated proteins and final product protein quality attributes. Experimental data are used for model parameter estimation and model validation. The operating conditions effect on cell culture and product compositions are investigated, including temperature, pH, ammonia concentration and other nutrients additions. Based on the correlations among operating conditions with product qualities, the process design space is defined. The overall process is optimized to achieve the required product specification. The purpose of this work is to apply mathematical modeling approaches to improve process understanding of mAbs production, achieve cell culture optimization and establish soft-sensors for variables that cannot be measured online experimentally.

1. Sha, S.; Huang, Z.; Wang, Z.; Yoon, S., Mechanistic modeling and applications for CHO cell culture development and production. *Current Opinion in Chemical Engineering* **2018,** *22*, 54-61.
2. Tziampazis, E.; Sambanis, A., Modeling of cell culture processes. *Cytotechnology* **1994,** *14* (3), 191-204.
3. Kontoravdi, C.; Jimenez del Val, I., Computational tools for predicting and controlling the glycosylation of biopharmaceuticals. *Current Opinion in Chemical Engineering* **2018,** *22*, 89-97.
4. Spahn, P. N.; Hansen, A. H.; Hansen, H. G.; Arnsdorf, J.; Kildegaard, H. F.; Lewis, N. E., A Markov chain model for N-linked protein glycosylation--towards a low-parameter tool for model-driven glycoengineering. *Metab Eng* **2016,** *33*, 52-66.
5. Jimenez del Val, I.; Nagy, J. M.; Kontoravdi, C., A dynamic mathematical model for monoclonal antibody N-linked glycosylation and nucleotide sugar donor transport within a maturing Golgi

apparatus. *Biotechnol Progr* **2011,** *27* (6), 1730-1743.
6. Jimenez del Val, I.; Fan, Y.; Weilguny, D., Dynamics of immature mAb glycoform secretion during CHO cell culture: An integrated modelling framework. *Biotechnology Journal* **2016,** *11* (5), 610-623.
7. Villiger, T. K.; Roulet, A.; Perilleux, A.; Stettler, M.; Broly, H.; Morbidelli, M.; Soos, M., Controlling the time evolution of mAb N-linked glycosylation, Part I: Microbioreactor experiments. *Biotechnol Progr* **2016,** *32* (5), 1123-1134.
8. Karst, D. J.; Scibona, E.; Serra, E.; Bielser, J. M.; Souquet, J.; Stettler, M.; Broly, H.; Soos, M.; Morbidelli, M.; Villiger, T. K., Modulation and modeling of monoclonal antibody N-linked glycosylation in mammalian cell perfusion reactors. *Biotechnol Bioeng* **2017,** *114* (9), 1978-1990.
9. Hutter, S.; Villiger, T. K.; Bruhlmann, D.; Stettler, M.; Broly, H.; Soos, M.; Gunawan, R., Glycosylation flux analysis reveals dynamic changes of intracellular glycosylation flux distribution in Chinese hamster ovary fed-batch cultures. *Metab Eng* **2017,** *43* (Pt A), 9-20.

556748 A Piecewise Parametric Model for Open Loop Screw Feeder Flow Rates

Screw feeders, a type of loss-in-weight feeder, serve as the crucial first unit in a pharmaceutical drug product continuous manufacturing (CM) process. They are responsible for feeding the API and excipient powders in a controlled manner. Therefore, there is industrial interest in an accurate and computationally efficient flow model of a screw feeder for use in process development [1]. Some proposed models have used statistical analysis [2], response surface methodologies [3], and semi-empirical differential equations [4]; however, these methods only model the mass flow under steady state or closed loop control, obfuscating the underlying dynamics of powder flow.

This work proposes a novel parametric piecewise-linear model to approximate the mass flow rate of powder discharged from a feeder under open loop operation. The piecewise model represents two flow regimes: a constant flow observed at higher powder masses in the hopper and a linearly declining flow observed at lower powder masses. First, a custom process model was developed in gPROMS™ (Process Systems Enterprise, London, UK). Then, open loop experiments were performed using six different powders, two different Coperion K-Tron screw feeders, and multiple different motor speeds. Using the feeder's operational data, parameter values for each run configuration were estimated. It was observed that for powders that mirrored our models assumptions, the model was accurate over the entirety of the run. For powders that did not display strictly constant mass flow in the first regime, the model would instead capture the average mass flow rate. Overall, this model is easily fit, computationally inexpensive, and reliably describes the general open loop flow behavior of a powder. It is suitable for use in the early stages of CM process development.

References

[1] S. García-Muñoz, A. Butterbaugh, I. Leavesley, L. F. Manley, D. Slade, and S. Bermingham. A flowsheet model for the development of a continuous process for pharmaceutical tablets: An industrial perspective. AIChE Journal, 64:511–525, 2017.

[2] W. E. Engisch and F. J. Muzzio. Method for characterization of loss-in-weight feeder equipment. Powder technology, 228:395–403, 2012.

[3] F. Boukouvala, F. J. Muzzio, and M. J. Ierapetritou. Design space of pharmaceutical processes using data-driven-based methods. Journal of Pharmaceutical Innovation, 5:119–137, 2010.

[4] F. Boukouvala, V. Niotis, R. Ramachandran, F. J. Muzzio, and M. G. Ierapetritou. An integrated approach for dynamic flowsheet modeling and sensitivity analysis of a continuous tablet manufacturing process. Computers & Chemical Engineering, 42:30–47, 2012.

557537 Hybrid Theoretical-Empirical Approach to Modelling of Twin Screw Feeders for Continuous Tablet Manufacturing

Screw feeders are popular equipment for the handling and metering of bulk solids in the pharmaceutical industry, especially in continuous tablet manufacturing. Predictive models are crucial to meet quality requirements, however a limited number of predictive process models have been developed so far in the scientific literature. Despite the increasing research in the last decade in the manufacturing of powder-based products, there is still a lack of knowledge on the physics governing the dynamic behaviour of the screw feeders for different material properties, screw designs and operating conditions. As a result, first-principles models for a complex system such as powder feeding have not been properly validated in both open and closed loop operations. Surrogate models, multivariate techniques and empirical correlations have often been used instead to tackle process design, optimisation and control applications.

In this work, a mathematical modelling of twin screw feeders, suitable for continuous drug manufacturing processes, has been developed using a combined physics-based- and data-driven approach. The aim of the model is to predict how the screw feeder design, the hopper fill level, the screw speed and the physical properties affect the dynamic behaviour of the open loop system. A First Order Plus Dead Time (FOPDT) model has been developed in MATLAB (MathWorks) where the vertical stress exerted by the powder in the hopper, estimated from the equilibrium of forces on slice elements of infinitesimal thickness, has been assumed to affect the powder densification within the screw feeder and, thus, the mass flow rate provided by the feeder. A method to predict the amplitude of the stochastic fluctuations in the mass flow rate observed in the process data has been suggested and included in the model. Model parameters have been estimated for a total of six different bulk solids (two non-cohesive powders, two cohesive ones and two in the transitional range, categorised according to their Hausner ratio values). Furthermore, two different screw sizes and two hopper geometries have been investigated. The model predictions are in good agreement with the experimental data sets. The model can be explored to investigate the mass flow – screw feed relationships and the optimal range of operating conditions. It also may be used for online control, to identify the original equipment manufacturer control system and to design alternative control strategies. Furthermore, the model can be exploited to optimise refill operations, when the process is not controlled and the system is running under volumetric mode (i.e. open loop, the screw speed is constant and the controller is temporarily off, as the weight in hopper is increasing).

571583 Continuous Twin-Screw Powder Feeding – a Crossroad for Continuous Process Development

Continuous Twin-Screw Powder Feeding – A Crossroad for Continuous Tablet Manufacturing

T.R. Hörmann[1,2,3], J. Kruisz[1,2], M. Beretta[1,2], E. Faulhammer[1,2], W.-K. Hsiao[1,2], J.G. Khinast[1,2,3]

[1] Research Center Pharmaceutical Engineering GmbH, Graz, Austria

[2] European Consortium for Continuous Pharmaceutical Manufacturing (ECCPM), 8010 Graz, Austria

[3] Institute for Process and Particle Engineering, Graz University of Technology, Graz, Austria

Abstract

Continuous powder feeding is of critical importance during continuous manufacturing (CM), since it is (next to in-process segregation) the most susceptible source of content uniformity fluctuations in the final drug product. While loss-in-weight twin-screw feeding technology is well developed and a large variety of equipment is available and pharmaceutical powders were extensively characterized, there is still room for further learnings on the processability of the pure powders and their respective bends. In order to make an informed decision on the optimal feeding strategy, i.e., split-feeding of pure materials vs. pre-blend feeding, critical material/blend attributes must be identified and considered in context to the available equipment.

This study investigates four pharmaceutical materials in different grades (in total nine powders) in four twin-screw feeders in different scales. The combinations of all powders and feeders were characterized in feeding experiments to elucidate the interaction of equipment and powder with regard to screw feeding capacity and powder densification and flow in the hopper. Multivariate data analysis was utilized to identify relevant material descriptors for powder feeding performance and to establish a model for their processability in different feeder equipment scales and screws. Thereby, the feeding equipment was represented by quantitative geometry descriptors in the statistical models. Moreover, qualitative observations, such as rat-holing, or powder bridging in the hopper were evaluated to gain deeper insight in events associated with a failure of the feeding process, that is a lack of material flow into the screws.

The aim for the statistical model is to predict, for which combinations of powder properties and feeding equipment geometries well-controlled continuous feeding can be achieved. Future investigations will expand this approach to pre-blend feeding, also involving mechanistic modeling methods.

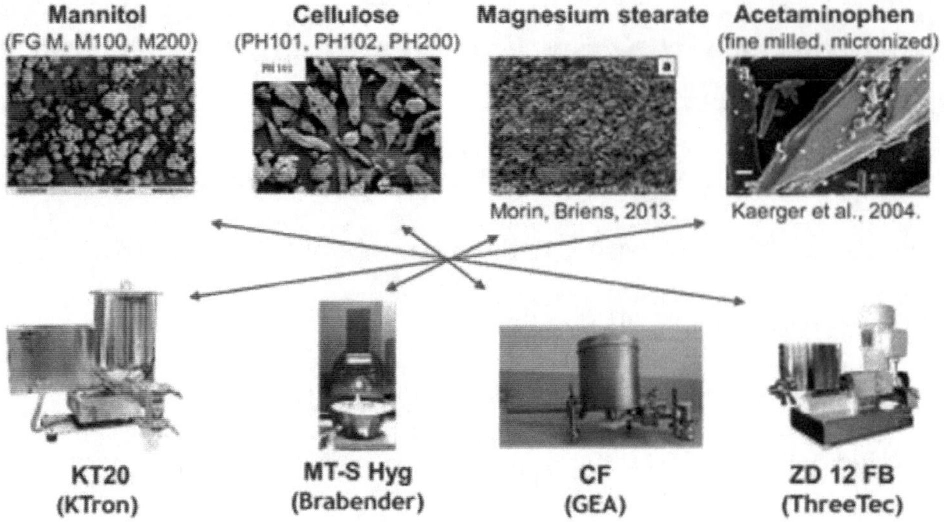

Figure 1. Tested selection of pharmaceutical powders and twin-screw feeders.

571453 Continuous Mixing Technology: Design Optimization with Discrete Element Simulations

Continuous Mixing Technology: Design Optimization with Discrete Element Simulations

Peter Toson[1], **Pankaj Doshi**[2], Eva Siegmann[1], Johannes Khinast[1,3], Daniel Blackwood[2], Ashwinkumar Jain[2], Alexandre Bonnassieux[2], Kai Lee[4], David Wilsdon[4] and James Kimber[4], Dalibor Jajcevic[1]

[1] Research Center Pharmaceutical Engineering, Inffeldgasse 13, 8010 Graz, Austria

[2] Worldwide Research and Development, Pfizer Inc. Groton CT 06340

[3] Graz University of Technology, Institute of Process and Particle Engineering, Inffeldgasse 13, 8010 Graz, Austria

[4] Pfizer Ltd. Discovery Park House IPC009, Sandwich, Kent. CT13 9ND

Keywords: Pharmaceutical Manufacturing; Continuous Mixing Technology; Process Design & Development

Mixing is a crucial processing step in a continuous pharmaceutical processing line. It is one of the three unit operations in a direct compression process (feeding, mixing, tablet pressing) and is responsible for content uniformity, lubrication of the powder and thus the tensile strength of the final tablet.

The operating space of a vertical continuous mixing device termed CMT (continuous mixing technology) has been analyzed with discrete element simulations. The advantage of the vertical design is that hold-up mass (and thus mean residence time MRT) and impeller speed (and thus the shear rate and stress) can be controlled independently from each other. In a previous study [1], process maps in terms of mixing quality, travel distance and extent of lubrication, and particle velocities have been created with DEM simulations and validated with tracer experiments. The process maps were created for operation at 10kg/h.

One challenge that remains is the scale up of the process to higher throughputs. The available operating space shrinks: Higher throughput means lower mean residence times. Higher impeller speeds and hold-up masses are required to offset the lower MRT, but both of these parameters have physical limits. In addition, higher impeller speeds come with higher centrifugal forces exerted on the powder bed, restricting the outlet flow and adding challenges to process control.

Instead of finding the optimal process conditions in the restricted operating space, the aim of this work is the design optimization of the CMT. Various impeller, screen, and baffle designs have been studied with DEM simulations. The DEM simulations revealed how the modifications to the geometry changed the particle dynamics inside CMT. This opened up the operating space, making high-throughput operation not only feasible, but also easy control. The best-performing virtual impeller prototype has been 3D printed and it was possible to confirm the predictions experimentally.

References:

[1] Toson et al., Int. J. Pharm. 552, 1–2 (2018), 288–300. doi:10.1016/j.ijpharm.2018.09.032.

560260 Optimization of Batch High Shear Granulation and Transfer to a Continuous Granulation Process: A Case Study of Metoprolol Succinate ER Tablets

While the most common implementation of continuous manufacturing (CM) has been for continuous direct compression, continuous wet granulation has been successfully established as a processing alternative in the continuous tableting lines. To date, no extended release products have yet been approved using CM technology. Continuous twin screw granulation utilizes a different mechanism than traditional batch granulation (e.g. high shear or planetary) and forms granules in a very short time (seconds). Twin screw granulation has the potential to produce granules with a broader size distribution compared to other wet granulation processes and may have different physical properties. Formulation properties, process parameters, and screw configuration have a complex influence on granule properties in twin screw granulation. This presentation will then highlight the critical material and process parameters to be controlled in high shear granulation process for the transfer to the continuous granulation process. A surrogate extended release formulation of metoprolol succinate was developed in house for this purpose. OFAT optimization of the formulation ingredients was done in a batch-mode. Optimization of the process parameters, namely impeller speed, granulation time, and spray rate, of the high shear granulator was done using a central composite design. Particle size distribution and bulk and flow properties of the granules and dissolution characteristics of the ER tablets were evaluated. The interactions among the effects of the input and intermediate factors on the in vitro performance of the finished product will be also discussed.

The identified critical process parameters of the batch high shear granulation were then translated into those of the twin screw granulation. The product transfer was done based on the granulation model developed from multivariate experiments, prior process knowledge, and a rational classification of formulation characteristics. Four external runs were performed on the twin screw granulator as a function of screw speed and shearing rate. Various variables of the granulation process will be also discussed in the presentation with an emphasis on their importance for "translating" a process between a batch process and continuous manufacturing line.

DISCLAIMER

This presentation reflects the views of the presenter and should not be construed to represent FDA's views or policies.

573952 Dampening Effect of Feedrate Variation By Downstream Unit Operations in Pharmaceutical Continuous Manufacturing

Purpose: Direct compression is the most straightforward continuous approach to tablet manufacturing and is appealing for its simplicity. The process begins with feeding individual raw materials into a continuous blender and finally a tablet press. Loss-in-weight feeders are typically used to maintain constant flow rate for both APIs and excipients. The materials then are blended to achieve an axial uniform distribution of the API and excipients. The material exiting the blender is then sent directly to a tablet press where the blend is compacted into tablets. Therefore, any perturbations to mass flow exiting the feeder can be transferred down to the subsequent unit operations. The blender is designed to reduce some of the inherent fluctuation due to back-mixing, but it is not currently known how much variation in flow can be filtered. Another location where mixing can occur is located in the feedframe of the tablet press. In this work, the aim is to study the dampening effect of blender and feedframe on incoming feed rate variability from the feeders. **Method:** A formulation including API, excipients and lubricant will be used in this study. All the materials are fed using loss-in-weight feeders at the designed mass flows. Controlled changes to the mass flow rate of the API feeder will be performed at different levels of amplitude and time to force a fluctuation in flow. To determine the impact of each the blender and the feedframe, a NIR probe will be fixed in the continuous line, scanning material exiting the blender and before it enters the feedframe. Tablets will be collected and scanned by offline NIR to determine the effect of the feedframe on the dampening of the perturbations. **Results and Conclusion:** A funnel plot is obtained from the results from NIR analysis and tablet content, which shows that variation in API concentration in the blends are significant reduced compared to that in feeding. Feedframe also contributes to the dampening effect of variation in tablet uniformity at different magnitude of disturbance in incoming feedrates.

576329 Investigation of Electric Capacitance-Based and X-Ray-Based Sensors for Real-Time Mass Flow Rate Monitoring in Continuous Tablet Manufacturing

The adoption of continuous pharmaceutical manufacturing is accelerating with six solid oral drug products produced via continuous tablet manufacturing approved by the FDA to date[1-3]. To continue the shift from batch operations to continuous operations to a wider range of products, using more complex routes such as wet and dry granulation, further advances in real-time process management (RTPM) are necessary. The key challenge in RTPM is to capture real-time data of critical process parameters (CPP) and critical quality attributes (CQA) of materials in a robust and accurate fashion so as to inform real-time manufacturing decisions.

The mass flow rate of particulate materials is an important CPP which affects the concentration and the content uniformity of the final product. Those attributes depend on the relative amounts of active pharmaceutical ingredients (API) and excipients which are combined into the powder blend. While loss-in-weight feeders can generally deliver granular materials at specified set points, mitigation of flow disturbances requires the implementation of the ratio control strategies [4] which depend on actual flow measurements. Moreover, mass flow measurements are important to establish whether the process is in a state of control and whether material losses or material accumulations are taking place in the process. This work is focused on investigating the use and comparative performance of Electrical Capacitance Volume Tomography (ECVT) sensor as well as an X-ray based sensor for real-time measurement of mass flow of powder/granular flows.

Our study draws on previous work reported in Ganesh et al [5] on the X-ray based sensor. The sensor, manufactured by Enurga Inc., (West Lafayette, IN) is based on X-ray attenuation and cross-correlation velocimetry [6]. When used in measuring powder blends composed of different proportions of acetaminophen (APAP), microcrystalline cellulose (MCC), lactose, magnesium stearate and silicon dioxide, the research indicated that the measured mass flow rate was within 5% of the actual mass flow rate in the pilot plant at Purdue University [5]. Although the measurement provided acceptable accuracy, the restriction where the X-ray sensor can be located by virtue of the requirement of bulky shielding has motivated the investigation of alternative technologies.

The ECVT sensor, offered by Tech4Imaging LLC, (Columbus, OH) is based on reconstruction of capacitance data [7]. It is a non-invasive sensing technology with fast data acquisition which makes it a good potential candidate for monitoring a continuous tablet manufacturing line. The ECVT sensors measure the dielectric properties of the particulate flow. A novel mapping between the change in capacitance and the displacement of materials affected by sensor sensitivity gradient is used to determine the displacement profile, and hence the velocity profile, which no longer depends on cross-correlation velocimetry [7]. Combined with volume fraction obtained from reconstruction of capacitance data, the velocity profile of solids is further transformed into real-time mass flow rate measurement.

In the first part of this study, we report on the performance of the sensor in a stand-alone system consisting of a loss-in-weight feeder, the ECVT senor and an independent gravimetric measurement. In the second part of this study, the ECVT sensor will be integrated into the tableting line at two locations – the exit of the API-excipient feeder-blender system and the entry to the tablet press. The performance of the ECVT sensor will be compared to that of the X-ray sensor for delivering the same functionality and the relative advantages and limitations of these technologies summarized.

Reference

1. Ned Pagliarulo. (2018) Pharma's slow embrace of continuous manufacturing. Available at: https://www.biopharmadive.com/news/pharmas-slow-embrace-of-continuous-manufacturing/532811
2. Pfizer Inc. (2018).S. FDA Approves DAURISMO™ (glasdegib) for Adult Patients with Newly-Diagnosed Acute Myeloid Leukemia (AML) for Whom Intensive Chemotherapy is Not an Option.

Available at: https://www.marketwatch.com/press-release/us-fda-approves-daurismotm-glasdegib-for-adult-patients-with-newly-diagnosed-acute-myeloid-leukemia-aml-for-whom-intensive-chemotherapy-is-not-an-option-2018-11-21

3. Lawrence De Belder, (2019) IFPAC Presentation. Information of Drug product is available at: https://www.janssen.com/canada/sites/www_janssen_com_canada/files/prod_files/live/tramacet_cpm.pdf
4. Lakerveld, R., Benyahia, B., Braatz, R. D., & Barton, P. I. (2013). Model-based design of a plant-wide control strategy for a continuous pharmaceutical plant. AIChE Journal, 59(10), 3671-3685.
5. Ganesh, S., Troscinski, R., Schmall, N., Lim, J., Nagy, Z., & Reklaitis, G. (2017). Application of X-ray sensors for in-line and noninvasive monitoring of mass flow rate in continuous tablet manufacturing. Journal of pharmaceutical sciences, 106(12), 3591-3603.
6. Yan, Y. (1996). Mass flow measurement of bulk solids in pneumatic pipelines. Measurement Science and Technology, 7(12), 1687.
7. Chowdhury, S., Marashdeh, Q. M., & Teixeira, F. L. (2016). Velocity profiling of multiphase flows using capacitive sensor sensitivity gradient. IEEE Sensors Journal, 16(23), 8365-8373.

555620 Pharmaceutical Process Modeling: Mechanistic Insights of the Impacts of Roller Compaction Process Variables and Material Properties on Product Attributes

Roller compaction (RC) is a dry granulation process that improves powder flowability prior to tableting by increasing material solid fraction (SF) followed by a milling step to produce larger agglomerates. Most of RC process work in pharmaceutical industry is performed empirically which is time consuming and expensive. Recent advances in material characterization, computational resources, and numerical simulation tools have provided the opportunity to develop fast and accurate predictive models for process development and scale-up. However, quantitative understanding of the impacts of RC major process variables, and input powder properties on product attributes is still limited due to complex material behavior. In this talk, a mechanistic model is presented which correlates the SF of compacted material with process variables, and powder compressibility and flow properties. Predictive capability of this model was demonstrated by comparing experimentally generated, and numerically calculated SF of various compacted pharmaceutical materials. Provided that the pertinent process variables and powder flow parameters are used in the equations, this model can be used to predict compacted material's SF at any RC scale and for any given pharmaceutical material. Alternatively, this approach can be used by both pharmaceutical industrial scientists and practitioners to significantly reduce the costs of process development and scale-up.

559646 System Model for Semi-Continuous Wet Granulation Pharmaceutical Tablet Manufacturing Process

In this presentation, system model is presented for GSK's semi-continuous wet granulation tablet manufacturing process. The model combines empirical and mechanistic understanding of all continuous unit operations, with focus on Active Pharmaceutical Ingredient (API) content (API weight/overall weight %). Special attention is given to quantifying impacts process parameters and disturbances have on API content, at the unit operation where process turns into batch.

566462 Assessing Performance of a Rotary Tablet Press Feeder Using the Discrete Element Method

Rotary tablet presses are routinely used in the pharmaceutical industry for the manufacture of tablets. The selection of appropriate feeder operational parameters and geometric configuration are important to ensure tablet quality measures are attained. Specifically, the feeder must deliver enough powder to each die to ensure complete filling while, at the same time, avoid applying too much stress or shear to the powder that may induce attrition or cause extensive lubrication of the granulation or blend. Should the process fail to meet these criteria, tablet weights could be too small or too variable and the tablet strength and dissolution could be adversely impacted. In this work, the performance of a commercial-scale rotary tablet press feed frame is investigated using the discrete element method. A range of process parameters and feeder configurations are examined with key simulation results compared with experimental data. Feeder performance metrics including the die fill weight and powder residence time are reported and suggest optimal operating regimes.

All authors are employees of AbbVie and may own AbbVie stock. AbbVie sponsored and funded the study; contributed to the design; participated in the collection, analysis, and interpretation of data, and in writing, reviewing, and approval of the final publication.

568543 Modeling Two-Dimensional Moisture Content and Size Evolution of Pharmaceutical Granules in a Semicontinuous Fluid Bed Dryer

In the transition to continuous manufacturing, the pharmaceutical industry is embedding mathematical modeling in its strategy for future process development and optimization. It hereby utilizes the benefits of modelling with respect to saving cost and materials of in-silico studies, and gaining and fixing process knowledge in the model structure and its calibrated parameters. As such, mechanistic modeling studies on the ConsiGma-25™ continuous wet granulation line have contributed to understanding of the process phenomena in function of the critical process parameters (CPP). The semicontinuous fluid bed dryer, configured to handle a continuous inflow of wet granules coming from a twin-screw wet granulator, has proven to be a complex part of the line. In a novel data collection in which granule moisture content was collected in conjunction with granule size, it was found that both properties are important with respect to the drying rate of the material. The 2D distribution dynamics are also influenced by the dryer's CPPs, i.e. filling time of the dryer cell and dryer inlet air temperature and flow (De Leersnyder et al., 2018).

A modeling study of this process was performed in (Ghijs et al., 2019). It comprised a model relating single granule drying behavior to the evolution in moisture content seen in the semicontinuous process, including the effect of granule size. The model of (Mortier et al., 2012) represented the drying phenomena on single granule scale, and was thus applied in a sequential addition mode in order to emulate the semicontinuous drying behavior. The model performed well when applied to the data from (De Leersnyder et al., 2018) collected at a dryer cell fill level of 1 kg of granules in the cell. However, lower levels could not yet be validated mechanistically. Therefore, in order to enhance model's predictive power for wider application, the work indicates that a more accurate inclusion of fluidization, air conditions and filling effects would benefit the model. This work presents these augmentations.

While the model of (Ghijs et al., 2019) assumed an ideal and constant fluidization behavior, the reality in the ConsiGma-25™ dryer is presumably different as air flow, particle size distribution and moisture content influence fluidization quality. The approach from (Peglow et al., 2007) is therefore incorporated into the model. It relates single particle drying kinetics to batch behavior by taking into account fluidization behavior conceptually, predicting the fraction of the drying air that brings the granules in suspension and the fraction that bypasses the granules. The single particle drying kinetics in this case still stem from the mechanistic model of (Mortier et al., 2012).
Other features of the model in (Peglow et al., 2007) are the mass and heat balances for exchange of water vapor and heat between granules and drying air. This updates the average relative humidity of the drying air in the fluid bed accordingly. That in turn influences the drying rate of the granules present in the bed. The heat and mass balances, which also include the heat transfer from the dryer to the environment, are moreover tailored to the ConsiGma-25™ dryer itself according to the work of (Mortier, Gernaey, De Beer, & Nopens, 2014).
A final model augmentation is the tracking of granules and the drying regimes they are subjected to throughout the drying process. Indeed, granules in the fluid bed chamber entered the system at different times due to the continuous filling of the cell. Moreover, those entering at a later time are moreover starting their drying curve in different conditions than these entering first in an empty cell. The different air conditions are captured with the heat and mass balances. The tracking of the granules per size fraction and moisture content occurs with a 2D population balance model (PBM). The latter models the moisture content distribution of the granules per size fraction. The range of moisture content of the granules is discretized into a number of levels, which serves as the grid for the PBM. The number of granules in each moisture content level is then tracked, and evolves due to the drying. The decrease in moisture content is modeled by a negative growth rate, as in the work of (Mortier, Gernaey, De Beer, & Nopens, 2013). Also the continuous inflow of wet granules into the dryer cell is introduced into the PBM as a source term in the moisture content level of the wet granules. This way, the entire population can be tracked for the semicontinuous operation.

It is concluded that a more accurate representation of semicontinuous drying of pharmaceutical granules has been achieved, capturing the influence of the CPPs of the drying process mechanistically. The model concept, implementation, simulation and calibration methodology will be presented.

References

De Leersnyder, F., Vanhoorne, V., Bekaert, H., Vercruysse, J., Ghijs, M., Bostijn, N., ... De Beer, T. (2018). Breakage and drying behaviour of granules in a continuous fluid bed dryer: Influence of process parameters and wet granule transfer. *European Journal of Pharmaceutical Sciences*, *115*, 223–232. http://doi.org/10.1016/j.ejps.2018.01.037

Ghijs, M., Schäfer, E., Kumar, A., Cappuyns, P., Van Assche, I., De Leersnyder, F., ... Nopens, I. (2019). Modeling of Semicontinuous Fluid Bed Drying of Pharmaceutical Granules With Respect to Granule Size. *Journal of Pharmaceutical Sciences*, 1–8. http://doi.org/10.1016/j.xphs.2019.01.013

Mortier, S. T. F. C., De Beer, T., Gernaey, K. V, Vercruysse, J., Fonteyne, M., Remon, J. P., ... Nopens, I. (2012). Mechanistic modelling of the drying behaviour of single pharmaceutical granules. *European Journal of Pharmaceutics and Biopharmaceutics*, *80*(3), 682–689. http://doi.org/10.1016/j.ejpb.2011.12.010

Mortier, S. T. F. C., Gernaey, K. V., De Beer, T., & Nopens, I. (2013). Development of a Population Balance Model of a pharmaceutical drying process and testing of solution methods. *Computers & Chemical Engineering*, *50*, 39–53. http://doi.org/10.1016/j.compchemeng.2012.11.005

Mortier, S. T. F. C., Gernaey, K. V., De Beer, T., & Nopens, I. (2014). Analysing drying unit performance in a continuous pharmaceutical manufacturing line by means of mass – Energy balances. *European Journal of Pharmaceutics and Biopharmaceutics*, *86*(3), 532–543. http://doi.org/10.1016/J.EJPB.2013.12.014

Peglow, M., Kumar, J., Heinrich, S., Warnecke, G., Tsotsas, E., Mörl, L., & Wolf, B. (2007). A generic population balance model for simultaneous agglomeration and drying in fluidized beds. *Chemical Engineering Science*, *62*(1–2), 513–532. http://doi.org/10.1016/j.ces.2006.09.042

568877 Cracking in Drying Films of Polymer Solutions

Thin coatings of polymer films find numerous industrial applications including tablet coatings in the pharmaceutical industry. The tablet may be coated for improving its appearance or taste, for increasing its stability or for achieving sustained or delayed release of the drug. Drying and the accompanying solidification process of the wet polymer film are often accompanied by shrinkage of the film. If the polymer film adheres strongly to the tablet surface, the latter prevents the film from shrinking leading to transverse tensile stress in the film. If the stress exceeds a critical value, the polymer film may develop cracks thereby compromising the integrity of the film.

In the present work, thin films of a polymer solution were cast on substrates of varying stiffness and investigated for cracking as a function of film thickness and substrate modulus. We measured the tensile stress and the maximum crack-free thickness. The measurements agree well with predictions of a model based on fracture mechanics of the drying film that balances the elastic energy released due to cracking to the increase in surface energy. These results provide guidelines to the formulator for designing crack-free coatings.

575048 Mechanistic Mass Transport Modeling of Ionizable Drug Dissolution Under *In Vivo*-Relevant Buffer and Hydrodynamical Conditions

There is a need to develop a biorelevant, predictive dissolution method that can be applied by pharmaceutical industry to facilitate marketing access for generic and novel drug products. In a recent clinical and computational study, the critical factors deriving the dissolution and absorption in the human GI tract have been more accurately determined. The primary goal of this study is understanding the rate-determining factors for ionizable drug dissolution from intubation studies, real-time magnetic resonance imaging (MRI) manometry, and computational fluid dynamic simulations (CFD), taking advantage of mechanistic mass transport modeling to quantify the GI tract hydrodynamical and environmental parameters, and finally incorporating the *in vivo* knowledge to an *in vitro* dissolution apparatus that can simulate the *in vivo* drug dissolution rate in the GI tract.

A mathematical mass-transport model was developed for ionizable drug dissolution in bicarbonate using the rules of conservation of mass and electric charge in addition to accounting for the diffusional times and reaction rate constants of the CO_2-H2CO_3 interconversion. The predictions made by the aforementioned model were compared to the experimental data generated by an intrinsic dissolution setup for three ionizable drugs: indomethacin, ibuprofen and haloperidol. A CFD approach was used to quantify the hydrodynamical conditions in commonly used USP II type *in vitro* dissolution device. Moreover, dissolution testing for suspended ibuprofen particles was performed under different hydrodynamical conditions in bicarbonate and phosphate buffers in a USP II type dissolution device. The particle-size distribution of ibuprofen particles was characterized using a laser diffraction particle size analyzer.

An agreement between the simulated and experimental dissolution rates obtained by intrinsic dissolution experiments lends credibility to use this model for a range of drugs. The sensitivity analysis with respect to the hydrodynamical conditions indicates that at high shear rates where the diffusion layer next to the solid-liquid interface is thin the interfacial pH is low. The low interfacial pH causes a lower dynamic solubility of acidic drugs at the solid-liquid interface and this led to a decrease in the dissolution flux. For ionizable particle dissolution in bicarbonate, comparing small particles with thinner diffusion layer to the large particles with thicker boundary layer, the flux is lower than that of the large particles. This is because the effective pKa governing the buffering action of bicarbonate in the boundary layer decreases as this layer gets thinner. This is due to the interconversion between H_2CO_3 and CO_2 not being fast enough to achieve equilibrium within the timeframes allowed by the diffusional processes in the boundary layer.

The mechanistic mass transport model for *in vitro* drug dissolution under different buffer and hydrodynamical conditions highlights the implications of using surrogate buffers such as phosphate for quality control tests. The gain in dissolution rate of ionizable drug that is obtained by the drug particles micronization in phosphate buffer is higher than in the *in vivo*-relevant bicarbonate buffer. The hydrodynamical conditions could potentially change the boundary layer thickness of particles and consequently the dissolution rate. Furthermore, in order to design an *in vivo*-relevant dissolution test both hydrodynamical and buffer conditions should match with the *in vivo* conditions to achieve better *in vivo-in vitro* correlation (IVIVC).

553726 Sticking Tendency Prediction of Pharmaceutical Powders

Sticking is one of the major issues in the pharmaceutical product development. Understanding what influences sticking in terms of process variables, material and microstructural properties and developing specialized techniques to correlate these variables with mechanical failures are practical interests of the pharmaceutical industry. In this talk, a nondestructive method to predict the sticking tendencies of pharmaceutical powders will be presented. Acceptance criteria of the *Sticking Tendency* has been proposed based on development and commercial manufacturing performance. These tools have been shown to correlate well with actual sticking performance of a diverse sample set. The developed predictive tool has potential to flag the risk of sticking issues and provide insight into the performance of common materials during formulation and process development.

561844 Cuprous Oxide Nanoparticles As Efficient Heterogeneous Photocatalyst for C-C Coupling Reactions

Carbon-carbon (C-C) coupling reactions are a diverse set of reactions that are widely used throughout industry. For instance, cross coupling reactions make up 25% of all pharmaceutical reactions. They are also used for many advanced materials such as metal-organic frameworks (MOFs) and OLED materials. These reactions are conventionally catalyzed using homogenous palladium (Pd) complex-based homogenous catalysts. Pd is expensive, toxic and rare-earth element.

To overcome this limitation, we propose heterogeneous cuprous oxide (Cu_2O) nanocatalysts to drive C-C homo-couplings as well as cross-couplings under relatively mild conditions using visible-light as energy input. With Cu_2O photocatalyst, we show improved rate of reaction and stability at a lower temperature than previously seen with conventional Pd catalysts. Furthermore, with a combination of spectroscopic techniques such as electrospray ionization mass spectrometry, UV-Vis extinction spectroscopy, flame atomic absorption spectroscopy and transmission electron microscopy, we demonstrate the stability of the catalyst as a function of time and temperature. The implications of this work is that Cu_2O can be an alternate catalyst for homogenous palladium.

577914 First-Principles Modeling of Hot-Carrier-Assisted Hydrogenation Reactions

Generation of highly energetic carriers (electrons and holes) in plasmonic nanostructures opens new pathways to trigger and enhance chemical transformations toward a certain desired product by harvesting solar energy. Here we intend to reveal, at the atomic level, the fundamental mechanism in the plasmonic hot-carrier-driven hydrogenation reactions. We employed density functional theory (DFT) calculations to explore the potential energy surfaces of hydrogenation of 4-nitrophenol and phenylacetylene. The reaction profiles with and without hot electrons at the different unoccupied molecular orbitals were calculated and compared to address the reaction mechanism. This work shed lights on the bond association process, which is a key elementary step in many practical industrial processes.

567446 Visible Light Driven Anti-Markovnikov Hydration of Styrene to 2-Phenylethanol in Continuous Flow Microreactor

Introduction

2-phenylethanol (2PE), which is an aromatic alcohol with a mild rose fragrance, is a key molecule in the fragrance and food industries. The traditional production method of 2PE involves multi-step chemical reactions, which require high temperature, strong acid, strong alkali and other harsh conditions. Furthermore, it is very difficult to separate the by-products of the reactions, leading to an increase in overall producing cost. Visible light catalyzed anti-Marko addition of olefin has the characteristics of less by-products and mild reaction conditions[1]. However, the required reaction time is up to ten or even dozens of hours. Like large number of other reported photocatalytic reactions, it is still very challenging to scale up this reaction in traditional batch reactors due to limited mass and photo transfer issues.

The continuous flow microreactor could be a promising alternative to circumvent the above common drawbacks of batch photoreactors because of its larger specific surface and shorter light path. Herein, for the first time, we attempt to synthesis 2PE through Anti-Marko hydration of styrene in continuous microreactor system with comparison of batch reactor. The effects of light source, reaction temperature, substrate concentration and catalyst concentration were examined experimentally so as to optimize the reaction conditions and better understand the reaction mechanism.

Experimental

In the present work, the continuous flow microreactor consisted of PFA capillary with 1mm in inner diameter (total reaction volume 4 mL) and blue LED irradiation. The schematic diagram of the reaction platform is shown in Figure 1. The raw materials are styrene and deionized water, the solvent is acetonitrile, and the catalyst is 9-trimethylene10-methyl acridine perchlorate and diphenyl disulfide. For each test, the prepared reactant mixture was pumped into micoreactor by a syringe pump, and the products were collected and analyzed by GC-MS. For comparison, batch tests were also conducted using 2 mL reactant mixture in lab test tube under blue LED irradiation.

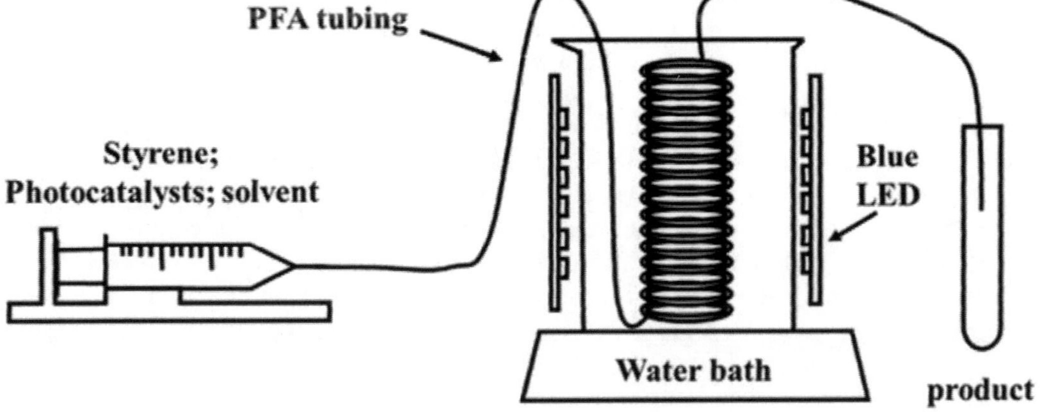

Figure 1. Schematic diagram of continuous flow microreactor platform

Results and discussion

Figure 2 compares the 2PE yield with different light sources and reactors. For batch reactor, replacing the plat LED by annular LED could shorten the reaction time from 24 hr to ~6 hr, indicating that the reaction rate was largely increased by increasing received radiation[2]. As for microreactor system, with annular LED, the 2PE yield reached maximum (62%) at 4 hr, then it slightly declined in 6 hr. Also, it can be seen from Figure 1 that the mircroreactor yielded much more 2PE than tube reactor in the first 3 hrs, which can be explained by more uniform light distribution in microchannels.

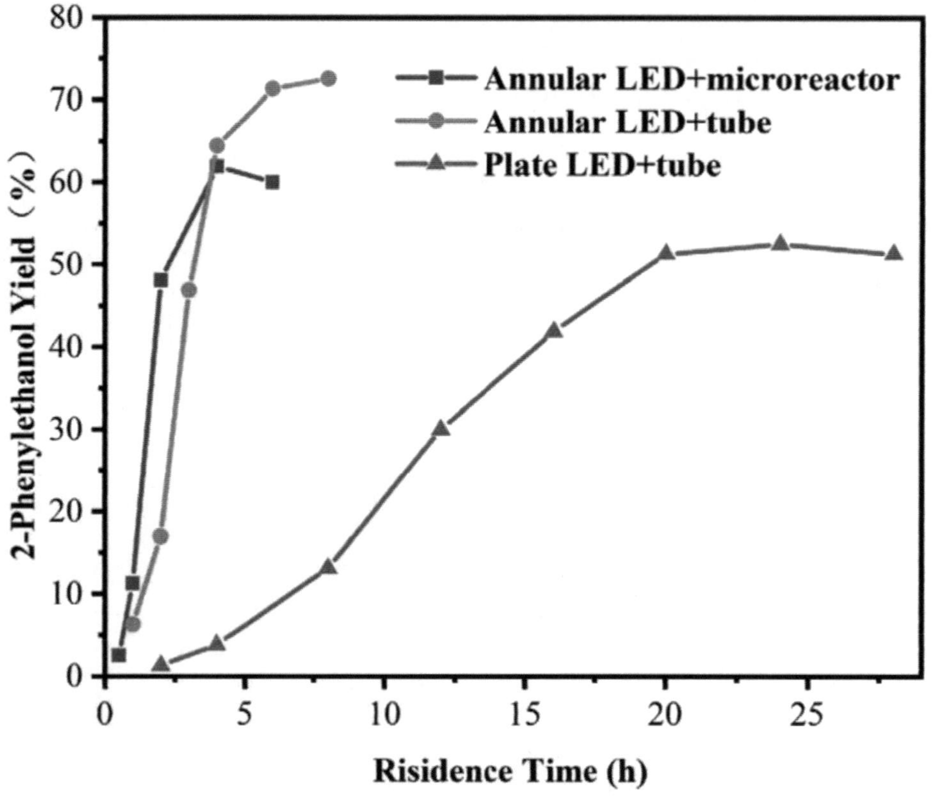

Figure 2. Effect of light sources and reaction system on yield of 2PE (Reaction conditions: styrene 0.15 M, Acr^+-Mes ClO_4^- 3 mol%, Ph-S-S-Ph 20 mol%, reaction temperature 25¡æ.)

With the same reaction time 1hr, the effect of reaction temperature on the 2PE yield is illustrated in Figure 3. As expected, the 2PE yield was significantly favored by increasing temperature from 25¡æ to 70¡æ for the both microreactor and batch systems,. The main reason is that the increasing temperature accelerates molecular thermal motion, and thus increases the reaction rate. Nevertheless, it should be noted that long-term operation of LEDs at high temperature would shorten their lifetime and also affect the emission wavelength. Moreover, it can be seen from Figure 3 that as compared to batch system, the 2PE yield was doubled in microreactor system at each reaction temperature, further confirming its advantages.

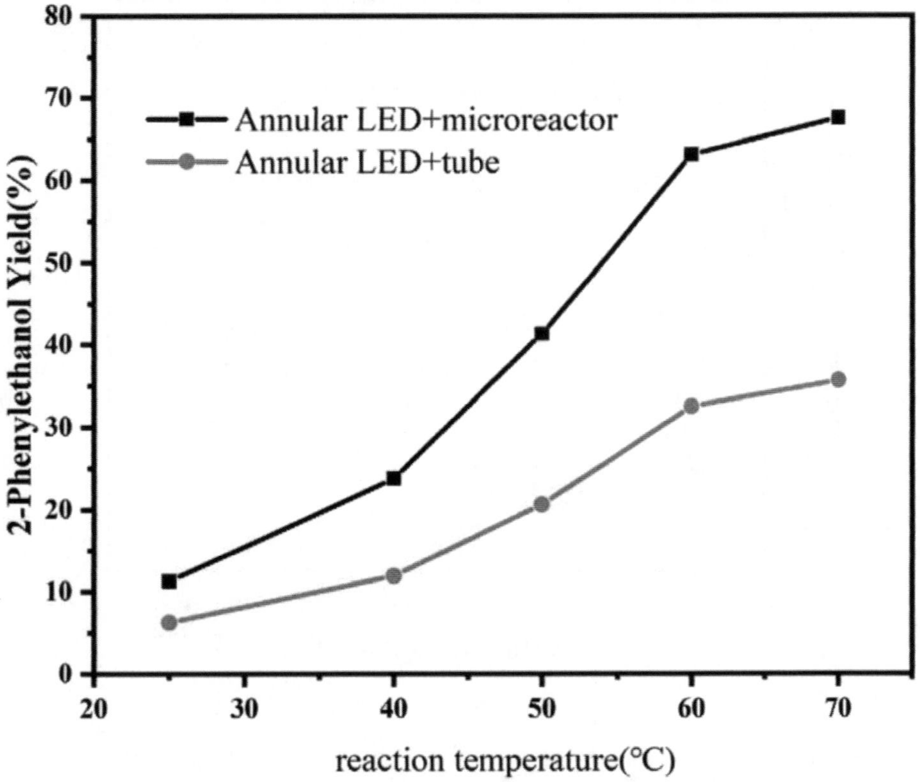

Figure 3. Effects of reaction temperature on the 2PE yield in microreactor system (Reaction conditions: styrene 0.15 M, Acr+-Mes ClO4- 3 mol%, Ph-S-S-Ph 20mol%, reaction time 1 hr.)

Figure 4 shows the trend of 2-phenylethanol yield for two different initial styrene concentrations. Increasing the styrene concentration from 0.15M to 0.3M led to a lower yield of 2PE. The peak of 1-phenyl-1,2,3,4-tetrahydronaphthalene was detected on GC-MS, and its area was larger when initial 0.3M styrene was applied. Therefore, we could conclude that the increasing styrene concentration will result in an increase of polymerization between styrene, which competes with the formation of 2PE. Moreover, we can notice that the 2PE yield remained almost constant after 1 hr or 1.5 hr for the both cases, indicating that the reaction equilibrium was reached.

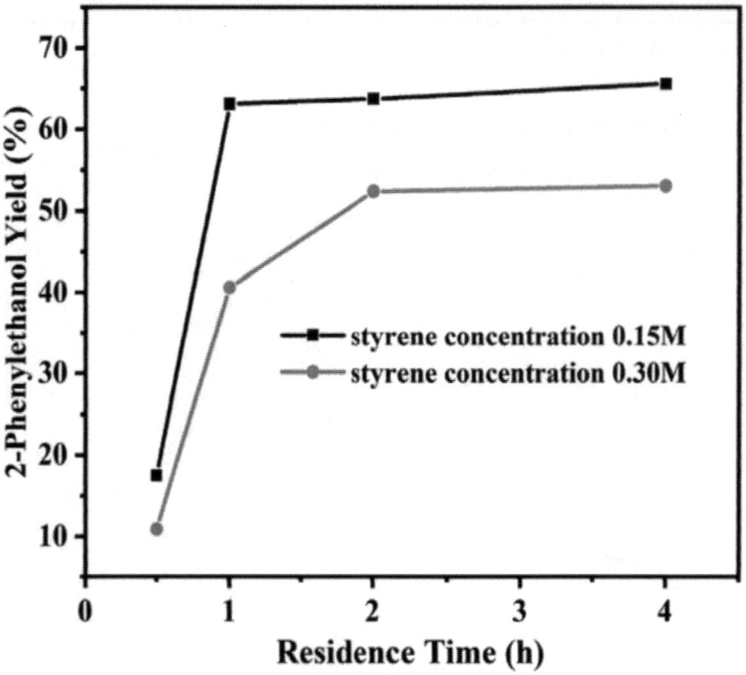

Figure 4. Effect of substrate concentration on the 2PE yield in microreactor system (Reaction conditions: styrene 0.15 M, Acr+-Mes ClO4- 3 mol%, Ph-S-S-Ph 20mol%, temperature 60¡æ.)

Conclusions

In this study, the continuous production of 2PE from Anti-Markovnikov hydration of styrene was realized for the first time by using PFA capillary microreactor. After the optimization of light source, reaction temperature and other reaction conditions, the reaction time was successfully reduced from the 24 hr to 1 hr. The increasing styrene concentration led to the polymerization of styrene carbocation intermediates, resulting in a lower selectivity of 2PE. Therefore, it is worthwhile to design a suitable structured microreactor so as to reduce the potential polymerization through enhanced mass transfer. The overall production capacity could then be further improved.

References

[1] X. Hu, G. Zhang, F. Bu, A. Lei, Visible-Light-Mediated Anti-Markovnikov Hydration of Olefins ACS Catalysis, (2017) 6.

[2] C.C. Le, M.K. Wismer, Z.C. Shi, R. Zhang, D.V. Conway, G. Li, P. Vachal, I.W. Davies, D.W.C. MacMillan, A General Small-Scale Reactor To Enable Standardization and Acceleration of Photocatalytic Reactions, ACS Cent Sci, 3 (2017) 647-653.

568672 Using Data-Rich Experimentation to Maximize Productivity of Photochemical Reactions upon Scale-up

Visible light promoted organic reactions access novel reaction pathways, enabling the synthesis of complex organic molecules using mild reagents under ambient reaction conditions. While photochemistry offers green solutions for manufacturing, scaling up reactions beyond the kilo scale remains a challenge due to the complex relationship between reaction conditions and the attenuation of light in finding the optimal operating conditions for productivity. For these reasons, it is important to have a firm grasp of the process kinetics and optical properties of the reactor. Herein we report a case study on the photocatalyzed cross coupling of aryl bromides with pyrrolidine where the interplay between reagent concentrations, pathlength, and light intensity were quickly elucidated. A design of experiments highlighted the optimal reaction conditions to maximize productivity by maximizing yield while minimizing residence time. Statistical or first principles kinetic models provided further insight into the impact of each parameter, and their interactions, on the reaction rate and final yield. These insights were applied in scaling this photoamination reaction and achieving productivities > 10 kg/day in a plug flow photoreactor.

576051 Development of a Continuous Stirred-Tank Reactor (CSTR) Platform for Handling Heterogeneous Visible-Light Photoredox Reactions

Visible-light photoredox has been demonstrated to be a powerful synthetic methodology to access pharmaceutical related compounds. However, most of the photoredox reactions involve insoluble starting materials or products which makes it challenging to be implemented in a continuous flow reactor. Here we present a new solid-handling platform for heterogeneous photoredox reaction systems in flow, which integrates a newly designed solid-feeding strategy and a continuous stirred-tank reactor (CSTR) cascade. The residence time distribution for single phase and solids was measured to characterize the hydrodynamics of the heterogeneous flow in the CSTR cascade. Silyl radical-mediated metallaphotoredox cross-electrophile coupling reactions was demonstrated with an inorganic base as the insoluble starting material. Multigram synthesis was achieved within 24 hours of stable operation without any human intervention.

564057 A Systematic Approach to the Design and Scale-up of Photochemical Reactors

Photochemistry allows efficient access to complex molecular scaffolds with high selectivity, and has been demonstrated to be a powerful tool in pharmaceutical chemistry applications. Determining reaction conditions and light intensity requirements for photochemical reactions, however, is challenging because light distribution through the reactor is highly dependent on reactor geometry and the absorbance of components. In scaling up these reactions, the effect is especially exacerbated as the ratio of illuminated area to reactor volume decreases.

Here we present a workflow for optimizing photochemical reaction parameters and reactor geometry to enable a robust and efficient scale-up in flow. The workflow starts with the characterization of the photophysical properties of the reaction components, and the irradiation parameters of the reactor and the light sources. In parallel, we utilize high-throughput experimentation (HTE) to determine reaction rates and to optimize reaction conditions. We incorporate the intrinsic reaction parameters and the radiation components to model the kinetics of the desired photochemical reaction for various reactor geometries and light sources. Through iterations of modeling and experimental work, the reaction conditions and residence time are selected to optimize productivity for a given photoreactor design.

Using this systematic approach and optimizing reactor geometry and light-source selection accordingly, we were able to achieve productivities exceeding those achieved in commercially-available laboratory scale flow reactors. We will present the application of our workflow to both a single component photocyclization reaction as well as to a more complicated visible-light photoredox catalysis reaction.

579335 A Scale-Down and Feedback-Control Based Methodology to Optimize Multi-Step Extraction for Botanical Preparation on the Commercial Scale

A systematical methodology from a commercial scale extraction apparatus to laboratory scale based on scale-down and feedback-control strategy was established to optimize the water-extraction process in the manufacturing of a botanical preparation, named as *Xutong injection* in China, on a commercial scale. The raw material of this preparation is *Salvia Miltiorrhiza.* Firstly, quality by design principle was used to achieve process characterization and process optimization of the water-extraction process in a scale-down of 2 L extractor. Several main quality markers of *Salvia Miltiorrhiza* were evaluated to determine the optimal formulation, including the extraction rate of phenolic acids as the effective components and the extraction rate of polysaccharides as the impurities. Secondly, the operation space of process parameters was optimized using a central composite design with response surface methodology as follows: *Salvia Miltiorrhiza* root length was11 cm, the ratio of solvent to raw material was 10 mL / g and each extraction time was 70-90 min for twice. Then, the optimized parameters were applied to make reproducibility test in a commercial scale of 4,000L extractor. In this phase, through in-line data collection, a feedback control strategy was implemented in the pilot production to achieve process real-time determination and control. Finally, the extracting solution manufactured in the commercial-scale process was compared with pre-existing extracting solution collected in the prior batches, which was verified of higher quality and process efficiency. More than 22% active ingredients were extracted than ever. The proposed methodology in this work, once applied to a commercial botanical preparation, is of significant value for continuous improvement in the aspect of reducing time and cost, decreasing batch-to-batch variability and realizing real time release.

571828 Scale-Independent Micronization: QbD Via a Specific Energy Model

It is well established in micronization theory that specific energy provides a holistic approach to evaluating and understanding the particle size reduction operation, and it can be used to fully characterize process performance and robustness. Utilizing a Fluid Energy spiral jet mill operating at ambient temperature, and nitrogen as the feed/grind gas, three operating parameters (Venturi gas pressure, grind gas pressure, and solid material feed rate) can be actively varied in the micronization process, each of which contribute to the calculated total specific energy. This study examined the application of a specific energy model to develop a scale-independent micronization process for commercial production. To develop the model, 34 development runs were performed across a 4" mill and an 8" mill where the Venturi gas pressure, grind gas pressure, and solid material feed rate were varied to investigate the impact of specific energy on the particle size of the micronized drug substance (i.e. output X_{90}). The data gathered at both scales adhere to the same curve, confirming the scale-independent nature of the specific energy model. With specific energy as the underlying mechanism for particle size reduction, the model was used to determine an appropriate registered range for specific energy to achieve the desired particle size distribution. In conclusion, predictive specific energy models can be implemented to create a scale-independent micronization design space; however, these product-specific models will only be applicable when used to predict performance using mills with similar internal geometries, operating principles and conditions, and processing gas.

573647 A Comparative Study of Surrogate Model Based Control Strategy for a Pharmaceutical Crystallization Process

Abstract

In pharmaceutical manufacturing, crystallization is the most preferred technique in downstream processing to recover crystal products. Crystal quality attributes such as crystal size distribution, purity, and polymorphic form, which are critical for not only following unit operations, but also therapeutic properties of the formulated product, are determined by the operating conditions of crystallizers. Therefore, control of a crystallization process is critically and essentially the control of quality attributes. Traditionally, crystallization processes are operated based on a pre-defined recipe of a process variable such as a temperature trajectory for a cooling crystallization system, and the quality of the crystal product is tested only at the end of the process. Quality by testing (QbT) approach leads to high profit loss since batch-to-batch variations of crystal quality attributes are unavoidable, due to the absence of control actions taken during the process in order to avoid effects of any disturbances available in the system through *e.g.* feedback signals coming from solid-state attributes or supersaturation. However, consistency of a process outcome is highly dependent on a robust process control strategy. In literature, emergence of the latest generation of process analytical technology (PAT) tools, mechanistic understanding and modeling of underlying mechanism of crystallization process and incentives by regulatory mechanism have accelerated the efforts on the development of advanced process control strategies in order to achieve desired product quality, so called quality by control (QbC). State of the art crystallization control strategies can be classified into two main groups as model free/data based and model based. As a model free method, direct nucleation control (DNC) is based on the principle of operating at a predefined set point of crystal number density to dominate crystal growth, while supersaturation control (SSC) as another model free method that aims to keep supersaturation level within metastable zone and constant over the experiment time. While the model free methods are simple and found many applications, they might require some investigations on the determination of a robust set point as well as some efforts on online sensor calibration. Model based control strategies utilize real time crystallization process simulations to predict the effects of inputs and disturbances on the process output, whose performance is strictly dependent on the complexity and accuracy of the implemented model as well as speed of solution time. However, often a simple but reliable control strategy is required in the pharmaceutical industries to achieve a fast and successful transition from laboratory to pilot scale crystallization operation. Therefore, in this work we aim to benefit from the simplicity of data based models, while maintaining the accuracy of the predictions in an acceptable range. To this end, we present a surrogate model based control strategy for a pharmaceutical batch cooling crystallization process. Two surrogate models namely Radial Basis Functions Network (RBF) and Polynomical Chaos Expansion (PCE) are utilized to follow a reference trajectory of a mean crystal size by manipulating and updating the temperature profile during process simulation. Performance of the proposed control strategy based on two surrogate models of RBF and PCE is evaluated in the presence of disturbances in the system such as initial concentration, temperature and seed specifications. Additionally, as a proof of concept, experimental applications of the proposed methods are supplemented.

Acknowledgment

We would like to thank the Danish Council for Independent Research (DFF) for financing the project with grant ID: DFF-6111600077B.

559274 Specifications By Design: Framework for Setting Specifications to Maximize Flexibility and Meet Design Requirements

Traditionally, specifications for API starting materials, intermediates and final API are set based on historical data, process capability and limit of quantitation/detection. The flexibility of the process is then determined based on using specification as constraints. Instead of the sequential approach, there is potential for a better outcome by integrating specification setting with process flexibility and design requirements. In this talk, we propose the framework of "Specifications by Design" where the objective is to set specifications for API starting materials, intermediates and final API to balance process flexibility with specification flexibility while meeting design requirements. We will illustrate the framework with few examples: (1) Setting residual solvent specifications for API to balance the flexibility of a key unit operation in the process and the flexibility of specification while meeting the design requirement of desired product performance (2) Setting a isomer impurity specification to balance specification flexibility and meet business design requirements of yield/API cost.

576015 Real-Time Concentration Monitoring Using a Compact Composite Sensor Array for Small-Scale Pharmaceutical Manufacturing Systems

Active pharmaceutical ingredients (APIs) are traditionally produced through time- and cost-intensive large-scale batch manufacturing.[1] The limitations of this approach have been revealed by the Food and Drug Administration (FDA) to be root causes for drug shortages, a major problem in today's pharmaceutical manufacturing.[2] A potential solution to overcome this threat are small-scale, end-to-end continuous manufacturing platforms that integrate synthesis, purification, and formulation within the same uninterrupted microfluidic reactor network.[3,4] This approach possesses a unique high demand on process analytical technology (PAT), for instance, to monitor and determine the strength of aqueous formulations (solution and suspension) *in situ*and in real-time as a critical quality attribute (CQA) prior to the release from the platforms, which are not comparable to lab-scale or industrial-scale needs.[5] The aim of this study is to demonstrate the application of a novel composite sensor array (UV-vis, NIR, Turbidity, Temperature) as a versatile, reliable, cost-effective, and small-scale PAT to measure accurately the concentration of aqueous pharmaceutical formulations (solution/suspension). Four APIs, warfarin sodium and lidocaine hydrochloride, formulated as solutions and acetaminophen and azithromycin formulated as suspensions were measured in their commercial aqueous forms using the composite sensor array. The results were compared with chemometrics data obtained from a state-of-the-art *in situ*Raman probe to provide a reference for the assessment of real-time monitoring. In addition, the concentrations of all formulations were analyzed by partial least squares (PLS) regression, validated and correlated with offline UV-vis spectrophotometer. Results show PLS regressions of $R^2 = 0.9958$ with a Bias = 0.061, which correlated well with offline measurements. The composite sensor array demonstrates potential as an *in situ*real-time PAT for concentration monitoring (solution and suspensions) in one single compact PAT device, as it is preferred in pharmaceutical manufacturing platforms to maintain small footprints.[3–5]

(1) Lee, S. L.; O'Connor, T. F.; Yang, X.; Cruz, C. N.; Chatterjee, S.; Madurawe, R. D.; Moore, C. M. V.; Yu, L. X.; Woodcock, J. Modernizing Pharmaceutical Manufacturing: From Batch to Continuous Production. *J. Pharm. Innov.***2015**, *10*(3), 191–199.

(2) FDA. Strategic Plan for Preventing and Mitigating Drug Shortages http://www.fda.gov/downloads/Drugs/DrugSafety/DrugShortages/UCM372566.pdf (accessed Oct 28, 2015).

(3) Adamo, A.; Beingessner, R. L.; Behnam, M.; Chen, J.; Jamison, T. F.; Jensen, K. F.; Monbaliu, J.-C. M.; Myerson, A. S.; Revalor, E. M.; Snead, D. R.; Stelzer, T.; Weeranoppanant, N.; Wong, S. Y.; Zhang, P. On-Demand Continuous-Flow Production of Pharmaceuticals in a Compact, Reconfigurable System. *Science (80-.).***2016**, *352*(6281), 61–67.

(4) Zhang, P.; Weeranoppanant, N.; Thomas, D. A.; Tahara, K.; Stelzer, T.; Russell, M. G.; O'Mahony, M.; Myerson, A. S.; Lin, H.; Kelly, L. P.; Jensen, K. F.; Jamison, T. F.; Dai, C.; Cui, Y.; Briggs, N.; Beingessner, R. L.; Adamo, A. Advanced Continuous Flow Platform for On-Demand Pharmaceutical Manufacturing. *Chem. - A Eur. J.***2018**, *24*(11), 2776 –2784.

(5) Stelzer, T.; Wong, S. Y.; Chen, J.; Myerson, A. S. Evaluation of PAT Methods for Potential Application in Small-Scale, Multi-Purpose Pharmaceutical Manufacturing Platforms. *Org. Process Res. Dev.***2016**, *20*(8), 1431–1438.

564477 Process Control Strategy for Synthetic Continuous Manufacturing

Continuous manufacturing enables a smaller manufacturing footprint, improved environmental sustainability, improved control strategy, and flexibility to meet changing demand among other benefits; however, the transition from batch to a continuous mode of operation can be resource intensive in terms of number of parameters that require evaluation for optimal process design and process characterization. These resource requirements can be reduced significantly by augmenting empirical design with integrated systems-based modeling.

A 'Virtual Plant for Synthetic Continuous Manufacturing' was developed on the grounds of a systems-based modelling platform. Model components were integrated to form the end-to-end process model, with which *in silico* studies (sensitivity analysis, uncertainty propagation, etc.) are performed to identify the key process parameters that will affect the critical quality attributes. Here we present the development and testing of an integrated advanced process control strategy enabled by the systems-based model. This approach is envisioned to support regulatory filings for CM processes. Development of a virtual plant for synthetic CM enables us to predict process performance and product quality in a real plant environment.

566877 Industry 4.0: Advanced Bi-Layer Control System for Continuous Pharmaceutical Manufacturing Pilot-Plant

The industry 4.0 standard continuous pharmaceutical manufacturing (CM) integrated with real time monitoring, advanced process control (APC), and a novel data management system is highly desired to consistently produce the better quality product with less time and resources. This modernized integrated system also open the possibilities of achieving Quality by Control (QbC) and real time release (RTR) paradigm [1-2]. However, there are different levels of challenges in implementation of industry 4.0 standard CM including advanced process control system. Therefore, this is still an open area of research.

In this work, a bi-layer advanced control strategy has been developed and implemented into CM pilot-plant using industry standard real time monitoring sensors, data management tool and control platform. In first layer, the CPP's and CQA's are controlled in real time using advanced model predictive control (MPC) system while in second layer, the none-confirming products are diverted in real time in waste to assure the final CQA's of qualified tablet lots. This bi-layer control strategy assures the final product quality, improves the production efficiency and minimizes the need of off line testing.

The critical control variables that have been controlled in first layer using model predictive control (MPC) system are drug concertation, powder level before tablet press, main and pre compression forces, tablet weight and hardness. A novel control strategy for powder level control in a chute placed in between blender and tablet press unit operation of continuous tablet manufacturing process has been developed, implemented and evaluated. A noninvasive technique based on change in electric field concept has been used for real time monitoring of powder level in continuous manufacturing pilot-plant. The sensor has been integrated with a control panel (DeltaV (Emerson)) through relays and charms. The control panel then send the signal (4-20 mA) to control platform where the control strategy, a program to operate the sensor and a program to convert the sensor signal into powder level have been implemented. Subsequently, an advanced model predictive controller for powder level have been implemented into our continuous pharmaceutical manufacturing pilot-plant facility. The APC has been used to control feeder, blender and tablet press as well.

A systematic framework including the methods and tools for real time diversion of tablets have been developed and implemented into continuous pharmaceutical manufacturing process as a second layer assurance of product quality. In CM, the drug concentration is measured in real time before the tablet compaction (chute &/or feed frame) using PAT sensor. The proposed control strategy then uses this inlet concentration to determine a signal for the diversion strategy that can accurately be used to reject tablets that are out of tolerance limits at the outlet of the tablet press. Two strategies, i.e. 'fixed window based strategy' and RTD based strategy have been developed, compared and evaluated. In fixed window approach, the tablet diversion is facilitated through knowledge of time delays from the point of detection to the point of the affect (tablet press outlet gate) in the system. The sensor that detects the concentration is connected to a comparator block which decides if the said concentration is within the specifications. If it is not within specification, the experimentally derived time delay is applied and post this the diversion begins. The diversion stops when a concentration within spec is detected and the another time delay is applied. In RTD based approach, the RTD is used to predict the outlet concentration from the inlet concentration. The predicted signal is then used to initiate the diversion. The first approach is simpler to implement but may lead to lower production efficiency. The second approach is based on more advanced technique and will ensure more efficiency but is relatively complex to implement. All the relevant data generated during continuous manufacturing has been systematically collected, stored and organized in a data hub (OSI PI) and cloud system as per industry 4.0 standard.

The objective of this presentation is to demonstrate the performance of integrated MPC and RTD based bi-layer control strategy together with advanced data management system implemented into our continuous pharmaceutical manufacturing pilot-plant facility.

References

1. Bhaskar, A., Barros, F. N., Singh, R. (2017). Development and implementation of an advanced model predictive control system into continuous pharmaceutical tablet compaction process. International Journal of Pharmaceutics, 534 (1-2), 159-178.
2. Bhaskar, A., Singh, R. (2018). Residence time distribution (RTD) based control system for continuous pharmaceutical manufacturing process. Journal of Pharmaceutical Innovation. DOI: 10.1007/s12247-018-9356-7.

566023 Avoiding Distribution Confusion: Using the Information in Particle Size Distributions

Particle size distributions of solid active pharmaceutical ingredients (APIs) govern the material properties important to bioavailability such as dissolution. To characterize API properties, rich data sets such as Morphologi G3 or surrogate measures such as laser light scattering are used to measure the full population of particles within a sample. However communicating this distribution often relies on histograms of the distribution or simple summary measures of volume weighted quantiles, e.g D10, D50, D90. These measures cannot adequately capture multi modal distributions and can be insufficient for setting material specifications.

By using concepts from nonparametric statistics and information theory, the information loss incurred by using these summary metrics and histogram binning can be quantified. A method is proposed for compressing a high resolution distribution into a summary measure that best captures the underlying particle characteristics. In addition to a theoretical study of information, an example of different summarization approaches applied to a PBPK model for a model API will be presented. The resulting approach of maximizing the value of particle characterization data is a useful methodology for communicating and modeling the behavior of particulate pharmaceutical compounds.

574784 Combining Liquid-Liquid Phase Separation and Mesoporous Carrier to Deliver Drug Product Intermediate

Combining Liquid-Liquid Phase Separation and Mesoporous Carrier to deliver Drug Product Intermediate

Moussa Boukerche, Sean Garner, and Nandkishor K. Nere

Center of Excellence for Isolation and Separation Technologies (CoExIST), Process Research and Development, AbbVie Inc

Liquid-Liquid Phase Separation (LLPS), typically known as oiling out is an undesired phenomenon when developing a robust and scaleable antisolvent crystallization process. LLPS can occur in ternary phase system composed of the molecule, a solvent and an anti-solvent. The solvent and anti-solvent alone are fully miscible but the presence of the solute can force the system to phase split and deliver a binary solution made of a continuous (solute rich) and a dispersed phase (solute lean). This phenomenon will impact the purity and physical properties of the molecule of interest and typically delivers a sticky amorphous phase not suitable for drug product development. In order to circumvent this issue and take advantage of LLPS, a phase diagram determination combined with the use of Mesoporous carrier (Neusilin US2 in this case) allows the preparation of Drug Product Intermediates (DPI) with tunable drug load from 0.1 to 65wt%. The DPI has the physical properties of the carrier and is not impacted by the physical properties of the Drug Substance (DS). This DPI has the potential to be readily suitable for direct compression, reduce the risk of variability of content uniformity, sticking behavior, polymorphic phase transition during DP processing. This approach is different than the Incipient Wetness Impregnation technique that is traditionally used to load mesoporous carrier. This presentation will describe the workflow of this new approach to quickly enable DPI by triggering LLPS purposefully to load mesoporous carrier with a defined drug load.

All authors are employees of AbbVie and may own AbbVie stock. AbbVie sponsored and funded the study; contributed to the design; participated in the collection, analysis, and interpretation of data, and in writing, reviewing, and approval of the final publication.

565588 Optimising the Nucleation of Polypeptides Using Nanotemplates: The Case Study of Insulin

Optimising the nucleation of polypeptides using nanotemplates: The case study of insulin

Frederik J. Link*[a], Jerry Y.Y. Heng[a]

[a]Department of Chemical Engineering, Imperial College London, South Kensington Campus, London SW7 2AZ, UK

*f.link18@imperial.ac.uk

Keywords: insulin, nanotemplates, crystallisation

Bio-active polypeptides have attracted interest in research because of their significant use in medical, biological and pharmaceutical industries. Thereby, the crystalline form offers significant pharmacokinetic advantages with regards to achieving higher bioavailability and better release control [1]. In the case of insulin, which is widely used for treating diabetes, the most common route of administration is by subcutaneous injections of either pure microcrystals or a mixture of microcrystals and amorphous insulin. Due to the slow dissolution of the crystals, the crystalline administration leads to a slow release of insulin into the blood stream [2]. However, one of the main limitation is the lack of controllability of the nucleation mechanism. However, nanotemplates are remarkable in controlling nucleation. Recent studies showed that by using small inorganic particle with a highly porous surface the induction time of nucleation of polypeptides can be significantly accelerated even if the supersaturation is low [3]. Furthermore, it is believed that nucleation is enhanced if the pore diameter has the same order of magnitude as the radius of gyration of the polypeptide [4]. This stems from the suggestion that nucleation in porous surfaces gets enhanced due to local increases in supersaturation which is controlled by the diffusibility of the protein into the pores. However, little is known about this mechanism and the impact of the chemical properties of the surface of the nanotemplates on controlling nucleation. It is believed that porous surfaces trigger nucleation due to the local increase in supersaturation. Therefore, investigating the non-covalent interaction between the surface patches of the nanotemplates and insulin will open up new routes for controlling the crystallisation of insulin.

Insulin has a molecular weight of 5.8 kDa and consist of 2 polypeptide-chains, A and B, with combined 51 amino acids. Furthermore, the two chains are inter-linked by two disulfide bonds and, additionally, chain A is intra-linked with an additional disulfide bond. Citrate buffer with a pH near the isoelectric point of insulin was used with zinc-sulphate as precipitant. Different concentrations were tested to identify the most promising ones. Since insulin is stored in a rhombohedral crystal configuration in combination with zinc-sulphate in the pancreas, zinc-sulphate was selected as the most appropriate precipitate. The impact of nanotemplates on crystallisability is investigated by mean of measuring induction time and crystal yield. Therefore, batch crystallisation experiments will be carried out in which the crystal yield and induction time will be tracked via measuring the concentration of insulin within the solution over time. To achieve this, the concentration of insulin is determined via UV-vis absorption in combination with Bradford's method [5]. Different types of nanotemplates will be investigated which differ in their surface chemistry and porosity, amount and size.

By using nanotemplates, the induction time of crystallising a complex molecule such like insulin can be reduced significantly even at low supersaturation. Although the nanotemplates do not have a notable influence on the crystal quality, multiple configurations can be observed which is due to the local availability of dissolved zinc-sulphate. When in the vicinity of zinc-sulphate insulin crystallises in its rhombohedral form, whereas otherwise it crystallises in a cubic shape.

Further work will be done to characterise the impact of the nanotemplates with regards to understanding the impact of the chemistry of the surface patches and whether the crystallisation can be enhanced solely due to local supersaturation or also because of specific non-covalent interaction between the surface patches and insulin. Additionally, the impact of the seeds on the solubility of insulin has to be investigated to evaluate the impact of the nanotemplates on the overall yield. All these findings should ultimately result in the ability of engineering appropriate nanotemplates to enhance the crystallisability of insulin. Further work will also be done in investigating the potential of scaling up from 1 ml to 10 and 100 ml scale.

References

1. Pechenov, S., et al., *Injectable controlled release formulations incorporating protein crystals.* J Control Release, 2004. **96**(1): p. 149-58.
2. Norrman, M. and G. Schluckebier, *Crystallographic characterization of two novel crystal forms of human insulin induced by chaotropic agents and a shift in pH.* BMC Struct Biol, 2007. **7**: p. 83.
3. Shah, U.V., et al., *Crystallization of Proteins at Ultralow Supersaturations Using Novel Three-Dimensional Nanotemplates.* Crystal Growth & Design, 2012. **12**(4): p. 1772-1777.
4. Shah, U.V., D.R. Williams, and J.Y.Y. Heng, *Selective Crystallization of Proteins Using Engineered Nanonucleants.* Crystal Growth & Design, 2012. **12**(3): p. 1362-1369.
5. M., B.M., *A Rapid and Sensitive Method for the Quantitation of Microgram Quantities of Protein Utilizing the Prinicple of Protein-Dye Binding.* Analytical Biochemistry, 1976. **72**: p. 248-254.

572187 Olanzapine Crystal Symmetry Originates in Preformed Centrosymmetric Solute Dimers

The symmetry of crystals is a central quality that determines their function and allure. Given the uncorrelated symmetries of a crystal and its constituent molecules, the symmetry breaking that begets a crystal phase has been thought to ensue during nucleation. Here we demonstrate that the one of the two symmetry element of olanzapine crystals, an inversion center, emerges in solute dimers extant in the solution prior to crystallization. Time resolved *in situ* atomic force microscopy monitoring of growth steps reveals that crystals grow nonclassically, exclusively by incorporation of dimers; this growth mode has been speculated and inferred from crystal and solute structures, but never experimentally observed. The growth rate of crystal layers exhibits a quadratic dependence on the solute concentration, characteristic of the second-order kinetics of dimer incorporation. Olanzapine dimers constitute a minority solution component. We show that growth by dimers is preferred not owing to greater activation entropy or faster dissociation of attached solvent molecules. The monomer and dimer adsorption energies on the crystal surface, evaluated in all-atom molecular dynamics simulations, suggest overwhelming accumulation of adsorbed dimers, which expedites their incorporation into growth sites. The proposed outlook on the cooperative creation of crystal symmetry illuminates analyses of structure selection, polymorph transformations, chiral separation, and crystal properties of natural and engineered crystals.

572060 Development of a Hybrid Quality-By-Control Framework for Optimal Crystallization Process Design

Strict regulatory standards are dictated by the FDA, EMEA and WHO for the active pharmaceutical ingredients (API) to meet high level quality criteria such as purity, size and shape of the API crystals or polymorphic structure. One of the main criteria is to achieve size uniformity as well as desired yield in industry. Despite of the efforts put in pharmaceutical industry, intensive fines formation and agglomeration of crystals are still common problems, often having a negative impact on the product crystal size distribution (CSD). A general practice to eliminate fines and agglomerates is to apply temperature cycles for internal fines dissolution and de-agglomeration.[1] For this purpose, quality-by-design (QbD) approaches can be applied for the design of the suitable temperature profile, but this requires intensive experimental investigation to explore the whole design space. Quality-by-Control (QbC) has been introduced more recently as a new novel alternative to QbD, which can be applied to control critical quality attributes of a certain process, whereby the desired operating curve like temperature profile is automatically determined by the feedback control algorithm rather than by excessive open-loop experimentation. In this way QbC can significantly reduce required process development time and material used.[2]

The work proposes a novel hybrid framework combining the model-free and model-based quality-by-control (mfQbC and mbQbC) paradigms for the optimal crystallization design with the aim of minimization of fines and agglomerates in the product. This developed framework describes the mfQbC based rapid design used in conjunction with direct nucleation control (DNC) and supersaturation control (SSC), along with (2) the mbQbC framework using a novel population balance model-based platform for the robust optimization of the operating trajectory of the crystallizer to achieve crystals with desired quality attributes.[3] This framework provides an improved alternative to the current industrial practice, by the combined use of feedback control and mathematical models for rapid design of robust crystallization processes. The implementation of the proposed framework enhanced the prediction of crystallization events and was able to achieve better control of the crystallization processes by providing controlled temperature cycles, for de-agglomeration and fines destruction. The information for parameter estimation and optimization calculations was collected from the experiments and the system diagnostics was done in mfQbC part, which also provided a rapid design of robust operating conditions, that were improved to achieve optimal operation subsequently in the mbQbC approach.

The advantage of the mfQbC is that it requires minimal preliminary system information, hence, it enables quick and efficient crystallization process design. On the other hand, the mf-QbC operation policy might be sub-optimal, which can be overcome by exploiting the predictive capabilities of the process models used by the mbQbC. In the mfQbC part, DNC experiments were run for two purposes: to determine the primary and secondary nucleation metastable zone widths (MSZW) and to obtain the temperature cycle profile that is required for producing the crystals with desired properties. DNC is an efficient approach to eliminate fines due to the application of temperature cycles. SSC was applied after DNC generates the MSZW information to obtain the temperature profile for only growth mechanism. The resulting temperature profile from SSC can also be easily implemented in industrial distributed control systems (DSCs). After mfQbC, using the mbQbC framework, a population balance model (PBM) was developed to simulate primary and secondary nucleation, growth, agglomeration and de-agglomeration mechanisms as well as dissolution. Model parameters were identified using experimental data for a model API crystallization with high agglomeration and secondary nucleation tendency. Using a model-based robust optimization, in which parametric uncertainty obtained from the model-identification step was also included, different temperature profiles were determined considering the dual objective of fine minimization and maximal de-agglomeration. The optimal temperature profiles obtained was validated experimentally to show the higher quality product that can be obtained using the this QbC framework.

References:

1. Wu, Z., Yang, S. & Wu, W. Application of temperature cycling for crystal quality control during

crystallization. *CrystEngComm* **18**, 2222–2238 (2016).
2. Yang, Y., Song, L. & Nagy, Z. K. Automated Direct Nucleation Control in Continuous Mixed Suspension Mixed Product Removal Cooling Crystallization. *Cryst. Growth Des.* **15**, 5839–5848 (2015).
3. Szilágyi, B., Borsos, Á., Pal, K. & Nagy, Z. K. Experimental implementation of a Quality-by-Control (QbC) framework using a mechanistic PBM-based nonlinear model predictive control involving chord length distribution measurement for the batch cooling crystallization of L-ascorbic acid. *Chem. Eng. Sci.* **195**, 335–346 (2019).

576759 Computation Fluid Dynamics (CFD) Guided Design of Single Use Bioreactors (SUBs) at Bench Scale

As single-use bioreactors (SUBs) become more prevalent in mammalian cell culture manufacturing, more robust equipment characterization and predictive scaling correlations are needed for better process understanding, scale-up, and technology transfers. During development of new bioreactor designs, modeling and characterization efforts occur in parallel to inform decisions. Over several iterations of design improvement, we found that small physical modifications can add up to incremental impacts on fluid behavior, making previous empirical correlations outdated. This is especially at bench scale and below; slight adjustments in physical structure due to product customization can impact overall power number and mixing.

In this study, we investigated a cell line specific sensitivity to being cultured in a single use vessel compared to similar glass bioreactor designs. Impacts on cell culture performance can be masked by variation from several sources, with the largest factor being inoculum source. It was hypothesized that minor deviations within tight manufacturing tolerances, especially at the bearings of bottom-mounted rotating assemblies, can result in higher strain rates through narrow gaps. Using Computational Fluid Dynamics (CFD) modeling, we discovered unexpected impacts to fluid flow behavior from adjustments such as baffle placement, baffle length, and the curvature of support structures for rotating parts.

568929 Physics-Based Modeling of Free Surface Dynamics Using Graphics Processing Units (GPUs)

Free-surface and two-phase systems are abundant in pharmaceutical equipment and processes, including vortex formation in agitated systems, nozzle sprays, tank filling, sloshing, spray coating atomization. Processes involving free surfaces are almost always time-varying, and involve transport processes that simultaneously span multiple length and time scales. This disparity in length and time scales makes first-principles modeling of free-surface and two-phase systems difficult.

In this work, we show how the Navier-Stokes equations with free surface boundary conditions can be solved efficiently on graphical processing units. We begin by introducing the concepts governing GPU-performance, as applied lattice-Boltzmann and free surface modeling. We then show how, given identical CFD algorithms, a single scientific GPU can execute simulations 100x faster than a single CPU. We then validate this approach by comparing predictions from simulations involving agitated tanks, shaker flasks, and sloshing tanks.

577222 Challenges in Transfer of a Shear Sensitive Lubricated Formulation from a Dosator-Type to a Dosing Disk-Type Capsule Filling Machine

According to current SUPAC Guidance, changes in equipment from the same class but different subclasses should be considered carefully and supported with appropriate scientific data to justify that the quality attribute(s) of the drug product are not adversely impacted. During transfer of a lubricated formulation from a dosator-type to a dosing disk-type encapsulation machine (same class but different subclasses), a significant decrease in both the rate and extent of dissolution, a critical quality attribute (CQA), was observed at the end of the encapsulation run, based on testing of samples pulled in a stratified fashion. Additionally, a reduction in plug hardness was observed towards the end of the encapsulation run. It is well known from the literature that the amount of shear energy (as a function of rate and time) induced on the blend can be much greater when a dosing disk-type encapsulation machine is used compared to a dosator-type encapsulation machine. This excess shear occurs within the feed hopper itself (via the rotating agitator) and in the powder bowl (via the rotating tamping pins and/or level scraper) and can reduce the hardness (tensile strength) of the plugs.

The lubrication sensitivity of the blend in the feed hopper of a dosing disk-type encapsulation machine and associated dissolution slowdown was first mimicked using a lab-scale Turbula blender, where the excess dispersion of the hydrophobic lubricant, magnesium stearate, was assessed. Next, a qualitative/semi-quantitative disintegration-based Process Analytical Technology (PAT) was developed and enabled prediction of the dissolution performance of capsules in parallel with the encapsulation unit operation. Subsequently, the feed hopper was re-designed by complete removal of agitator (main source of shear), and finally granule characterization (from a prior roller compaction step and via factorial DoE) was optimized to ensure proper flow and compactability of the blend. These approaches led to a successful transfer of the new process and ensured continuity of drug product CQAs to meet or exceed all regulatory commitments, specifically those around the Quality Target Product Profile (QTPP).

557997 A Quality By Design Approach to Address the Challenges of Direct Compression Scale-up and Technology Transfer between Rotary Tablet Presses

A quality by design approach to address the challenges of direct compression scale-up and technology transfer between rotary tablet presses

Raghu V. G. Peddapatla[1], Gerard Sheridan[2], Conor Slevin[2], Shrikant Swaminathan[3], Ivan Browning[4], Clare O' Reilly[4], David Egan[2], Stephen Sheehan*[4], Abina M. Crean[1,5]

[1]School of Pharmacy, University College Cork, Cork, Ireland

[2]Bernal Institute, University of Limerick, Limerick, Ireland

[3]Alkermes Inc, Waltham, MA, USA.

[4]Alkermes Pharma Ireland Limited, Athlone, Ireland

[5]Synthesis and Solid State Pharmaceutical Centre, Ireland

* Alkermes Pharma Ireland Limited, Athlone, Ireland, Tel: + 353906495412,

email: Stephen.Sheehan@alkermes.com

INTRODUCTION

The key outcome of implementing quality by design (QbD) is attainment of process understanding, i.e. understanding the relationships between product critical quality attributes (CQAs), process parameters and formulation attributes.[1] During the process development stage a degree of process understanding can be attained using pilot scale equipment. However, differences in equipment designs utilised at pilot scale and production scale makes extrapolation of process understanding from pilot to production scale challenging. When scaling-up a process, or transferring a process between equipment models, different processing parameters can be encountered which need to be considered for thorough process understanding and hence QbD.

The concept of a design space is integral to QbD. Design space is a multidimensional space that encompasses combinations of material attributes and processing parameters to provide assurance of processing conditions that can achieve product CQAs. However, due to the restricted processing parameters that can be explored on certain pilot scale equipment, it is difficult to generate a single design space that will encompass both pilot and production scale equipment during development. This is particularly evident for tablet presses where feeder designs and press speeds can vary greatly between equipment models.[2] Therefore, to create an accurate design space for a product processed on a particular production tablet press additional experimental runs are required using that production press. Each design space generated is to some degree limited by the equipment on which the experimental runs were conducted.

The aim of this study is to develop a generic understanding of how direct compression formulations with differing material attributes are compacted on selected pilot and production scale tablet presses. Six placebo formulations with differing flow and compaction properties were designed and compacted on one pilot scale and two production scale presses. For each press an experimental design space was generated encompassing both formulations attributes (e.g. flow and compaction behaviour). The objectives of this study were two fold. Firstly, to enhance understanding of the impact of transferring formulations between press models, specifically how formulation attributes and press parameters influence compaction and flow behaviour. Secondly, to benchmark the attributes of commercial formulations to the placebo formulations and thereby predict their behaviour on different press models. The overall goal of this work is to create a knowledge base on how to optimize a future OSD active formulation on each tablet press. From a commercial process development perspective this information will be vital to reduce a new products time to market.

MATERIALS AND METHODS

Tablet presses

The following rotary presses were utilised in this study KG RoTab (KGPharma, Scharbeutz, Germany), Fette 1200i (Fette Compacting, Schwarzenbek Germany) and Modul™ P (GEA, Belgium).

Formulations

Six placebo formulations were used in this study designed to achieve differing flow and compaction properties, shown in Table 1.

Blend preparation and characterisation

Formulation ingredients were dispensed and fed through a 450µm Sweco sieve to remove any agglomerates. Except lubricant, all other formulation components (diluents, flow aid and disintegrant) were added into a 100L IBC and blended for 18 minutes at 18 rpm. 1%w/w of lubricant was also passed through the 425µm sieve before being added to the other components in IBC. Blending was then continued for a further 3 minutes.

Design of Experiments (DoE)

To investigate the impact of formulation on flow and compaction behaviour on each press, two separate DoEs were designed using JMP statistical software. Based on process parameters investigated and levels, a factorial experimental design was used with 4 centre points (Table 1).

A 2^4 factorial DoE was used to study the compaction behaviour for formulations 1-3 (Table 1). The effect of formulation, pre-compression force, main compression force and press speed on the tablet CQA's (weight variability, thickness, breaking force, porosity, friability and disintegration time (DT)) was investigated.

A 2^3 factorial DoE was used to study the flow behaviour for formulations 4-6 (Table 1). The effect of formulation, feeder speed and press speed on tablet CQA's.

Table 1 Formulations with different compaction and flow behaviours used in the study

Formulation	Formulation behaviour	Factorial DoE	Factors	Total runs (including centre points)
1	Predominantly Plastic	2^4	-Formulation -Pre-Compression Force -Main Compression Force -Press Speed	16
2	Predominantly Elastic			
3	Predominantly Brittle			
4	Good Flow	2^3	-Formulation -Feeder Speed -Press Speed	12
5	Medium Flow			
6	Passable Flow			

Blend characterisation

Yield pressure and elastic recovery were used to differentiate the different compaction mechanisms within formulations 1-3. Freeman FT4 powder rheometry was used to assess the flow behaviours of formulations 4-6.

Blend compaction

All tablet blend formulations were compressed at target compression forces of 1.2 to 21 kN, which corresponded to compression pressures of 20 MPa to 360 MPa using the shield shape punches on all three tablet presses. Compression profile was further used to optimise the main compression forces for compression and flow DoE.

RESULTS AND DISCUSSION

Formulation characterisation

Yield pressures and tablet elastic recovery were determined to characterise the compaction mechanism (plastic or brittle or elastic) in formulations 1-3. Table 2 shows the yield pressures of the formulations on two tablet presses. On both tablet presses, formulation 1 showed a lower yield pressure, indicating higher plasticity compared to formulations 2 and 3. Tablets 24 hr thickness recovery, showed higher elastic recovery in formulations 2 (Table 2). Formulation 3 contained a higher percentage of an excipient with known brittle behaviour.

Table 2 Yield pressures and elastic recovery of formulations 1-3 compacted on Modul P and KG RoTab.

Formulation	Yield Pressure (MPa)		Percentage thickness recovery (%) on Modul P
	Modul P Tablet Press	KG RoTab Tablet Press	
Formulation - 1	121.9	88.4	0.26
Formulation - 2	222.2	204.0	0.75

Formulation - 3	196.0	217.3	0.00

FT4 flow analysis of formulations 4-6 showed that formulation 4 is good flowing compared to formulation 5 (medium flow) and formulation 6 (passable flow), shown in Figure-1.

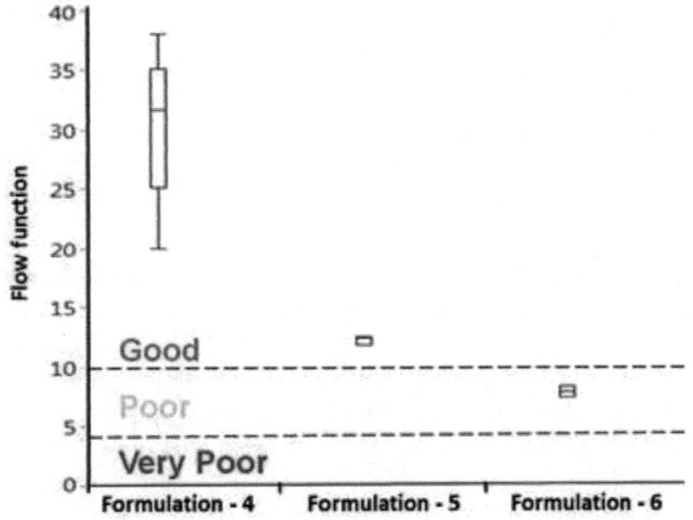

Figure 1 Flow function of formulations 4-6 analysed by FT4 powder flow rheometer.

Breaking force and disintegration time variation by model factors on Modul P

For each tablet press and each formulation a model was developed using JMP software from which a design space could be constructed. A representative output of the model is shown in Figure 2. The effect of process parameters on tablet breaking force and DT was assessed by model factors for formulations 1-3. Press speed, formulation type and main compression force showed a significant effect on the tablet breaking force and DT for the formulations, shown in Figure 2AB. Pre compression force did not show significant effect on breaking force and DT.

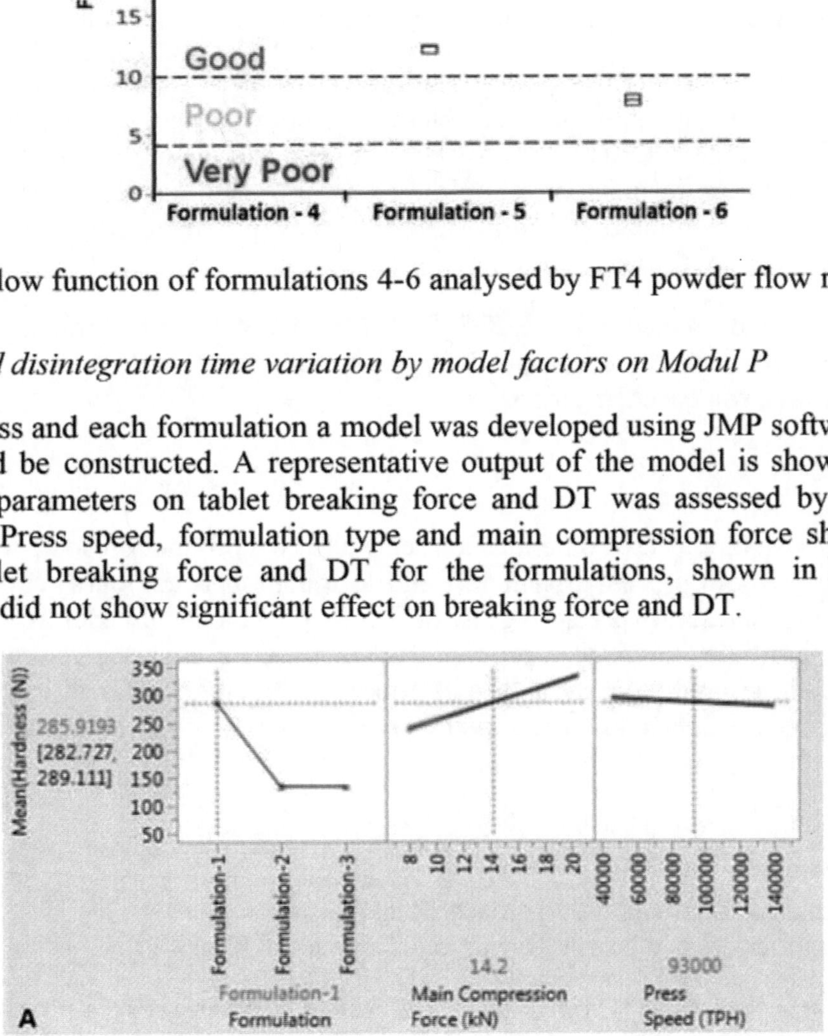

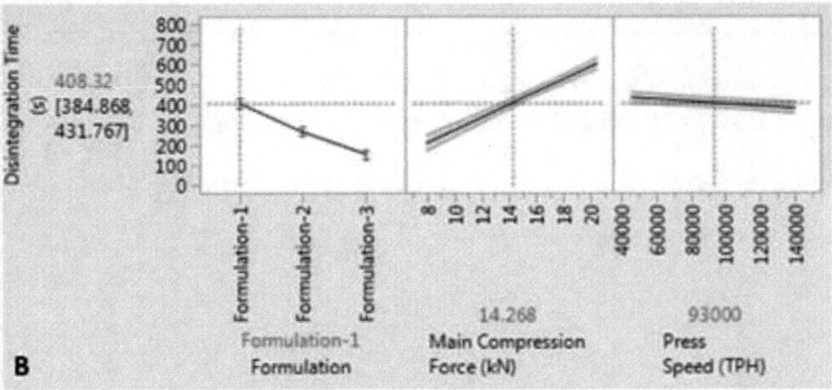

Figure 2 Prediction profiler for mean breaking force and disintegration time for formulation 1-3 on Modul P. (A) Mean breaking force (B) Disintegration time

This model can be used to predict the quality attribute of tablet for specific type of formulation, on specific tablet press using prediction profiler (Figure 2). For example, at 14.2kN compression force and 93000 tablets per hour (TPH) press speed, the expected quality for formulations can be predicted. At these settings formulation-1 produced tablets with greater breaking force and DT (Figure 2B) compared to formulation 2 and 3, where they produced tablets with similar breaking force (Figure 2A).

Similar models will be presented for the KG RoTab and Fette 1200i presses. Examples will be presented illustrating how these models can be applied to support scale-up and technology transfer of formulations between rotary tablet presses.

CONCLUSIONS

Experimental designs were executed on one pilot-scale and two production scale rotary tablet presses studying the impact of both process parameters and formulation behaviours (compaction and flow) on tablet quality attributes. A predictive model was developed to identify significant factors impacting on tablet attributes to create a process design space for each tablet press. For future OSD active formulations these predictive models will allow for a seamless transition in scale-up and/or technology transfer between each tablet press, thus reducing a new product's time to market.

BIBLIOGRAPHY

1. FDA, 2009. Guidance for Industry: Q10 Pharmaceutical Quality System. (Accessed on April 04, 2019) https://www.fda.gov/downloads/Drugs/GuidanceComplianceRegulatoryInformation/Guidances/UCM073517.pdf.

2. Narang AS, Rao VM, Guo H, Lu J, Desai DS. Effect of force feeder on tablet strength during compression. Int J Pharm. 2010;401(1-2):7-15. doi:10.1016/j.ijpharm.2010.08.027

564257 Scale-up of Roller Compaction Processes Using a Geometric Scaling Model

The key indicator of successful scaling of the roller compaction process of a drug product formulation is achieving similar ribbon density, which is the primary attribute measured that effects downstream processes, subsequent material properties of granules, and tablet critical quality attributes. The ribbon density is controlled by equipment parameters such as the roll force and gap. Prediction of the required roll force at targeted scale is key to enable efficient scale up of roller compaction processes. A novel methodology for the scale up of roller compaction processes has been developed that leverages a geometric scaling model. A relationship between roll forces and roll diameters of different scales is achieved through a calibration process. Case studies will be presented demonstrating the application of the geometric scaling model coupled with an equipment correlation term to predict scale-up of roller compaction processes.

571764 Local Residence Time Distributions for Hot Melt Extrusion: Making a Black Box Concept Mechanistic

Local Residence Time Distributions for Hot Melt Extrusion: Making a black box concept mechanistic

Hannes Bauer*, Josip Matić*, Johannes Khinast*,**

*Research Centre Pharmaceutical Engineering GmbH, Inffeldgasse 13/III, 8010 Graz, Austria

**Institute for Process and Particle Engineering, Graz University of Technology, Inffeldgasse 13/III, 8010 Graz, Austria

Email for correspondence: khinast@tugraz.at

Introduction

Hot Melt Extrusion (HME) is a continuous manufacturing process used in the pharmaceutical industry to produce amorphous solid dispersions (or solutions) of poorly soluble active pharmaceutical ingredients (APIs). Process setup and scale-up are still a significant issues for the temperature and shear sensitive formulations used in the pharmaceutical industry.

Besides the high temperatures reached during the process, it is generally agreed upon that the Residence Time Distribution (RTD) significantly impacts the product properties. However, RTD theory was derived for ideal systems with uniform characteristics, such as plug flow reactors (PFR) or continuously stirred tanks (CSTRs). The characteristics in a Twin-Screw extruder vary considerably, along the screw axis, and across the screw cross section. Meaningful kinetic models which aim to predict product formation (or degradation) need to account for that.

Methodology

Smoothed Particle Hydrodynamics[1] (SPH) has an excellent track record in predicting the flow in extruder elements.[2,3] The methods meshless nature allows for a convenient treatment of rotating moving boundaries such as extruder screws. In SPH the fluid consists of numerous flowing particles, which carry the fluid flow information. Therefore, the method is uniquely suited for *in-silico* tracer experiments. Such tracer experiments are useful for determining local RTDs in extruder screw elements. Three zones of similar behavior can be distinguished in the cross section of the elements: (1) high shear screw-barrel gap zone; (2) medium shear screw-screw intermeshing zone and (3) low shear screw channel zone. Following the trajectories of fluid elements we find them passing through these zones. Hence, we determine the local residence time in these zones as well as the number of transitions (number of passes) from one zone to the other.[4,5]

Results

Instead of taking the classical black box point of view, which is usually associated with RTD models, we take a closer and more detailed look at twin-screw extruder elements. We derive local RTDs in the gap zone, the intermeshing region and the channel. We discuss the number of passes from one zone into another and we investigate how the total time the melt spends in a specific zone relates to the overall RTD. This approach is the basis for future mechanistic product property prediction of HME products.

Literature

1. J. Monaghan, "Smoothed particle hydrodynamics," *Reports Prog. Phys.*, vol. 68, no. 8, pp. 1703–1759, Aug. 2005.
2. Eitzlmayr and J. Khinast, "Co-rotating twin-screw extruders: Detailed analysis of conveying

elements based on smoothed particle hydrodynamics. Part 1: Hydrodynamics," *Chem. Eng. Sci.*, vol. 134, pp. 880–886, 2015.
3. Eitzlmayr, J. Matić, and J. Khinast, "Analysis of Flow and Mixing in Screw Elements of Corotating Twin-Screw Extruders via SPH," *AIChE J.*, 2017.
4. Mann, U. , Rubinovitch, M. and Crosby, E. J. (1979), Characterization and analysis of continuous recycle systems. *AIChE J.*, 25: 873-882.
5. Mann, U. and Rubinovitch, M. (1981), Characterization and analysis of continuous recycle systems: Part II. Cascade. *AIChE J.*, 27: 829-836.

571567 Simulation Aided Pharmaceutical HME Scale-up

Simulation Aided Pharmaceutical HME Scale-up

Josip Matić*, Abdelhamid Mostafa*, Carolina Alva*, Johannes Khinast*,**

*Research Centre Pharmaceutical Engineering GmbH, Inffeldgasse 13/III, 8010 Graz, Austria

**Institute for Process and Particle Engineering, Graz University of Technology, Inffeldgasse 13/III, 8010 Graz, Austria

Email for correspondence: khinast@tugraz.at

Introduction

Hot melt extrusion (HME) is a continuous manufacturing process primarily facilitated by co-rotating intermeshing twin-screw extruders (TSE). The process is mostly used in order to produce amorphous solid dispersions of poorly soluble active pharmaceutical ingredients (APIs) by dispersing them in polymer carriers. The process setup is modular, allowing the process to be tailor-made for the processed formulation. This might however be problematic in the sense of choosing the appropriate and efficient setup for an unknown formulation. Besides the process setup challenge the process scale-up is a considerable issue during the different product development stages (from formulation development to pilot plant/clinical study to production scale). The current state-of-the-art is based on utilizing different extruder similarity based approaches (simple 0D models), developed mostly in the polymer industry and heavily coupled to extensive experimental efforts. The immediate restriction in the pharmaceutical industry is the lack and the expense of APIs in the early development stages in addition to the connected costs and efforts in handling the process in a GMP environment. Moreover, the 0D models do not provide any insights into the process, as they are not predictive and do not guaranty the most important aspect in the process transfer and scale-up: the equivalence of product quality at different scales. Hence, understanding why certain process settings may lead to a certain product quality is the key for ensuring an easy process setup and a reliable subsequent process scale-up.

Methodology

As a response to the challenges, simulations have been increasingly used as means for process understanding. In general, HME simulations have been used for understanding individual extruder screw elements, the HME process as a whole and to relate HME to other unit operations in a continuous manufacturing line. The most notable among them is the individual extruder screw investigation using the Lagrangian based Smoothed Particle Hydrodynamics (SPH) approach for understanding the melt flow and mixing in fully and/or partially filled extruder screw elements, which applies for both Newtonian and non-Newtonian fluids [1]–[5]. The knowledge gained is applied in a less complex and faster 1D HME code, allowing for the accurate representation of the process as a whole[6], [7]. These approaches have a focus on understanding the process, either by investigating the melt flow in detail and/or by investigating the influence of the process settings on the process state variables like melt temperature and RTD. Based on the knowledge gained from process simulations, experiments and product investigations, the next challenge is the development of tools for the *in-silico* prediction of the product performance. Hence, relating the process state variables (melt temperature, SMEC, RTD...) with the product performance is the next step towards understanding the HME process.

Results

In our last talk we looked at the degradation of Famotidine-Eudragit RL formulation as a function of different HME process parameters of the 12mm TSE from Leistritz. The API degradation was taken as a measure of the product quality after extrusion and the process states that lead to the degradation were analyzed in detail using the 1D HME simulation code developed in-house. As a next step, a number of scale-up approaches were tested experimentally, for a process transfer from the Leistritz 12mm to 18mm TSE. The goal was reaching the same API degradation state formed in the original process. The different resulting HME processes were analyzed using the 1D HME software in order to explain variations of the product quality in comparison to the original process, keeping in mind the temperature peaks and mean residence times archived in the different processing zones. *In silico* based process settings were then obtained and tested with the goal of achieving the same product quality. This lays the foundation for automated process setup and scale-up with a prescribed product quality in mind.

Literature

[1] J. J. Monaghan, "Smoothed particle hydrodynamics," *Reports Prog. Phys.*, vol. 68, no. 8, pp. 1703–1759, Aug. 2005.

[2] K. Kohlgrüber and W. Wiedmann, *Co-Rotating Twin-Screw Extruders*. Munich: Carl Hanser Verlag, 2008.

[3] A. Eitzlmayr and J. Khinast, "Co-rotating twin-screw extruders: Detailed analysis of conveying elements based on smoothed particle hydrodynamics. Part 1: Hydrodynamics," *Chem. Eng. Sci.*, vol. 134, pp. 880–886, 2015.

[4] A. Eitzlmayr, J. Matić, and J. Khinast, "Analysis of Flow and Mixing in Screw Elements of Corotating Twin-Screw Extruders via SPH," *AIChE J.*, 2017.

[5] R. Baumgartner, J. Matić, S. Schrank, S. Laske, J. G. Khinast, and E. Roblegg, "NANEX: Process Design and Optimization," *Internaitonal J. Pharm.*, vol. 506, pp. 35–45, 2016.

[6] A. Eitzlmayr *et al.*, "Mechanistic modeling of modular co-rotating twin-screw extruders," *Int. J. Pharm.*, vol. 474, no. 1–2, pp. 157–176, 2014.

[7] J. Matić, A. Witschnigg, M. Zagler, S. Eder, and J. Khinast, "A novel in silico scale-up approach for hot melt extrusion processes," *Chem. Eng. Sci.*

576463 Possible Synergistic Effects of Binary Polymers in Enhancing Release of a Poorly Soluble Drug from Amorphous Solid Dispersions

Over the last few decades, majority of the new chemical entities (NCEs) in the drug discovery pipeline are poorly-water soluble [1], and their slow dissolution turns out to be the rate-limiting step during intestinal absorption. Formulating a drug in its amorphous form could significantly enhance the dissolution rate as it has higher apparent solubility [2]. Despite its potential benefit, the application of pure amorphous form is limited due to its inherent thermodynamic instability. Polymers are often incorporated to form amorphous solid dispersions (ASDs) with the intent of providing stability during processing, storage, and dissolution. They play an important role in inhibiting the recrystallization in both the matrix and the dissolution medium; thus, they allow for supersaturation generation during dissolution.

Selection of a proper polymer/surfactant system for ASD formulations is essential to achieving storage stability and dissolution enhancement. It is well-known that most polymers used in ASDs are relatively hydrophobic and dissolve slowly, which leads to slow drug release from ASDs. To resolve this issue, surface-active agents such as surfactants have been used alone or in combination with polymers. The formulations with surfactants exhibited improved dissolution and physical stability compared to those w/o surfactants [3, 4]. However, detrimental effects of surfactants have also been reported [5, 6]. Considering these conflicting findings, a binary polymer system can be an alternative to surfactants in ASDs. Tian and Taylor [7] produced ASD of celecoxib (CXB) and compared the effectiveness of binary polymer system with that of the single polymer. Polyvinylpyrrolidone (PVP), hydroxypropyl cellulose (HPMC), and HPMC acetate succinate (HPMCAS) were used as polymers. Combining cellulosic polymers with PVP enabled improved CXB release and prevented recrystallization more effectively than single polymers. Recently, other studies have also shown the effectiveness of polymer combination in the dissolution performance of ASD and recrystallization inhibition from supersaturated solution over single polymer [8–10]. Only a few studies have investigated binary polymer system in ASDs systematically. In these studies, supersaturation generation and recrystallization inhibition properties of the binary polymer systems have been the main focus; however, the wetting effectiveness of the individual polymer and their combination has not received much attention. It is hypothesized that wettability enhancement of the hydrophobic drug, as modulated by the polymers, could also affect drug release rate from the ASDs. This aspect should be considered besides the impact of polymers on supersaturation generation and maintenance, and their recrystallization inhibition property.

The aim of this study is to examine the effectiveness of binary polymers vs. single polymer in the production of spray-dried ASD of a poorly water-soluble drug, Griseofulvin (GF). GF has an aqueous solubility of 14.5 mg/L at 37 °C. Hydroxypropyl cellulose (HPC), Soluplus® (Sol), and Vinylpyrrolidone–Vinyl acetate (Kollidone VA64) were used as polymers. For all formulations, drug:polymer ratio was fixed at 1:3. Combination of VA64–HPC, Sol–HPC, and Sol–VA64 were used for the binary polymer systems. Acetone–deionized water mixture was used to dissolve GF–Sol, GF–HPC, and GF–HPC–Sol formulation. On the other hand, mixture of acetone–ethanol–deionized water was used to dissolve GF–VA64, GF–VA64–HPC, and GF–Sol–VA64 formulation. Completely dissolved drug–polymer (s) solutions were spray-dried to evaporate the solvents and produce ASD. X-ray powder diffraction (XRPD), modified Washburn experiments, drug desupersaturation experiments, and *in vitro* USP II dissolution testing were used for the characterization of the spray-dried powders. Dissolution tests were performed with a USP II apparatus by dispersing 100 mg GF in 1000 mL deionized water at 37 °C.

XRPD results suggest that the spray-drying of GF–polymer(s) solutions produced ASDs for all polymers used. The dissolution tests on ASDs with a single polymer reveal that relatively low supersaturation ratios (as percentages), i.e., 42% and 33%, were achieved respectively byVA64 and HPC, whereas notable supersaturation (163%) was generated by Sol at 210 min albeit at a slower GF release rate. This could be explained by fast recrystallization of GF in the dissolution medium and/or ASD matrix when VA64 or HPC was used alone, and Sol appeared to be an effective recrystallization inhibitor. The solubility parameter difference for the GF–polymer pairs were estimated as follows: 7.2, 7.5, and 11.8 $MPa^{0.5}$, respectively, for GF–Sol, GF–VA64, and GF–HPC. The poor miscibility of GF with HPC could explain the observed lower extent of supersaturation and faster recrystallization than those for Sol. The desupersaturation experiments showed that HPC and VA64 cannot maintain supersaturation; hence, they are poor recrystallization inhibitors in the solution. While Sol appears to be a good recrystallization inhibitor, due to slow release of GF from the Sol matrix, the supersaturation in the dissolution medium built up slowly. The faster GF release from VA64 and HPC matrices than for Sol matrix can be explained by the more hydrophilic nature of the former, as quantified by the higher wetting effectiveness factors calculated via the modified Washburn method, and faster erosion of the former polymers compared to the Sol.

With some exceptions, the dissolution behavior of the ASDs could be elucidated based on the performance of ASDs with single polymers. Despite higher supersaturation ratios attained at 210 min (58%–77% for different mass ratios of VA64–HPC) compared with the single polymers, VA64–HPC combination was still inferior to Sol alone in terms of recrystallization inhibition. This is of no surprise as neither VA64 nor HPC inhibited recrystallization and maintained supersaturation—as independently confirmed by desupersaturation experiments. When Sol was combined with either HPC or VA64, more hydrophilic polymers, the supersaturation levels were higher than those obtained with VA64–HPC. This result shows the criticality of having at least a major polymer, (Sol here), that has good miscibility with the drug and that can inhibit recrystallization in the polymer matrix and maintain supersaturation in the dissolution medium. When the more hydrophilic polymer (HPC or VA64) was added to Sol, the dissolution rate increased as compared with that of Sol alone. For Sol–HPC, a decrease in Sol:HPC mass ratio led to faster GF release, but lower extent of supersaturation. This observation can be explained by more effective wettability enhancement, but poor recrystallization inhibition capability of HPC both in the matrix (due to poor GF–HPC miscibility leading to phase separation) and in the dissolution medium. On the other hand, for Sol–VA64, a decrease in Sol:VA64 mass ratio led to faster GF release. The supersaturation at 210 min was lower for Sol:VA64 mass ratios of 1:1 and 1:5 than that for Sol alone, but surprisingly both the GF release rate and the supersaturation ratio at 210 min (222% vs. 163%) were higher for Sol:VA64 mass ratio of 5:1 than those for Sol alone and VA64 alone. Apparently, a synergistic effect of Sol–VA64 combination exists when Sol was used in excess of VA64.

This study has demonstrated that the use of a binary polymer system in drug ASDs could result in synergistic positive effects on the drug release rate and extent of supersaturation. The synergy occurs when both polymers are miscible with the drug, the major polymer (Sol) is an excellent recrystallization stabilizer in the dissolution medium, and the minor polymer (VA64), being more hydrophilic, could bring significant wettability enhancement to the hydrophobic drug (GF). These findings could help formulators in rational design of immediate-release drug ASDs with binary polymers, obviating the need for surfactants.

References

[1] S.P. Chaudhari, R.P. Dugar, Application of surfactants in solid dispersion technology for improving solubility of poorly water soluble drugs, J. Drug Delivery Sci. Technol., 41 (2017) 68–77.

[2] B.C. Hancock, M. Parks, What is the true solubility advantage for amorphous pharmaceuticals?, Pharm. Res., 17 (2000) 397–404.

[3] C. Goddeeris, T. Willems, K. Houthoofd, J. Martens, G. Van den Mooter, Dissolution enhancement of the anti-HIV drug UC 781 by formulation in a ternary solid dispersion with TPGS 1000 and Eudragit E100, Eur. J. Pharm. Biopharm., 70 (2008) 861–868.

[4] B. Lang, J.W. McGinity, R.O. Williams III, Dissolution Enhancement of Itraconazole by Hot-Melt Extrusion Alone and the Combination of Hot-Melt Extrusion and Rapid Freezing— Effect of Formulation and Processing Variables, Mol. Pharm., 11 (2013) 186–196.

[5] C. Liu, Z. Chen, Y. Chen, J. Lu, Y. Li, S. Wang, G. Wu, F. Qian, Improving oral bioavailability of sorafenib by optimizing the "Spring" and "Parachute" based on molecular interaction mechanisms, Mol. Pharm., 13 (2016) 599–608.

[6] D.P. Medarević, K. Kachrimanis, M. Mitrić, J. Djuriš, Z. Djurić, S. Ibrić, Dissolution rate enhancement and physicochemical characterization of carbamazepine-poloxamer solid dispersions, Pharm. Dev. Technol., 21 (2016) 268–276.

[7] T. Xie, L.S. Taylor, Dissolution performance of high drug loading celecoxib amorphous solid dispersions formulated with polymer combinations, Pharm. Res., 33 (2016) 739–750.

[8] G.B. Park, H. Yoon, J.W. Bae, Y.U. Kim, D.Y. Jeon, J.E. Song, D. Lee, G. Khang, Release behavior of cilostazol according to the fabrication methods and ratio of HPMC/PVP, Macromol. Res., 21 (2013) 971–976.

[9] M.T. Davis, C.B. Potter, M. Mohammadpour, A.B. Albadarin, G.M. Walker, Design of spray dried ternary solid dispersions comprising itraconazole, soluplus and HPMCP: Effect of constituent compositions, Int. J. Pharm., 519 (2017) 365–372.

[10] N. Ohyagi, K. Ueda, K. Higashi, K. Yamamoto, K. Kawakami, K. Moribe, Synergetic role of hypromellose and methacrylic acid copolymer in the dissolution improvement of amorphous solid dispersions, J. Pharm. Sci., 106 (2017) 1042–1050.

577640 Modeling Fluidized Bed Impregnation of Active Pharmaceutical Ingredients Onto Porous Excipients

Impregnation of active pharmaceutical ingredients (APIs) onto porous excipients offers several advantages to solid dosage pharmaceutical preparation over conventional formulation methods. Previous work has successfully introduced fluidized bed (FB) impregnation of APIs onto porous carriers as a robust manufacturing process with numerous benefits (easy to implement, excellent blend uniformity and dissolution kinetics, stabilization of amorphous APIs). This study aims to develop a simple, multi-scale modeling approach to FB impregnation. The model consists of three main parts: 1) an energy/mass balance on the FB and an evaporative flux calculation, 2) an impregnation model for a single particle and 3) a drying model for a single particle. The model allows the calculation of some important process parameters (e.g. product temperature, maximum liquid spray rate) for given drying conditions. These are important process parameters needed to achieve successful impregnation. The model accounts for convection, diffusion and crystallization within the particle. The main characteristic is the 2D modeling of the liquid and API distributions within the particle during impregnation and drying. Several cases are presented to illustrate the effect of some material properties (fast crystallization, slow diffusion, low porosity) on the final API profile. Simulation results regarding API profiles of impregnation are in a very good agreement with experimental measurements carried out in our previous work.

560467 The Interaction of Solvent and Hypromellose Acetate Succinate Relate to Pharmaceutical Spray Dried Dispersion

An increasing number of new chemical entities suffer from poor solubility in aqueous media making drug delivery via oral solid dosage form challenging. To address this issue, formulations are increasingly relying on the use of amorphous solid dispersions of the active with an excipient. The excipient serves the function of stabilizing the active as amorphous in the solid state and acts as a nucleation inhibitor upon dissolution. The use of spray drying is one of the leading technologies for this application space. Similarly, hypromellose acetate succinate (HPMCAS) is one of the leading excipients used in spray drying. Although extensively used in spray drying, little is understood about the affinity of organic solvents for the HPMCAS polymer. The affinity of the spray solvent for HPMCAS can impact drying rate, secondary drying rate (removal of residual solvent in a secondary batch process), and the glass transition temperature of the spray dried dispersion. The objective of this study was to understand the desorption/adsorption of organic solvents to HPMCAS.

In this study, Dynamic Vapor Sorption (DVS-Resolution) was used to measure the methanol and acetone vapor sorption isotherms of the samples at 25 °C. HPMCAS was provided by The DuPont Nutrition & Health as AFFINISOL™ HPMCAS and was used as received. Anhydrous reagent grade of acetone and methanol were used in DVS study. HPMCAS polymers of various grades were analyzed over a relative partial pressure (P/P_0) range of 0-90%. The vapor program begins by exposing the sample to dry air to establish the dry mass. Then, the vapor concentration was increased in 10% P/Po steps to 90% P/Po followed by a final step at 95% P/Po. The vapor concentration is then decreased in a similar manner to accomplish a full sorption/desorption cycle. *In-situ* video images were collected at the end of each P/Po stage. Fickian, 1-D diffusion coefficients were determined at each step change in partial pressure. A spherical geometry was assumed for the particle shape and the EQPC d50 values were used as the average particle diameter. To make simple comparisons for the different vapors and samples, the diffusion values across the entire partial pressure range were averaged[1].

It was found that HPMCAS 912G has the highest residual solvent after exposing to acetone and methanol up to a partial pressure of 95% even though HPMCAS 716G has highest equilibrium absorption content during the sorption process. The residual acetone in HPMCAS is significantly higher than the residual methanol in HPMCAS. To understand such behavior, various fitting models were used and diffusion coefficients were calculated. It appeared that the methanol vapor has higher diffusion coefficients compared to acetone vapor of the same sample. This can be explained by the higher affinity of acetone with acetate group than that of methanol. In comparing the three typical HPMCAS grades, diffusion rates for acetone are as follows: HPMCAS 126G > HPMCAS 716G > HPMCAS 912G, while the diffusion rates for methanol are as follows: HPMCAS 126G > HPMCAS 912G > HPMCAS 716G. This can also be explained by the higher affinity of acetone and methanol from higher concentration of acetate of HPMCAS. Analysis using various fitting model indicated that HPMCAS has a much stronger interaction with acetone than with methanol and water. The acetone content to saturate the monolayer is larger than that of methanol.

In conclusion, HPMCAS has higher acetone uptake and retention compared to methanol and water. The methanol vapor has higher diffusion coefficients compared to acetone vapor on the same grade of HPMCAS. The understanding of the interaction between organic solvent and HPMCAS excipients will help formulation scientist to improve spray dry dispersion process.

References

1. Dynamic Vapour Sorption for Organic Solvent Sorption by Surface Measurement Systems. https://azom.com/article.aspx?ArticleID=5181

560620 Effect of Preparative Methods on the Polymorph Found in Crystalline Solid-Dispersions of Flufenamic Acid

A solid dispersion is a strategy that embeds active pharmaceutical ingredients (APIs) in a polymeric carrier: (a) to improve the solubility of poorly water-soluble APIs and more recently (b) as an alternative solid dosage formulation approach for continuous manufacturing (e.g. hot melt extrusion). Three main solid dispersion preparation techniques were employed: *(i)* fusion, *(ii)* solvent-evaporation, and *(iii)* solvent fusion methods. Although these techniques are also widely recommended as a pretreatment step for phase diagram determination, very little examples exist in the literature concerning the effect of these methods on the resulting polymorphic form of the API once the solid dispersion is obtained. Polymorphism is a critical quality attribute that allows molecules to exhibit multiple crystalline phases, which may affect the physicochemical properties (e.g. solubility, bioavailability, melting point). In this study, the influence of these preparation methods was studied for flufenamic acid (FFA), a polymorphic API, and polyethylene glycol (PEG 10,000), as polymeric carrier. The physical mixtures and solid dispersions were characterized by powder X-ray diffraction (PXRD), Fourier transform infrared spectroscopy (FTIR), differential scanning calorimetry (DSC), and hot stage microscopy (HSM). The results show that the fusion method, applying fast and slow cooling profiles, lead to concomitant formation of FFA forms I and III. On the other hand, the solvent evaporation and solvent fusion methods, applying slow and fast evaporation, both lead to FFA form III. Moreover, the results show that the preparation methods have a significant influence on the phase diagrams determined, which are often used to justify the thermodynamic design space of manufacturing processes. Consequently, when employing compounds prone to display polymorphism, choosing a solid dispersion preparation method that enables polymorphic control is crucial to ensure *(i)* the accurate determination of the phase diagram and *(ii)* the quality attributes of the formulation in crystalline solid dispersions.

565418 Methodology to Rapidly Screen Lyophilization Formulations

Formulation screening can take significant time and resource to identify viable excipient combinations. The proposed methodology allows for hundreds of formulations to be screened within a single lyophilization cycle to identify viable formulations. Manufacturability of the formulations was determined by evaluating the primary drying glass transition temperature (Tg') and commercial equipment limitations. The lyophilization cycle was determined via an automated end point using a pressure differential between the capacitance manometer and the Pirani gauge. A scoring criterion was then adopted to evaluate the appearance, dispersibility, and other quality critical attributes. The methodology resulted in a reduction in development time from months to less than two weeks.

562484 Towards Modelling the Morphology of Particles Obtained from Spray Dried Droplets

TOWARDS MODELLING THE MORPHOLOGY OF PARTICLES OBTAINED FROM SPRAY DRIED DROPLETS

Hassan Abdullahi [1], Christopher L. Burcham, [2] Thomas Vetter [1*]

[1] School of Chemical Engineering and Analytical Science, University of Manchester, Manchester, UK

[2] Eli Lilly & Company, Indianapolis, USA

Spray drying is used to form particles from droplets (clear liquid or a suspension of particles) in a hot gas stream. There is recent and renewed interest in this technology within the pharmaceutical industry as it presents a way to co-formulate active pharmaceutical (APIs) with excipients [6]. By manipulating drying conditions (gas temperature, relative humidity, ratio of gas flow rate to liquid/suspension flow rate) spray drying also allows tuning the external shape, apparent density, etc. of the particulate product. Figure 1 shows an overview of possible morphologies that can be obtained from such a process [3].

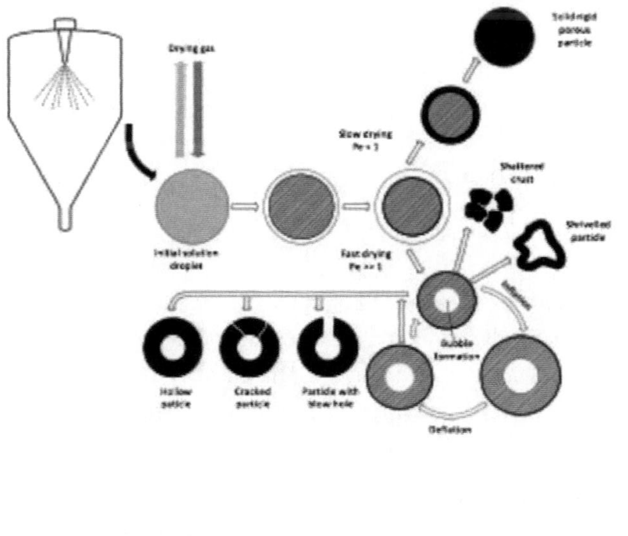

Figure 1: The morphological evolution of a droplet during drying. The dimensionless Peclet number (the ratio of convective transport to diffusive transport) is used to describe the various drying routes.

While there has been progress in describing the formation of such complex structures during drying in recent years [1][5][6], there is scope to develop a robust mechanistic model coupling the phenomena of evaporation, mass transfer and particle formation. Such a model is useful in the prediction of particle morphologies and ultimately product properties. The aim of this work is to introduce such a mechanistic model based on population balance equations that allows describing several of the above morphologies. As a first step, we describe a situation where droplets are saturated with liquid. The droplet is heated to the wet-bulb temperature where shrinkage occurs at a near constant rate and particles are formed within the droplet as illustrated in Figure 1 (first drying stage)[4]. In the second stage, shrinkage stops as a locked shell is formed around the droplet. The evaporation rate falls continuously as the thickness of the shell increases [2][3]. In the end, the balance of various forces determines the type of particle formed [3]. An accurate description of such a process requires detailed formulation of internal and external mass and energy conservation equations.

By taking a differential control volume for the homogeneous droplet (Figure 2) and assuming a radially symmetrical droplet, we can describe the nucleation/growth of particles as well as the advection/diffusion of liquid and solids within the droplet during drying.

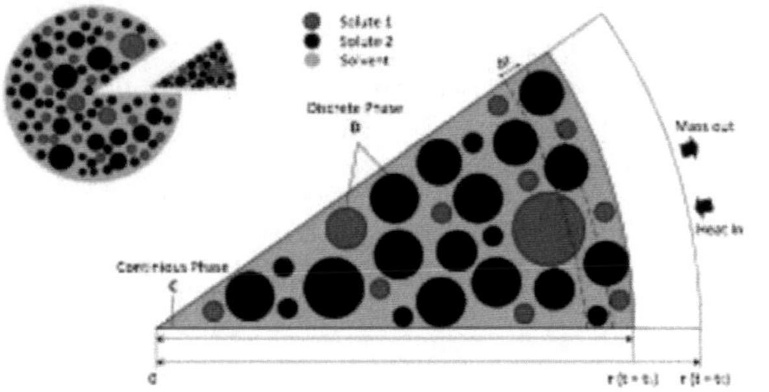

Figure 2: A conceptual control volume of a droplet drying during the first drying stage. Energy is transported to the droplet from the drying gas while mass is transported away from the surface.

A material balance for the liquid phase is introduced (Equation 1), where the change in liquid concentration **c** of component i (for i = 1Én-1) within the droplet is described by the rate of advection and diffusion of solute molecules, the rate of nucleation of particles and the growth rate of particles within the droplet respectively. The number density distribution of particles $\mathbf{N_i}$ and temperature distribution are described similarly by Equation 2 and 3 respectively [4][5].

$$\frac{\partial c_i}{\partial t} = -\frac{1}{r^2}\frac{\partial}{\partial r}\left(r^2 u c_i\right) + \frac{1}{r^2}\frac{\partial}{\partial r}\left(r^2 \sum_{j=1}^{n-1} D_{ij}^{(C)} \frac{\partial c_j}{\partial r}\right) - J_i \delta(L_{nuc}) k_v \rho_i (1-\varepsilon) L_{nuc}^3 - \rho_i k_v \frac{\partial}{\partial t}\int_0^\infty N L^3 dL$$

$$\frac{\partial N_i}{\partial t} = -\frac{1}{r^2}\frac{\partial}{\partial r}\left(r^2 u^{(D)} N_i\right) + \frac{1}{r^2}\frac{\partial}{\partial r}\left(r^2 \sum_{j=1}^{n-1} D_{ij}^{(D)} \frac{\partial N_j}{\partial r}\right) - \frac{\partial (G_i N_i)}{\partial L}$$

$$\frac{\partial}{\partial t}\left(\rho C_{pl} T_d\right)^{(C)} = -\frac{1}{r^2}\frac{\partial}{\partial r}\left(r^2 u \rho C_p T_d\right)^{(C)} + \frac{1}{r^2}\frac{\partial}{\partial r}\left(r^2 \lambda^{(C)} \frac{\partial T_d}{\partial r}\right) - \sum_{j}^{n-1} \rho_j^{(D)} \Delta h_{crys,j} k_v \frac{\partial}{\partial t} \int_0^{\infty} N_j L^3 dL$$

The equations are complemented by appropriate boundary conditions. In the second stage, the equations are similar and are only adapted to describe the presence of a shell region. Solving the above equations yields data for the evolution of particle properties such as size, shell lock time and porosity. Results from simulations are validated using an extensive data set obtained from the controlled drying of single droplets containing varying compositions of mannitol and paracetamol in a solvent using acoustic levitation. In addition, Scanning electron microscope (SEM) imaging yields external morphology data and X-ray-computed tomography allows visualisation of the internal structure of individual particles and hence yields comparable data of shell thickness. This in fact represents the first time such a shell thickness analysis is applied to single droplet drying. In the end, we demonstrate key fundamental understanding of droplet drying and particle formation through combined mechanistic modelling and detailed experimental methodologies. The resultant model can be applied to the simulation of typical spray drying processes when combined with knowledge of the flow pattern in the processing vessel.

References

1) Liang, H., Minoshima, H.,Matsushima, K., Shinohara, K. Basic model of spray drying granulation. Journal of Chemical Engineering of Japan 2001, 34 (4), 472–478.

2) Maurice, U., Mezhericher, M., Levy, A. and Borde, I. (2015). Drying of Droplets Containing Insoluble Nanoscale Particles: Second Drying Stage. *Drying Technology*, 33(15-16), pp.1837-1848.

3) Mezhericher, M., Levy, A. and Borde, I. (2011). Modelling the morphological evolution of nanosuspension droplet in constant-rate drying stage. *Chemical Engineering Science*, 66(5), pp.884-896.

4) Mezhericher, M., Naumann, M., Peglow, M., Levy, A., Tsotsas, E. and Borde, I. (2012). Continuous species transport and population balance models for first drying stage of nanosuspension droplets. *Chemical Engineering Journal*, 210, pp.120-135.

5) Seydel, P., Blšmer, J. and Bertling, J. (2006). Modeling Particle Formation at Spray Drying Using Population Balances. *Drying Technology*, 24(2), pp.137-146.

6) Vehring, R. (2007). Pharmaceutical Particle Engineering via Spray Drying. *Pharmaceutical Research*, 25(5), pp.999-1022.

571697 Predicting the Stability of Biotherapeutics in Liquid and Solid Formulations

A key barrier in the manufacturing of new biotherapeutics is the formation of protein aggregates, which can lead to increased production costs, loss of the biological function, and immunogenic responses from patients. To detect the presence of aggregates and ensure protein stability upon storage, efficient pre-formulation screening methods are required. Many of these biotherapeutics are lyophilized (freeze-dried) to avoid or slow down the degradation of the product; however the stability for freeze-dried cakes is highly critical too as damaged freeze-dried products cannot be delivered to patients. Therefore, it is of great interest to investigate methods that are capable of predicting changes in mechanical and physical stability of fragile freeze dried cakes.

Self-Interaction Chromatography (SIC) has been shown to be an accurate and automated high-throughput technique to predict protein stability. In this study the efficiency of SIC as a screening tool in the biopharmaceutical industry, was investigated. Common formulation conditions were investigated using a therapeutic mAb and BSA. Additionally, the stabilising effects of NaCl, Tween 80®, L-arginine and trehalose were examined in the range of 0-250 mM, 0-0.1% v/v and 0-200 mM, respectively. The protein-protein interactions for each formulation condition were quantified by calculating the osmotic second virial coefficient (B_{22}), a fundamental physico-chemical property that describes protein-protein interactions in solution. These results were later compared with the Z-average sizes obtained with Dynamic Light Scattering (DLS). A number of these conditions were lyophilized afterwards to see if the conditions that had shown to provide the greatest stability for the proteins in solution also were suitable for use in solid formulation, as freeze-dried cakes.

SIC showed an increase in the stability of the mAb with the ionic strength and arginine concentration, but no major effect with the addition of polysorbate. BSA showed repulsive protein-protein interactions in all the range of concentrations analysed, as expected due to its high stability in solution, which could be increased with the addition of stabilisers, such as trehalose. Even in the freeze-dried cakes higher concentration of trehalose led to an increased Young's modulus, and therefore higher mechanical stability.

Based on these results it could be seen that a protein-protein interaction technique such as SIC could be used as an early predictor for formulation behaviour both in liquid and solid states.

569669 Two-Phase Optimal Design of Pharmaceutical Separation: Sampling-Based Uncertainty Analysis and Reinforcement Learning

Pharmaceutical industry has been consistently growing up because of diverse factors such as the increased size of the global aging population, drug affordability, government policies, and supply-side factors. It has reached for a number of challenges including a changing health care landscape, expiring patents and generic competition, pricing pressures, and a persistent economic slowdown. One of the feasible solutions to overcome the current situation is to develop an optimal and smart separation process of pharmaceuticals in order to improve the performance and maintain the system flexibility under the uncertainty. Liquid-liquid extraction (LLE), which is also known as solvent extraction, is the main process that is much employed during the downstream separation. In the LLE process, a feed stream of two or more components is faced with a second liquid stream containing solvent. Therefore, the use of proper solvents and the implementation of an optimal operating design for the LLE process are inevitably important issues to keep obtaining high yield of pharmaceuticals. This research aims to construct an advanced LLE framework including two-phase strategy. The first phase describes the optimal design of the LLE process based on thermodynamics and transport properties and uncertainty analysis. The second phase presents a smart operating planning by reinforcement learning according to results from the first stage. In the reinforcement learning part, a deep Q-learning algorithm to control the LLE process is investigated. Each stage in the purpose of this study is underpinned by several internal steps as the follows. For the first step in the first stage, electrolyte non-random two-liquid segment activity coefficient (eNRTL-SAC) model is employed to estimate segment parameters and activity coefficients of a target pharmaceutical. eNRTL-SAC model is reminiscent of NRTL-SAC model and additionally uses an electrolyte conceptual segment parameter. Segment parameters and activity coefficients are explicitly utilized to perform solubility modelling. The next step is that liquid-liquid extraction column as a separation process is considered and distribution coefficients are calculated based on results of thermodynamics properties; thereafter, uncertainty analysis is conducted to identify optimal solvents and operating conditions. Reinforcement learning in the second stage technically aims to optimize an operating structure using the results from the first stage. For the first step in the second stage, an agent and an environment have to be precisely defined, each of which corresponds to the controller and the LLE process respectively in this research. The agent in the second step takes an action to the environment whereby the environment operates the designed process model according to the action. The environment in the last step discharges states and rewards to the agent to be updated by deep learning algorithms. Actions, states, and rewards can be constructed in various ways as the relationship in between the agent and the environment should be considered.

Property process modelling and uncertainty analysis

Screening of an optimal solvent or solvents mixture is an inherent procedure to formulate an efficient pharmaceutical process. Solubility modelling is most often used as a criterion of solvent screening and requires a knowledge of a proper thermodynamic model. As aforementioned, eNRTL-SAC model is applied to this study, whereby regressed segments parameters as well as activity coefficients are able to be disclosed. Activity coefficients are determined by regressed segment parameters and a distribution coefficient, which is also known as a partition coefficient and means the ratio of concentration of a compound in a mixture at equilibrium, is calculated by activity coefficients. Extraction factor consists of the distribution coefficient, the flow rate of the feed stream in the LLE process, and the flow rate of the solvent stream in the LLE process, and Kremser-Souders-Brown theoretical stage equation directly uses the distribution coefficient and the extraction factor to estimate the number of stages in the extraction column that ultimately affects the operating costs. Parameter uncertainties and an influence thereof are implemented by uncertainty analysis in order to give a decision maker the guideline of an optimal process design. Sampling-based model solution prior to uncertainty analysis is conducted as means to explore and identify the optimal process variables thereby Monte Carlo method is exploited to illustrate the influence of parameter uncertainties.

Off-policy control with deep Q-leaning

General reinforcement learning basically encompasses five components, each of which is state, state transition probability matrix, action, reward, and discount factor. Action is determined by a policy included in the agent. Conventional objective function in reinforcement learning is to maximize a reward function according to the policy. Therefore, understanding of the optimal policy is the main purpose in reinforcement learning. Off-policy strategy is associated with the concept of exploration and exploitation. In the case of on-policy strategy, a specific policy can be upgraded through agent training, on the other hand, it has the weakness of exploring solutions available in the vicinity of the current policy. At the expense of consuming computational costs, off-policy can flexibly explore distinct candidates and approach much closer to the global solution. The agent is designed by deep learning algorithms to suggest the optimal action every single episode. State information and value of reward from the environment play a role of input variables of the agent and suitable hyperparameters of the deep learning algorithm have to be set up to improve the efficiency of the agent.

This study purposes to design a novel extraction process in pharmaceutical industry based on uncertainty analysis and reinforcement learning. Thermodynamics property modelling, solubility modelling, solvent screening, and uncertainty analysis are consecutively examined to design the feasible LLE process of pharmaceuticals. Afterwards reinforcement learning is considered to suggest the optimal operating planning based on the results from uncertainty analysis. Moreover, the proposed two-phase framework of the LLE process would be extended to other processes in downstream process.

573815 Computer-Aided Design of Solvent Blends for the Cooling and Anti-Solvent Crystallisation of Active Pharmaceutical Ingredients

The majority of pharmaceutical products are delivered in the solid form, such as tablets or via aerosols. As such, crystallisation remains the process of choice for the manufacturing of active pharmaceutical ingredients (APIs) – typically via solvent-based techniques. Thermodynamically, solvent choice will affect the solubility, and thus crystal yield of the API, but the effects of this decision on transport properties, process and product safety, and the final crystal morphology – which can impact downstream processing and ultimately the *in vivo* efficacy of the drug – are all driving forces for selection. For novel drug molecules, solvent selection is often performed via time-consuming and expensive experiments [1], limiting the full range of solvent mixtures and process conditions from being explored. Consequently, fast development of manufacturing processes that achieve high standards in terms of productivity, quality and environmental performance can be difficult.

Over the last decades, Computer Aided Molecular Design (CAMD) has been developed to overcome such difficulties, with the ultimate aim of guiding experiments towards optimal candidate molecules [2,3]. To reduce the number of potential solutions in such designs, a decomposition-based approach has been proposed [4], whereby smaller subproblems are posed and then solved successively. By fixing process conditions and utilising a single crystallisation technique, this has been applied to crystallisation design [5]. Indeed, most existing methodologies focus on the selection of single solvents, utilising a specific crystallisation technique, whereas more integrated problems, in which cooling and anti-solvent effects are combined and solvent mixtures are considered, have not yet received significant attention. This is likely due to the complexity of formulating and resolving a mixed-integer optimisation problem to represent these design choices; such problems result in challenging non-convex design spaces for the continuous variables, in addition to the combinatorial solution space.

More recently, the use of Generalised Disjunctive Programming (GDP) [6] within the Computer Aided Mixture/blend Design (CAM^{b}D) framework has been proposed to design optimal solvent mixtures that maximise solubility of an API [7,8]; optimal mixtures were shown to outperform pure compounds. This strategy can be used to use to solve more integrated problems, enabling the user to explore the relative merits of cooling crystallisation, anti-solvent crystallisation and combined approaches concomitantly and solvent mixtures can be considered throughout [9], but this type of approach has not yet received significant attention. We investigate the formulation and solution of such an integrated crystallisation design problem, whereby the optimal process temperature, selection of solvent and anti-solvent molecules, and design of the solvent mixture composition, can be identified simultaneously – resulting in a mixed integer nonlinear problem (MINLP). Based on an equilibrium model, the formulation considers solvent use and crystal yield, but can be extended to account for other factors. To calculate the relevant thermodynamic properties, we use the SAFT-γ Mie equation of state, which has been shown to provide high quality predictions of solubility [10].

This design formulation is implemented in gPROMS and successfully applied to a case study on the crystallisation of lovastatin, showing that this more general approach to crystallisation design can be used effectively to optimise desired metrics, exceeding solutions that would otherwise be achieved by decomposition-based methods. Furthermore, it is shown that coupling of cooling and anti-solvent crystallisation can achieve greater crystal yields, whilst reducing the mass of solvent required for the process, further emphasising the usefulness of this approach, and of solvent mixtures in general.

[1] Brown CJ, McGlone T, Yerdelen S, Srirambhatla V, Mabbott F, Gurung R, et al. Enabling precision manufacturing of active pharmaceutical ingredients: workflow for seeded cooling continuous crystallisations. Molecular Systems Design & Engineering. 2018.

[2] Achenie L, Venkatasubramanian V, Gani R, editors. Computer aided molecular design: theory and

practice. Elsevier. Amsterdam, The Netherlands; 2003.

[3] Gani R. Chemical product design: challenges and opportunities. Computers & Chemical Engineering. 2004;28(12):2441-57.

[4] Karunanithi AT, Achenie LE, Gani R. A new decomposition-based computer-aided molecular/mixture design methodology for the design of optimal solvents and solvent mixtures. Industrial & engineering chemistry research. 2005;44(13):4785-97.

[5] Karunanithi AT, Achenie LE, Gani R. A computer-aided molecular design framework for crystallization solvent design. *Chemical Engineering Science*. 2006;61(4):1247-1260

[6] Raman R, Grossmann IE. Modelling and computational techniques for logic based integer programming. Computers & Chemical Engineering. 1994;18(7):563-78.

[7] Jonuzaj S, Akula PT, Kleniati P, Adjiman CS. The formulation of optimal mixtures with generalized disjunctive programming: A solvent design case study. AIChE J. 2016;62(5):1616-1633

[8] Jonuzaj S, Adjiman CS. Designing optimal mixtures using generalized disjunctive programming: Hull relaxations. Chemical Engineering Science. 2017; 159 106-130.

[9] Watson OL, Galindo A, Jackson G, Adjiman CS. Computer-aided design of solvent blends for the cooling and anti-solvent crystallisation of ibuprofen. European Society of Computer-Aided Process Engineering. 2019, submitted

[10] Hutacharoen P, Dufal S, Papaioannou V, Shanker RM, Adjiman CS, Jackson G, Galindo A. Predicting the solvation of organic compounds in aqueous environments: from alkanes and alcohols to pharmaceuticals. Industrial & Engineering Chemistry Research. 2017;56(38):10856-76.

568701 Two-Dimensional Moisture Content and Size Evolution during Fluid Bed Drying As Part of a Continuous Pharmaceutical Manufacturing Process Using NIR-CI

The pharmaceutical industry is currently making the transition from batch to continuous manufacturing for the latter method's benefits of e.g. process flexibility, less waste of materials and improved control of the product. For the manufacturing of oral solid dosages, the innovative ConsiGma-25™ continuous wet granulation line applies twin-screw granulation for production of granules for qualitative tableting. To enable downstream processing, the line uses a semi-continuous fluid bed dryer which reduces the excess liquid in the wet granules coming from the twin-screw wet granulation unit. This process is robust, yet features a multitude of factors that make the dryer system complex. Firstly, the dryer distributes a continuous stream of wet granules over its six drying cells. The cells are sequentially filled for a fixed amount of time, after which the granules are dried and discharged. Due to the gradual filling of each cell, the wet granules start drying at different points in time and the drying regime changes over time. Furthermore, critical process parameters such as the inlet air flow rate, inlet air temperature and cell filling time influence the fluidization and evaporation behavior of the material in the dryer cell. Finally, also the granule size has a major influence on the drying behavior (De Leersnyder et al., 2018).

The study of this complex process, with the aim of Design Space determination in the Quality-by-Design paradigm of the pharmaceutical manufacturing industry (ICH, 2009), is therefore best performed by means of mechanistic modelling. In this way the various interacting phenomena such as fluidization behavior, single particle-level drying and conditions of the drying air can be captured in order to predict the drying behavior of the key product. In models with many interacting factors such as fluid bed drying, sufficiently detailed data is necessary to calibrate and validate the model, to ensure that the final model is the correct representation of the process. As such, these drying models do predict a moisture content distribution in the material (Ghijs et al., 2019; Peglow et al., 2007). These arise from fluidization behavior, the continuous filling of a dryer cell and the heterogeneity in granule size encountered in pharmaceutical granules. Yet in practice, the only available product data is the average moisture content of the granules, meaning that the model cannot compare its output to reality in full extent. This makes that some parts of the model might be insufficiently validated with the reality, and therefore are not able to extrapolate to other process conditions or systems. With two-dimensional moisture content and size distribution data, these models instead would be able to compare their output one-to-one to experimental data, and more components of the model (effect of fluidization, granule size, air conditions) can be validated, which enhances its predictive power.

Therefore, in this study an at-line analytical method was developed to simultaneously determine the moisture content as well as size of individual pharmaceutical granules to construct a 2D distribution. Spatially distributed moisture content measurements of the granules were obtained by dispersing the granules on a conveyor belt (ENP, Hjälteby, Sweden) with a Retsch DR 100 vibrating feeder (Retsch, Haan, Germany), which passed the granules underneath a hyperspectral camera: a near-infrared chemical imaging (NIR-CI) device (VLNIR with OLES56 lens, Specim, Finland). Method validation experiments were performed by measuring multiple samples of granules first with the NIR-CI set-up, immediately followed by a reference LOD instrument (Mettler LP16, Mettler-Toledo, Zaventem, Belgium). By drying wet granules (i) in open air, (ii) in a 50°C oven and (iii) in a fluid bed dryer (Glatt GPCG1), granules with different levels of moisture content were targeted for method validation. All hyperspectral data was then processed in MATLAB (MathWorks®, Natick, Massachusets, USA) using the HYPER-tools package (Mobaraki & Amigo, 2018). The obtained spectra were processed in order to analyze the granules captured on the image. To relate the granules' spectra to the actual moisture content, the average spectrum of the granules on each image was obtained and compared to the LOD measurement of the sample. A partial least squares (PLS) model was used to derive the moisture content from the spectra. Good model predictions were obtained), and thus the spectra could accurately indicate the actual moisture content, and this even for granules larger than 3 mm.

In the same setup, the size of each granule on the image is determined, and by means of the PLS-model the moisture content of each granule is found, resulting in a two-dimensional moisture content and size distribution. These distributions capture the spatial distribution of the moisture in the key product. By analyzing the distribution at different drying times and process conditions, highly accurate information is obtained on the evolution of moisture content of the material. In conjunction with a 2D population balance model, this will allow to gain more in-depth knowledge of the drying process. The latter is planned as next step and is out of scope of this abstract which focuses on the data collection.

References

De Leersnyder, F., Vanhoorne, V., Bekaert, H., Vercruysse, J., Ghijs, M., Bostijn, N., ... De Beer, T. (2018). Breakage and drying behaviour of granules in a continuous fluid bed dryer: Influence of process parameters and wet granule transfer. *European Journal of Pharmaceutical Sciences*, *115*, 223–232. http://doi.org/10.1016/j.ejps.2018.01.037

Ghijs, M., Schäfer, E., Kumar, A., Cappuyns, P., Van Assche, I., De Leersnyder, F., ... Nopens, I. (2019). Modeling of Semicontinuous Fluid Bed Drying of Pharmaceutical Granules With Respect to Granule Size. *Journal of Pharmaceutical Sciences*, 1–8. http://doi.org/10.1016/j.xphs.2019.01.013

ICH. (2009). *International Conference on Harmonisation (ICH) of Technical Requirement for Registration of Pharmaceuticals for Human Use, Pharmaceutical Development, Q8 (R2).*

Mobaraki, N., & Amigo, J. M. (2018). HYPER-Tools. A graphical user-friendly interface for hyperspectral image analysis. *Chemometrics and Intelligent Laboratory Systems*, *172*(October 2017), 174–187. http://doi.org/10.1016/j.chemolab.2017.11.003

Peglow, M., Kumar, J., Heinrich, S., Warnecke, G., Tsotsas, E., Mörl, L., & Wolf, B. (2007). A generic population balance model for simultaneous agglomeration and drying in fluidized beds. *Chemical Engineering Science*, *62*(1–2), 513–532. http://doi.org/10.1016/j.ces.2006.09.042

571347 Continuous Mixing Technology: Design Space Exploration with Discrete Element Simulations

Continuous Mixing Technology: Design Space Exploration with Discrete Element Simulations

Peter Toson[1], Pankaj Doshi[2], Eva Siegmann[1], Johannes Khinast[1,3], Daniel Blackwood[2], Ashwinkumar Jain[2], Alexandre Bonnassieux[2], Dalibor Jajcevic[1]

[1] Research Center Pharmaceutical Engineering, Inffeldgasse 13, 8010 Graz, Austria

[2] Worldwide Research and Development, Pfizer Inc. Groton CT 06340

[3] Graz University of Technology, Institute of Process and Particle Engineering, Inffeldgasse 13, 8010 Graz, Austria

Keywords: Pharmaceutical Manufacturing; Continuous Mixing Technology; Process Design & Development

Continuous manufacturing has many advantages over batch processing, including smaller footprints, faster turnaround times, and better control of process parameters and product quality. A minimal setup for a continuous tableting line consists only of three unit operations (direct compression): feeding of the individual components, mixing and tableting. Many critical quality attributes (CQAs) of the final tablet depend on the mixing step: the mixing quality directly influences the content uniformity of the tablet, the extent of lubrication influences the tensile strength of the final tablet.

The aim of this work is an in-depth analysis of a vertical continuous mixing device termed Continuous Mixing Technology (CMT, see Fig. 1). The vertical design allows to control the impeller speed (and thus the shear rates) independently from the hold-up mass (and thus mean residence time). The same number of blade passes can be generated with different impeller speed and hold-up mass combinations. Thus, the vertical design of the CMT opens up more degrees of freedoms when it comes to process design compared to the traditional horizontal mixing devices, where the screw speed directly influences both shear rate and mean residence time.

Discrete Element Method (DEM) simulations have been performed using the commercial software package XPS (Extended Particle System). The simulations model production-scale processes with a hold-up mass of up to 1000g and mass throughputs ranging from 30kg/h to 40kg/h with 5.7million particles. The model has been validated with tracer experiments in a previous study [1].

A wide range operating conditions have been examined with DEM simulations. Simulation results elucidate the powder mixing dynamics as captured by residence time distribution throughout the operating space. It helps in identifying most optimal operating space. Thus virtual design space exploration with DEM simulations has been proving to be a valuable tool in process development.

References:

[1] Toson et al., Int. J. Pharm. 552, 1–2 (2018), 288–300. doi:10.1016/j.ijpharm.2018.09.032.

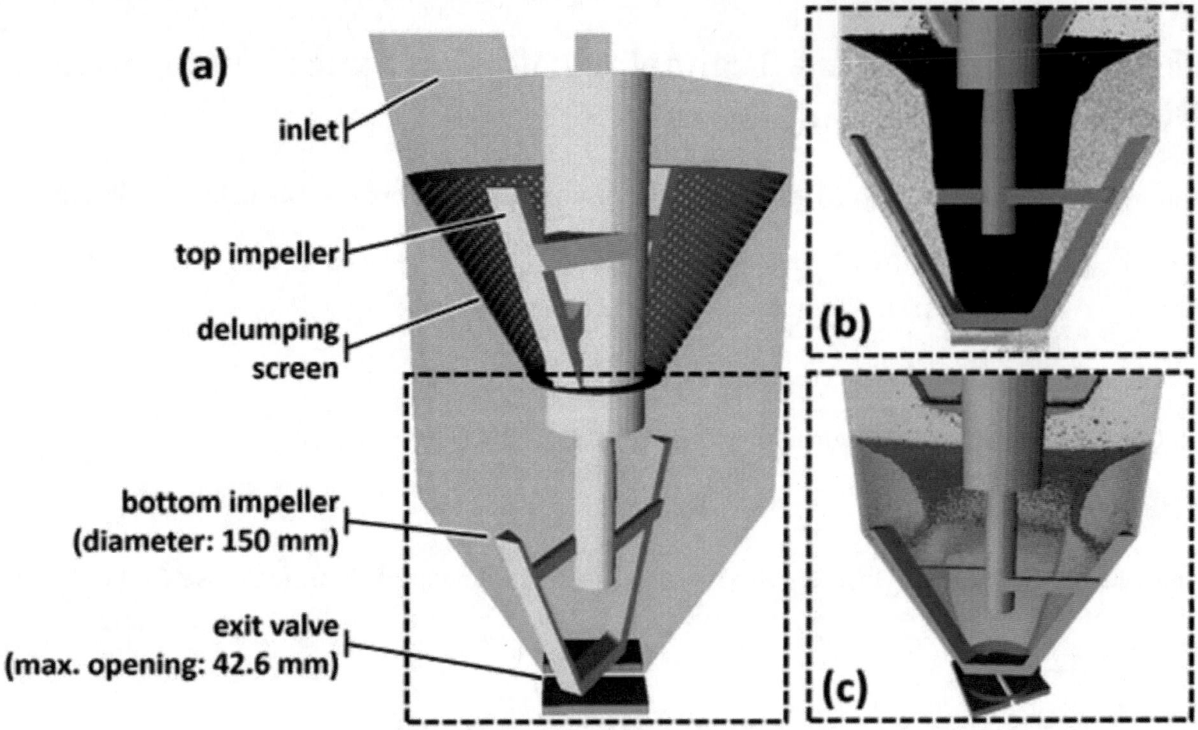

Figure 1. (a) CMT overview. (b) A cut through the CMT to reveal the powder bed shape in the mixing zone at 350 rpm. (c) Tracer particles (red) used to determine the residence time distribution. Snapshot taken 2s after insertion of the tracer impulse.

573794 Validation of Combined PBM-RTD Technique to Simulate Continuous Wet Granulation Systems for Pharmaceutical Manufacturing Processes

Wet granulation is a key unit operation in the pharmaceutical industry employed in the production of solid oral dosage formulations through the continuous manufacturing route. In an effort to predict the outlet compositions of the particles of a continuous granulator, the study of residence time distributions (RTDs) is crucial in order to predict the effects of variabilities in the inlet parameters such as feeder fluctuations and blend inhomogeneity on the outlet granule particle sizes and compositions. Population Balance Models (PBMs) are developed and extensively employed in modeling wet granulation systems in order to track the particle size distributions (PSDs) and porosities of the particles [1].

An initial experimental study on a twin screw granulator was performed using a low API dose blend across a DOE consisting of variation in powder feed rate, liquid content and screw rotation speed. The RTD pulse experiments were carried out on the DOE and exit concentration was measured in real time using an inline near infrared probe (NIR). The residence time profile was fitted to a theoretical model consisting of tank-in-series, plug flow fractions, parallel streams, etc. Moreover, empirical mathematical correlations were obtained for the mean residence time (MRT) with respect to the operating DOE conditions. These studies and results were in line with previously published results in literature [2, 3].

A granulation PBM with RTD mechanics incorporated to it would track the flow of key species such as API thereby quantifying the processes and attributes of the outlet granules completely. The model would provide the researcher with the potency of the API across granule sizes. The goal of this proposed study would be to experimentally validate the developed models. The results of the study would help in developing real-time control strategies that would ensure granules with critical quality attributes (CQAs) consistent over time during the production of the granules.

[1] A. Chaudhury, 'Mechanistic modeling, simulation and optimization of wet granulation processes', Rutgers University - Graduate School - New Brunswick, 2015.

[2] A. Kumar et al., 'Conceptual framework for model-based analysis of residence time distribution in twin-screw granulation', European Journal of Pharmaceutical Sciences, vol. 71, pp. 25–34, Apr. 2015.

[3] H. Y. Ismail, M. Singh, S. Darwish, M. Kuhs, S. Shirazian, D. M. Croker, M. Khraisheh, A. B. Albadarin, and G. M. Walker, 'Developing ann-kriging hybrid model based on process parameters for prediction of mean residence time distribution in twin-screw wet granulation,' Powder Technology, vol. 343, pp. 568-577, 2019.

566521 Computational Modeling of Liposome Formation in a Continuous Jet Flow Process Using CFD and MD

Purpose: Liposomes are spherical vesicles composed of a bilayer and/or a concentric series of multiple bilayers separated by aqueous compartments formed by amphipathic molecules such as phospholipids that are commonly used as carriers for drug delivery purposes. The current study focused on the continuous liposome processing system developed at UConn, where two liquid flows are mixed under a controlled manner through a coaxial jet in co-flow arrangement, of which the center flow consists of an ethanol solution of lipids and the exterior flow is an aqueous solution. It was suspected that the intermolecular forces among various components (e.g., ethanol-lipid-water) along with the unique fluid dynamics (e.g., jet flow) caused lipid aggregation and subsequent liposome formation. However, the underlying mechanism as well as the process dynamics are yet fully understood. Accordingly, we have carried out a multi-scale computational study of liposome formation in coaxial turbulent jet flow to probe the underlying mechanism and to quantitatively predict the formation of liposomes.

Methods: Both computational fluid dynamics (CFD), as a macro-scale simulation, and coarse-grained molecular dynamics (CG-MD), as a micro-scale investigation, have been conducted to not only reveal the detailed mechanism of liposome formation, but also implement multi-scale case studies for the process. CFD simulations were verified by comparison of flow pattern as well as formation temperature with experiments.

Results: The CG-MD simulations revealed that MARTINI force field (FF) could not capture a realistic behavior of lipids and cholesterol in ethanol solution due to lipids aggregation, which caused deviation from both the experimental results and all-atom MD simulations. Further optimization of MARTINI FF with reference to all-atom MD simulations was performed using a versatile object-oriented toolkit for coarse-graining applications (VOTCA). The multi-scale simulations were found to be most accurate when compared to the experimental data and trends. The CFD simulation results suggested that including heat of mixing in the energy equation would be desired to obtain a formation temperature comparable with the experiments.

Conclusions: Default MARTINI potential energy parameters require optimization with reference to an all-atom MD simulation for properly modeling lipids in ethanol solution. Discrete and continuum computational modeling of liposome formation were found to be two viable complementary approaches which cover micro-scale and macro-scale, respectively.

Acknowledgements: FDA Grant# 1U01FD005773-01.

Disclaimer: This article reflects the views of the authors and should not be construed to represent FDA's views or policies.

579336 A Big Data Analytics Workflow for Pharmaceutical Manufacturing Industry

Pharmaceutical manufacturing is a complex process, consisting of blending, granulation, tableting and other unit operations. The aim of pharmaceutical manufacturing is to reliably produce good quality drugs with high efficiency and low cost. However, due to lack of process understanding and control strategies, it is hard to avoid drug quality defects and production efficiency decreasing. Nowadays, many pharmaceutical manufacturing unit operations have been equipped with lots of sensors, controllers and even process analytical instruments, and this has enabled the rapid growth of pharmaceutical manufacturing industry big data. The vast data resource hides the process information and characteristics, so mining knowledge behind the manufacturing data is of vital importance.

In this study, we proposed a big data analytics workflow for pharmaceutical industry manufacturing to mine process knowledge, which helped to improve drug quality and production efficiency. The workflow contains five steps: problem definition, data acquisition, data preprocessing, data modelling and model application. The workflow provides an idea for pharmaceutical enterprises to collect, organize and analyze manufacturing data systematically. Meanwhile, a data analytics case study of a double effect evaporation process for herbal medicines was used for further illustration. In the case, the production efficiency of a double effect evaporator was observed to decrease during a long time. To handle this problem, the data of sensors, valves and instructions from 172 batches were collected. After data preprocessing, data modelling task was carried out. Correlation analysis was first applied to study the relationships among each single process variables. Next, multivariate data analysis methods were used to evaluate the profiles of all the batches: different process phases were identified using hierarchical clustering analysis according to the dissimilarities among the batches; the root causes of the phase transformations were also investigated through principal component analysis. The models above could be applied for on-line process monitoring, thus helping to enhance process understanding and to support manufacturing decision.

574446 An Efficient Data-Based Methodology to Identify the Design Space of Continuous Pharmaceutical Manufacturing Processes

Flowsheet modeling[1,2] is an increasingly accepted tool to simulate and improve continuous processes for solid oral dosage manufacturing. Advances in continuous manufacturing have led to increased focus on developing accurate and detailed models for the unit operations. Flowsheet models developed using the unit models to simulate dynamic behavior of the continuous process are computationally expensive. Strategies are needed to overcome this limitation and fully realize the potential of these models. Specifically, the models can be utilized to enable decision making during detection of off-spec product, risk assessment and process optimization.

Tools such as sensitivity analysis are established[3], and can be applied to flowsheet models to help identify the critical process parameters. While this helps in identifying variables where further research efforts can be focused on, the resulting problem is still high dimensional for implementing advanced analyses such as identification of design space. A design space is characterized by the range of variables within which the process satisfies equipment, quality and production constraints. Traditional methods to identify the design space require high sampling cost or closed form constraints, which limits its use with flowsheet models. Hence, effective strategies are required to handle problems typically encountered in pharmaceutical processes.

Current work uses surrogate based feasibility analysis method[4] that builds surrogate models to approximate the feasibility function that characterizes the maximum constraint violation. This strategy uses a modified expected improvement function to identify samples close to the feasible region boundaries and unexplored regions. Previous work published used kriging[4] and radial basis function[5] as the surrogate models, which suffered limitations for high dimensional problems. Specifically for flowsheet models, the design space was identified using a set of two dimensional problems, which ignores interactions between the variables. In this work, a novel artificial neural network (ANN) based methodology is used to identify the design space. An ANN based surrogate model is built to approximate the feasibility function through carefully identifying samples using an adaptive sampling strategy. The unexplored regions are identified using the modified expected improvement function and a variance estimator. The variance of an unexplored sample is estimated using a statistical technique known as Jackknifing[6,7]. The developed approach is utilized to effectively identify the design space for an integrated direct compaction as well as wet granulation line that simulates a plant scale operation.

References

1. Boukouvala F, Niotis V, Ramachandran R, Muzzio FJ, Ierapetritou MG. An integrated approach for dynamic flowsheet modeling and sensitivity analysis of a continuous tablet manufacturing process. *Computers & Chemical Engineering*. 2012;42:30-47.
2. Yoon S, Galbraith S, Cha B, Liu H. Flowsheet modeling of a continuous direct compression process. 2018:121-139.
3. Saltelli A, Annoni P, Azzini I, Campolongo F, Ratto M, Tarantola S. Variance based sensitivity analysis of model output. Design and estimator for the total sensitivity index. *Computer Physics Communications*. 2010;181(2):259-270.
4. Boukouvala F, Ierapetritou MG. Derivative-free optimization for expensive constrained problems using a novel expected improvement objective function. *Aiche J*. 2014;60(7):2462-2474.
5. Wang Z, Ierapetritou M. A novel feasibility analysis method for black-box processes using a radial basis function adaptive sampling approach. *Aiche J*. 2017;63(2):532-550.
6. Eason J, Cremaschi S. Adaptive sequential sampling for surrogate model generation with artificial neural networks. *Computers & Chemical Engineering*. 2014;68:220-232.
7. Jin Y, Li J, Du W, Qian F. Adaptive Sampling for Surrogate Modelling with Artificial Neural

Network and its Application in an Industrial Cracking Furnace. *The Canadian Journal of Chemical Engineering*. 2016;94(2):262-272.

565756 Condition Based Maintenance for Sensor Network Reliability in Continuous Oral Solid Dose Manufacturing

The advent of Industry 4.0 (Smart Manufacturing) in the 2010s, is a manifestation of cultural and mindset change across multiple industries towards the acceptance of digital tools in process design and operations, resulting in highly automated and data-rich systems to realize lean manufacturing. These process intelligence tools leverage cross-industry advances in manufacturing methods, data analytics, fault management, and implementation technologies, for sustainable process operations, safety, quality risk assessment, and enhanced customer relations. In pharmaceutical systems, such digitalization tools facilitate the implementation of continuous manufacturing and real-time release testing by tracking product quality in real-time, thereby enabling the release of life critical medication with reduced off-line quality testing compared to existing practices.

Continuous manufacturing of drug product is the outcome of targeted process intensification with systematic integration of product and process knowledge, instrumentation and automation systems, quality control protocols and real-time process management for a solids processing system handling pharmaceutical materials. Success in the technology necessitates appropriate infrastructure for Information and Operations Technology integration.

At the core of such integrated systems is sensor network analytics, consisting of reliable and accurate measurements from field devices combined with mature process understanding to provide good quality estimates of unmeasured process variables [1]. In continuous tablet manufacturing, four leading causes of instrumentation faults are observed, which affect monitoring robustness. These include communication failure, instrument configuration, calibration and instrument age. Such instrumentation failures could potentially cause a systemic failure of the CM process.

This work focuses on leveraging systems engineering methods for real-time operations management, in particular, condition-based maintenance to improve operational reliability by mitigating sensor network failures in drug product CM. We discuss the following: (i) an integrated data-driven and model-based framework for systematic sensor network fault monitoring, (ii) the configuration and implementation of condition-based maintenance in a tablet processing pilot plant at Purdue University [2].

The proposed maintenance management approach is comprised of three workflows which are linked in an integrated data-driven and model-based framework. First, sensor network analytics methods of statistical process monitoring, data reconciliation, and gross error detection are leveraged for real-time process data validation, ensuring measurement accuracy, detecting sensor faults, and estimating unmeasured variables. Proof of concept of these approaches have recently been demonstrated by the research group [3,4]. Second, gross errors are identified for their root cause for appropriate maintenance activities, such as corrective, preventive and predictive maintenance. Third, to reconnect a maintained sensor, guidelines such as ASTM D6122, E2968 are leveraged and customized for the process.

The integrated framework is implemented at Purdue University in a pilot scale continuous direct compression facility. An analytically redundant sensor network using process equipment sensors, inline and at-line sensors are networked with automation systems such as Emerson DeltaV, OSIsoft PI System and Applied Materials Smart Factory Rx. A hierarchical systems integration architecture closely following ISA-95 is set up to advance the practice of Quality-by-Design in continuous manufacturing towards Quality-by-Control and Industry 4.0 [4].

References:

[1] Bagajewicz MJ. Smart Process Plants: Software and Hardware Solutions for Accurate Data and Profitable Operations. McGraw-Hill, 2009

[2] S. Ganesh, G. Reklaitis et al. Leveraging the digital infrastructure for real-time operations management in continuous pharmaceutical manufacturing: Condition-based maintenance for sensor network robustness

(in preparation).

[3] M. Moreno, S. Ganesh, Y. Shah, Q. Su, M. Gonzalez, Z. Nagy, G. Reklaitis. Sensor Network Robustness using Model-based Data Reconciliation for Continuous Tablet Manufacturing. J Pharm Sci (2019).

[4] Q. Su, S. Ganesh, M. Moreno, Y. Bommireddy, M. Gonzalez, G. Reklaitis, Z. Nagy. A perspective on Quality-by-Control (QbC) in pharmaceutical continuous manufacturing. Comput Chem Eng (2019).

562229 Sensor Fusion with Irregular Sampling and Varying Measurement Delays

In multi-sensor fusion, several sources of information are combined in order to increase the estimation quality for the quantity of interest. This activity finds many applications from tactical missile defense to self-driving cars and the estimation of variables difficult to measure such as concentrations in chemical processes (Gao & Harris 2002, Kumar et al. 2007, Safari et al. 2014). In these situations, it is common to resource to laboratory analysis that provide more accurate measurements, but usually at slower sampling rates and with significant delays. Therefore, soft sensors and online analyzers are often introduced in the process to provide more frequent and updated measurements (Fatehi & Huang 2017).

Since Bar-Shalom & Campo (1986) introduced track-to-track fusion (TTF), several methods have been proposed based on the use of Kalman filter (KF) for sensor fusion. TTF makes use of a KF to track each sensor and fuses the sensors' state estimates into a new state estimate using a static linear estimation equation (Chen et al. 2003). Gao & Harris (2002) reviewed measurement fusion processes and proposed a modified track-to-track fusion (MTF). However, these methodologies rely on first-principle models and do not capitalize on prior knowledge about the relative quality of the sensors. Therefore, in this work we present a KF based sensor fusion scheme that integrates data collected from a dynamic soft sensor, an online analyzer and a laboratory analyzer. Fusion occurs in the absence of a first-principle model and accommodates a less reliable (but more frequent) online analyzer recurring to laboratory measurements. This scheme has the capability for managing asynchronous sampling rates and detecting faulty measurements while reducing false alarms and missed identifications.

Two fusion schemes are proposed and outlined: a more classic tracked Bayesian fusion scheme (TBF) and a novel modification of the track-to-track algorithm, designated bias-corrected track-to-track fusion (BCTTF). When comparing these two schemes, the new BCTTF has shown to be better in terms of prediction performance and alarm identification sensitivity. This algorithm produces also a less noisy signal. The analysis of the figures of merit leads us to recommend the use of BCTTF as a fusion algorithm under multi-rate sensor fusion conditions.

The presented case study arises from a real world chemical plant and comprises two trayed distillation columns for product purification in a Dow production facility. Process measurements include pressures, temperatures and flow rates across the columns, as well as, reflux ratio and online analyzer readings from upstream reactors. Quality variables regard the concentrations of impurity components, say A and B. These are sampled every hour by an online analyzer and a soft sensor. Furthermore, every 12 hours a laboratory sample is grabbed for analysis (Lu & Chiang 2018). By applying the BCTTF algorithm and analyzing the results, we could confirm its good performance and infer that the laboratory sampling rate can be reduced to 1 sample every 48 hours without significant loss of prediction power. This allows for a meaningful cost reduction in laboratory material and hours of work. Finally, an Emergency Laboratory Sampling Plan is outlined to cope with possible faults that may put the online analyzer out of service and to have better control over the process during an abnormal operation time period.

References

Bar-Shalom Y, Campo L. The effect of the common process noise on the two-sensor fused-track covariance. *IEEE Transactions on Aerospace and Electronic Systems*. 1986;22:803-805.

Chen H, Kirubarajan T, Bar-Shalom Y. Performance limits of track-to-track fusion vs. centralized estimation: Theory and application. *IEEE Transactions on Aerospace and Electronic Systems*. 2003;39:386-398.

Fatehi A, Huang B. Kalman filtering approach to multi-rate information fusion in the presence of irregular sampling rate and variable measurement delay. *Journal of Process Control*. 2017;53:15-25.

Gao J, Harris C. Some remarks on Kalman filters for the multisensor fusion. *Information Fusion*.

2002;3:191-201.

Kumar M, Garg DP, Zachery RA. A method for judicious fusion of inconsistent multiple sensor data. *IEEE Sensors Journal*. 2007;7:723-733.

Lu B, Chiang L. Semi-supervised online soft sensor maintenance experiences in the chemical industry. *Journal of Process Control*. 2018;67:23-34.

Safari S, Shabani F, Simon D. Multirate multisensor data fusion for linear systems using Kalman filters and a neural network. *Aerospace Science and Technology*. 2014;39:465-471.

562176 Knowledge-Based Interactive Preprocessing of Process Data

Zhong-Yan Liu[1], Min Hong[1], Min-Yi Peng[1], Yi Heng[2,3,4]*, (1) School of Chemical Engineering and Technology, Sun Yat-sen University, Zhuhai, China, (2) School of Data and Computer Science, Sun Yat-sen University, Guangzhou, China, (3) Guangdong Province Key Laboratory of Computational Science, Guangzhou, China, (4) National Supercomputer Center in Guangzhou (NSCC-GZ), Guangzhou, China

*** Corresponding author. E-mail address: hengyi@mail.sysu.edu.cn (Yi Heng)**

Keyword: Data Preprocessing, Workflow Support, Knowledge Base

How to preprocess data with advanced analysis algorithms and leverage it into applications is a crucial task in practice from a holistic viewpoint [1]. To this end, distributed data processing has been proposed to deal with a wide range of industrial applications [2]. On the one hand, Jashapara et al. considered well-established information technologies such as databases and data mining for discovering knowledge [3-5]. On the other hand, the workflow used for processing real-life data can be created by tools, methods, and standards that provide better support in the domain of knowledge management, modeling, simulation, validation and other aspects of designing, testing, and commissioning [6-8]. Meanwhile, the amount of process data obtained from engineering has been significantly increasing. Personalized and efficient data preprocessing encountered challenges to develop next-generation intelligent systems, such as maintaining the quality, accessibility and traceability of data, reviewing and reusing knowledge as well as providing useful workflows for practical applications.

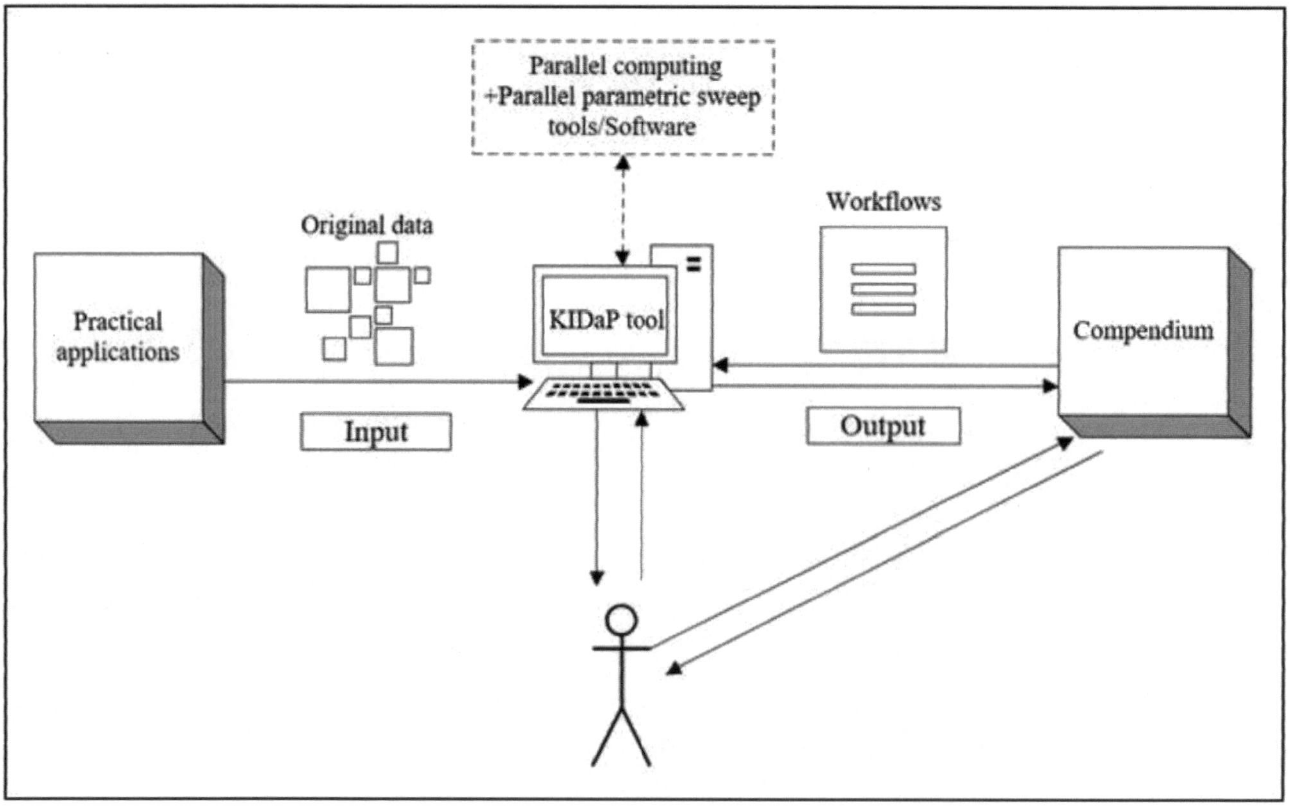

Figure 1. Overview of the KIDaP management framework

In this work, taking an application instance of boiling experiments as an example, we consider a type of management framework that enables knowledge-based big data preprocessing (KIDaP) for practical tasks in a systematic and efficient way (see Fig. 1). The KIDaP management framework was proposed by Heng, Theissen and Soemers [9]. It consists of two main components: a compendium of workflows that summarizes algorithms and methods for massive data processing and a knowledge-based interactive open source software tool KIDaP that has direct access to the compendium. Software user interface is shown in Fig. 2.

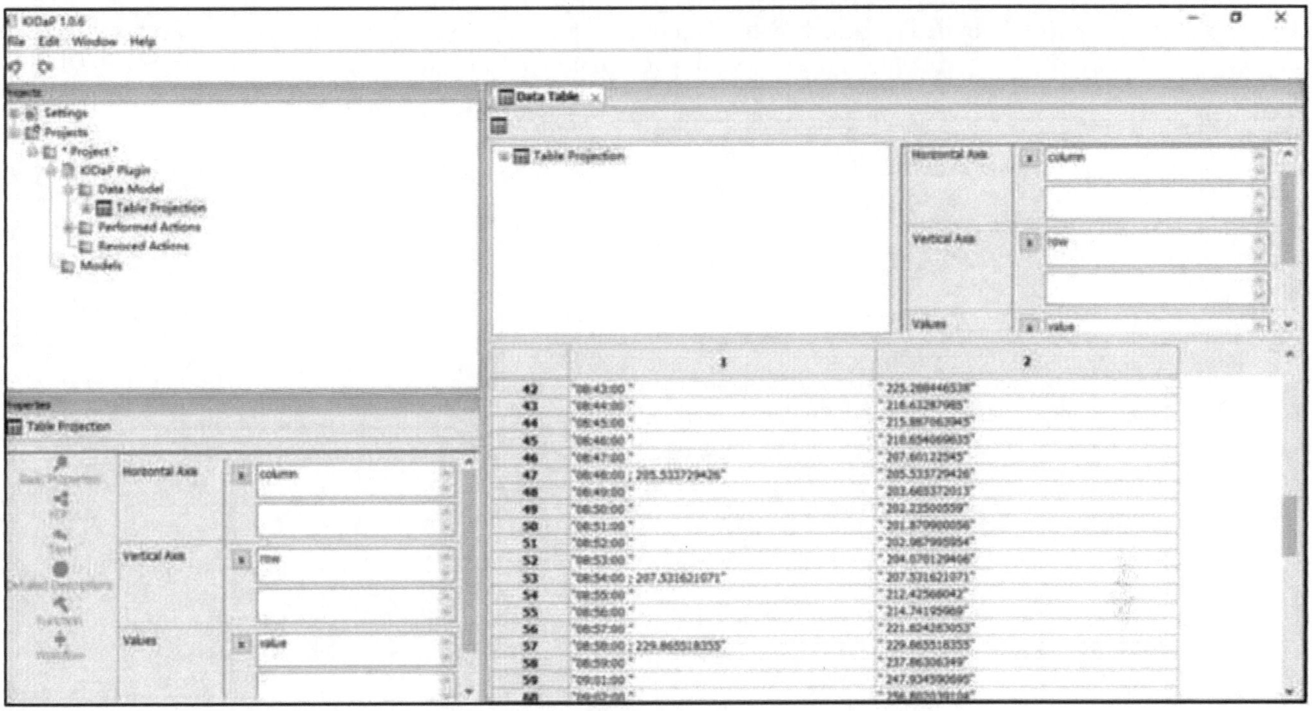

Figure. 2. Software user interface of KIDaP

This knowledge-based management framework is supposed to be used to guide the user to automate repetitive tasks with workflow support, which is open for future extensions in other fields of science and engineering. By using the software KIDaP, users can specify the input format of their data, design and execute a workflow. The use of different processing tools within the platform is available while the designed workflow and related single steps can be automatically recorded so that the data and the actions can be traced easily at a later time.

The data preprocessing task considered in this work arises from a pool boiling experiment conducted by our collaboration partner. It is known that boiling heat transfer has many applications and can be enhanced by micro/nanostructured surfaces. In our collaboration partners' work, wang et al. fabricated micro/nano bi-porous surfaces, taking into account both optimal cavity size and wettability modification [10]. Firstly, a micro/nano bi-porous copper surface with abundant micro pores whose diameters are fabricated in the optimal range for cavity size according to previous study from other researchers [11] is considered. Secondly, a prior development by wettability modification was made to further enhance the boiling heat transfer on the micro/nano bi-porous copper surface. Boiling heat transfer performance of the samples fabricated in these two steps were tested with water at atmosphere pressure respectively [10]. Through the performance test, original experimental temperature time series data are obtained. They manually processed these massive data step by step in order to obtain the boiling performance curve. According to the data analysis results, the mechanism of the boiling heat transfer can be further analyzed.

Here, we are going to use the proposed knowledge-based management framework to preprocess the original data. As shown in Fig. 3, the data processing work-process is divided into five tasks, of which task 4 can be optionally selected. Each task applies a different algorithm or a single step, so that it can process data efficiently. At first, we import the original data (*.CSV) into the KIDaP tool and perform task 1 (prepare data structure) by using the KIDaP built-in functionalities. Then, task 2 (data screening) is executed with an equal-width random selection. In task 3 (data transformation), a series of numerical algorithms (such as Fourier 1-D conduction equation) are used to analyze the data. Task 4 (data integration) is optional, and will be performed when necessary. And then, task 5 (data visualization) will be executed as a final step. The drawing module is called to plot the data, for further analysis and study of the boiling heat transfer mechanism. Compared with manual processing of data, the proposed knowledge-based management framework is supposed to solve the problems of excessive data volume, numerous data processing steps, efficiently process and record the workflows as well as implement parallel parametric sweep, etc. As a long-term goal, the proposed knowledge-based management framework is supposed to provide a universal solution strategy for this kind of experimental data processing problems in the future.

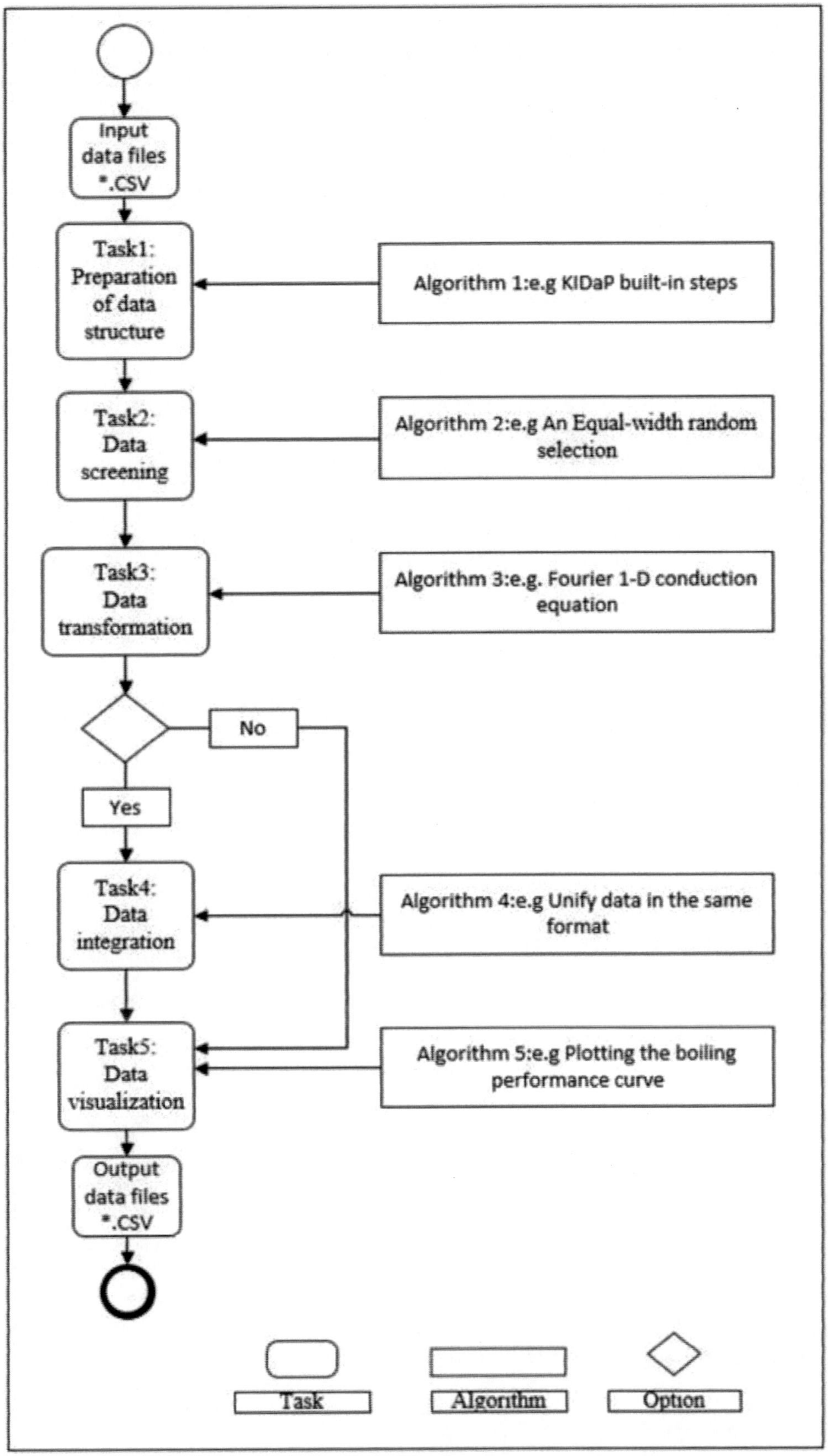

Fig.3. Suggested workflow based on the KIDaP management framework for the example arising from pool boiling experiment

In conclusion, the proposed data preprocessing approach is supposed to provide workflow support and guidance for the experimenters in a more convenient and flexible way. In case of complex work-processes and huge amount of industrial data, the in-depth research of the software framework KIDaP is also supposed to meet the industrial requirement and move one step forward to the development of future intelligent data preprocessing systems, which will be focused on the method of deep learning, the support of the docking of multiple platforms and the basis of the foundation for solving complex sequential as well as parallel problems. In future work, more industrial big data-preprocessing problems arising from different engineering fields are suggested to be tackled by KIDaP based on the proposed management framework. Furthermore, KIDaP will be continuously extended to support intelligent data preprocessing with workflow support so as to satisfy the practical requirement.

Acknowledgments

The authors particularly thank our collaboration partners Dr. Marcus Soemers and Manfred Theissen for providing the open source software KIDaP and the corresponding work-process compendium. Dr. Ya-qiao Wang and Hao-ran Lu are appreciated for giving the measurement data from their pool boiling experiments.

References

[1] G.F., Wang., X.T., Tian., J.H., Geng., et., al. A Process Innovation Knowledge Management Framework and its Application. Advanced Materials Research, 655-657 (2013) 2299-2306.

[2] P., Gaj., A., Malinowski., T., Sauter., et., al. Guest editorial: Distributed Data Processing in Industrial Applications. IEEE Transactions on Industrial Informatics, 11(3) (2015) 737-740.

[3] A., Jashapara. Knowledge Management, An Integrated Approach. Prentice Hall, (2004) 73-108.

[4] E., Žižmond., M., Novak. Controversies of Technology Convergence within The European Union. Industrial Management & Data Systems, 107(5) (2007) 618-635.

[5] S., Natek., D., Lesjak. Improving Knowledge Management by Integrating Hei Process and Data Models. Journal of Computer Information Systems, 53(4) (2015) 81-86.

[6] Z.P., Zhang., S.M., Jasimuddin. Knowledge Market in Organization: Incentive Alignment and IT Support. Industrial Management & Data Systems, 112(7) (2012) 1101-1122.

[7] Y., Wang., S., Yu., T., Xu. A User Requirement Driven Framework for Collaborative Design Knowledge Management. Advanced Engineering Informatics, 33 (2017) 16-28.

[8] D., Mishra., S., Aydin., A., Mishra., et., al. Knowledge Management in Requirement Elicitation: Situational Methods View. Computer Standards & Interfaces, 56 (2018) 49-61.

[9] Y., Heng., M., Soemers., M., Theißen. Wissensbasierte Aufbereitung von Prozessdaten mit Workflow-Support. Mitteldeutsche Mitteilungen, 1 (2012) 8-9.

[10] Y.Q., Wang., J.L., Luo., D.C., Mo., et., al. Wettability Modification to further Enhance the Pool Boiling Performance of the Micro Nano Bi-porous Copper Surface Structure. International Journal of Heat and Mass Transfer, 119 (2018) 333-342.

[11] C., Li., Z., Wang., P.I., Wang., et., al. Nanostructured Copper Interfaces for Enhanced Boiling. Small, 4(8) (2008) 1084-1088.

570790 Data-Driven Predictive Monitoring and Operation Support for Change-over Processes in Biopharmaceutical Drug Product Manufacturing

In biopharmaceutical drug product manufacturing, product-change-over is a crucial but time-consuming process. This step is necessary to establish and maintain a sterile production environment, which is essential to guarantee the product quality. During the change-over process inside the isolator, both the exterior and interior surface of the production machinery require intensive cleaning/disinfection, achieved through hydrogen peroxide decontamination, and clean-in-place (CIP)/sterilization-in-place (SIP), respectively. Numerous sensors are available to constantly monitor the process and provide real-time measurements of process variables, whereby the stored sensor data is only used for backtracking actions. In our work, we exploit the available data to develop a system for predictive monitoring of the process and early failure detection.

The data is initially prepared for exploitation through several steps. First, a dimensionality reduction step through principal component analysis (PCA) is conducted followed by a data cleaning and classification step, where the runs are classified into failed and successful. The boundaries of successful runs are then defined and are further referred to as "Golden Zone." Machine learning algorithms such as Random Forest and k Nearest Neighbor are used to predict the performance of a run following a few minutes of operation. The quality of the prediction is assessed, and the distance to the "Golden Zone" boundaries is measured to guide the decision-making process by the operators in case abnormalities are detected.

The application of this predictive monitoring approach for early failure detection to the historical data of a hydrogen peroxide decontamination process of an industrial filling line at Hoffmann La Roche in Switzerland yielded a 50% reduction in the number of repeated runs due to failures leading to significant time and financial savings.

A complementary root cause analysis is also conducted to deepen the understanding of the complex system where a limited understanding of the causes of failure is provided. Pattern recognition within the analyzed data is used to identify different sources of errors and to recognize the full impact of certain abnormalities or failures (such as leaks) on the system operation and performance.

We further explore opportunities to improve the machine learning algorithms to achieve the earliest possible failure detection points, i.e., to maximize the time distance from the predicted failure to the occurrence of a recorded alarm. This could be especially helpful in the case of hydrogen peroxide decontamination where failures can be detected before the introduction of the hydrogen peroxide to avoid the need to repeat the aeration process which is the most time-consuming step in the procedure.

574765 A Decision Support Tool and Closed-Loop Startup Framework for a Hydrogen Plant

Hydrogen is one of the most extensively used chemical components, with wide range of applications in petroleum and chemical industries. An economic way to commercially produce hydrogen is to utilize steam methane reforming [1]. The process forms an intricate network of material and energy due presence of vast number of material and heat flows. Startup of a hydrogen plant involves implementation of a predefined recipe that consists of a series of events taking place over a finite duration of time. Specifically, it involves continuous actions such as ramping the natural gas fuel to reach a sufficient reformer exit temperature, regulation of nitrogen flow in the reactor tubes, circulation of natural gas feed and steam etc. Further, discrete decisions such as starting up the PSA, recirculating the tail gas from the PSA and ramping up to full capacity in several stages are also part of the startup recipe. Based on the decisions that are taken at various stages in the startup, the startup time can vary significantly. Startup is typically categorized as cold startup and warm startup, based on the initial state of the plant from which the plant is required to be driven to nominal operation. The cold startup will be central to the discussion in this work, where the plant is started from completely shutdown conditions.

Existing startup recipes have been built heuristically from past successful startups, with the possibility of optimizing the startup procedure remaining unexplored. Further, due to the uncertainty associated with the startup duration, the availability of the product cannot be predicted accurately. Motivated by these considerations, in this work we first developed a decision support tool (DST) that can assist the operator during startup to predict the completion of startup. One key recognition that we make is that due to the nature of startup/shutdown phase, they can be treated as a batch like operation with variable lengths, where the objective is to drive the process from one state to the other. Thus, batch subspace identification approach [2] is used for data-driven modeling of the hydrogen plant [3]. Development of a data-driven LTI model, that can handle variable batch duration, paves the way for reliable prediction of the startup evolution and its utilization in a tractable optimization formulation. The DST in real time predicts the timeline of the startup by utilizing the data-driven model and state-trajectories from historical batches. This paper will present the proposed algorithm and results from plant startup.

Secondly, we addressed the problem of synthesis of an optimal startup procedure for the hydrogen production plant. A key requirement for operating procedure synthesis (OPS) is a good model for the startup process. There exist several modeling approaches that are well suited for small scale processes (see [5] for an excellent review) but difficult to implement to large chemical units. Out of these approaches, dynamic simulation based strategies are most prominent. In these approaches, dynamic simulations are used to select the best startup scenarios. In the present work, startup is optimized such that it improves not only the process economics but also its completion time using dynamic optimization. The optimization problem for the present application is an optimal control problem where the objective is to reach the desired terminal state in minimum time, while adhering to various constraints (such as the bounds on reformer exit temperature [6]) and making discrete decisions along the way. To this end, first a high-fidelity test bed model of the entire plant is developed in UniSim, capable of simulating the startup and shutdown phase, with appropriate adaptation of the plant SOP. Note however that such detailed first principles models pose computational challenges when directly embedded in optimization problems, and thus existing SOP synthesis approaches utilize heuristics to determine optimal profile [7-10]. Thus, in this work, several simulated startups are performed to generate the training data. Then, an LTI data-driven model of the process using batch subspace identification [2-4] is identified. The identified data-driven model is subsequently utilized within an optimization framework to synthesize the startup SOP. Through closed-loop startup, we are able to achieve startup in significantly less time. Results from the closed-loop startup and the corresponding optimization problem will be presented.

[1] Kroschwitz, J., Seidel, A., 2006. Kirk-Othmer Encyclopedia of Chemical Technology, No. V. 20 in Encyclopedia of Chemical Technology. Wiley.

[2] Corbett, B., Mhaskar, P. Subspace identification for data-driven modelling and quality control of batch

processes, AIChE Journal 62 (2016) 1581-1601.

[3] Garg, A., Corbett, B., Mhaskar, P., Hu, G., Flores-Cerrillo, J. (2017), High Fidelity Model Development and Subspace Identification of a Hydrogen Plant Startup Dynamics, Computers and Chemical Engineering, 106 183-190.

[4] Garg, A., Mhaskar, P. (2017), Subspace Identification Based Modeling and Control of Batch Particulate Processes, Industrial and Engineering Chemistry Research, 56 (26) 7491-7502.

[5] Batres, R., Soutter, J., Asprey, S.P., Chung, P., 2002. Operating procedure synthesis: science or art. Knowl. Eng. Rev. 17 (3).

[6] Du, M., Mhaskar, P., Zhu, Y., Flores-Cerrillo, J., 2014. Safe-parking of a hydrogen production unit. Ind. Eng. Chem. Res. 53 (19), 8147–8154.

[7] Scenna, N., Ruiz, C., Benz, S., 1998. Dynamic simulation of start-up procedures of reactive distillation columns. Comput. Chem. Eng. 22, S719–S722.

[8] Reepmeyer, F., Repke, J.-U., Wozny, G., 2004. Time optimal start-up strategies for reactive distillation columns. Chem. Eng. Sci. 59 (20), 4339–4347.

[9] Yang, X., Xu, Q., Li, K., Sagar, C.D., 2010. Dynamic simulation and optimization for the start-up operation of an ethylene oxide plant. Ind. Eng. Chem. Res. 49 (9), 4360–4371.

[10] Zhao, Y., Zhang, J., Qiu, T., Zhao, J., Xu, Q., 2014. Flare minimization during start-ups of an integrated cryogenic separation system via dynamic simulation. Ind. Eng. Chem. Res. 53 (4), 1553–1562.

575372 Using Channel State Information for Estimating Moisture Content in Wood Via 5 GHz Wi-Fi

Using Channel State Information for Estimating Moisture Content in Wood via 5 GHz Wi-Fi

Kerul Suthar, Forest Dewberry, Zhihua Jiang, Q. Peter He

Moisture content monitoring is a very important factor in manufacturing for forest industry as well as pulp and paper industry. For a manufacturing process that uses wood as a raw material, it is important to have real-time knowledge about the moisture content of wood so that the process can be optimized and/or controlled to achieve satisfactory product quality. Besides the manufacturing industry, the energy exploitation, storage suitability, and delivery price also depend on the moisture content of wood as well. Both destructive and non-destructive methods have been developed for moisture detection and/or estimation in wood. Destructive methods such as oven drying are accurate, but they also lead to destructive changes in the properties of wood. In addition, they are usually off-line and time consuming and therefore cannot be used for online applications due to the time constraint [1]. On the other hand, several non-destructive techniques that are less time consuming and require less man power have been developed, which include capacitance method, resistance method, microwave method, NIR method, etc. But these techniques are relatively complex and costly to install and operate [2].

To address the above-mentioned limitations of existing techniques, we propose a non-destructive and economic approach based on 5 GHz Wi-Fi, and use Channel State Information (CSI) to predict the moisture content in wood chips. We extract CSI by modifying the open source device drivers for Intel Wi-Fi link 5300 Network Interface Card based on CSITool [3]. CSI contains information about the channel in form of individual data sub-carriers capturing indoor channel features like effect of scattering, fading and power decay with distance. Modern Wi-Fi systems are equipped with Orthogonal Frequency Division Multiplexing (OFDM), dividing the data in multiple orthogonal subcarrier groups which solves the issue of selective frequency fading [4] . CSI has been used for indoor localization, device free sensing including fall detection, activity recognition and heart rate monitoring [5]. In addition, CSI and phase difference data have been successfully used for multi-class moisture detection in wheat [5].

In this work, we collect CSI and phase difference data using IWL5300 NIC by configuring the transmitter and receiver in injection and monitor mode respectively. We use Lenovo ThinkPad systems equipped with Linux based OS 14.02 and kernel version 4.2 due to the version specific selectivity of CSI tool. Both systems are equipped with IWL5300 NIC with modified driver and firmware for data collection. Our work includes CSI data collection, preprocessing, outlier detection, offline training and online testing. The CSI data is collected for wood chips at different moisture levels for offline training of models. The statistics pattern analysis (SPA) framework [6,7] is adapted in this work to build a multivariate statistical model for predicting the moisture content in wood. The most significant difference between SPA based approach and other multivariate statistical methods, such as partial least squares (PLS), is that instead of extracting correlations between the variables and moisture content measurements, we extract correlations between the statistics of the variables and the moisture content measurements. The proposed method is compared with other multivariate statistical and machine learning methods, including PLS, long short-term memory (LSTM) network, etc. We also discuss some challenges of the project and potential solutions to address them.

References:

1. Govett, R., Mace, T., & Bowe, S. (2010). A practical guide for the determination of moisture content of woody biomass. *University of Wisconsin*, 20.

2. John Bogart, Kett US - Learn the six methods for determining moisture: Advantages, Deficiencies and who uses each method.

3. Halperin, D., Hu, W., Sheth, A., & Wetherall, D. (2011). Tool release: Gathering 802.11 n traces

with channel state information. *ACM SIGCOMM Computer Communication Review*, *41*(1), 53-53.

4. Wang, X., Mao, S., Pandey, S., & Agrawal, P. (2014). CA2T: Cooperative antenna arrays technique for pinpoint indoor localization. *Procedia Computer Science*, *34*, 392-399.

5. Yang, W., Wang, X., Cao, S., Wang, H., & Mao, S. (2018, July). Multi-class wheat moisture detection with 5GHz Wi-Fi: A deep LSTM approach. In *2018 27th International Conference on Computer Communication and Networks (ICCCN)*(pp. 1-9). IEEE.

6. Wang and Q. P. He, "Multivariate process monitoring based on statistics pattern analysis," Ind. Eng. Chem. Res., vol. 49, pp. 7858–7869, 2010.

7. P. He and J. Wang, "Statistics Pattern Analysis - A New Process Monitoring Framework and Its Application to Semiconductor Batch Processes," AIChE J., vol. 57, pp. 107–121, 2011.

571273 Identification of Repeating Sequential Alarms in Historical Operation Data of Chemical Plants

The advance of distributed control systems (DCS) in the chemical industry has made it possible to inexpensively and easily install numerous alarms in DCS. While most alarms help operators detect an abnormality and identify its cause, some are unnecessary. A poor alarm system might cause sequential alarms, which are a collection of numerous alarms that almost always occur simultaneously with specific time lags within a short period of time. Sequential alarms reduce the ability of operators to cope with plant abnormalities because critical alarms are buried under many unnecessary ones. It is very important to identify sequential alarms in a plant operation data for safe plant operations.

Nishiguchi and Takai (2010) proposed an identifying method of sequential alarms in plant operation data using event correlation analysis. Event correlation analysis is a method of extracting knowledge that detects statistical similarities between discrete events. The method can identify sequential alarms from the operation data of chemical plant, but it occasionally failed to detect similarities between two physically related sequential alarms when deletions, substitutions, and/or transpositions occurred in the alarm sequence. A modified Smith-Waterman algorithm was proposed to calculate similarities of alarm flood sequences in plant operation data (Cheng *et al.*, 2013). Experts can make a thoroughly analysis, such as root cause, based on the clustered patterns of alarm floods. However it is not the method that directly finds out sequential alarms hidden in the plant operation data. Wang and Noda (2017) proposed an identification method of sequential alarms by applying the dot matrix method to a plant operation data. The dot matrix method is one of sequence alignment methods for identifying similar regions in DNA or RNA, which may be a consequence of functional, structural, or evolutionary relationships between the sequences (Mount, 2004). The method can identify sequential alarms from the operation data of chemical plant, but it occasionally failed to detect similarities between two physically related sequential alarms because the time information when alarms occurred is not for evaluating similarities between them.

In this research, we propose a new identification method of repeating sequential alarms in historical operation data of a chemical plants. In the method, time stamped alarm data is converted into a set of windows containing adjacent alarms. All combinations of windows are compared, and repeating similar windows are identified on the basis of similarities between them. The alarms in each repeating window comprise a sequential alarm. Application of this method to simulated operation data for an azeotropic distillation column demonstrated that it can identify sequential alarms in noisy plant-operation data. By classifying such alarms into small numbers of sequential alarms, an engineer is able to effectively reduce the number of unnecessary alarms.

References

1. Cheng, Y., Izadi, I., and Chen, T., Pattern matching of alarm flood sequences by a modified Smith-Waterman algorithm, *Chemical Engineering Research and Design*, 91, 1085-1094 (2013)
2. Nishiguchi, J., and Takai, T., IPL2 and 3 Performance Improvement Method for Process Safety using Event Correlation Analysis, *Computers & Chemical Engineering*, 34(12), 2007-2013 (2010)
3. Wang, Z., and Noda, M., Identification of Repeated Sequential Alarms in Noisy Plant Operation Data Using Dot Matrix Method with Sliding Window, *Journal of Chemical Engineering of Japan*, 50(6), 445-449 (2017)
4. Mount, D., W., Bioinformatics Sequence and Genome Analysis Second Edition, Cold Spring Harbor Laboratory Press, Cold Spring Harbor, New York (2004)

573955 Big Data for Enhanced Reduced Order Modeling: Application to Hydraulic Fracturing

Many chemical processes undergo non-linear dynamics which are usually described by high-dimensional high-fidelity models whose complexities limit their usage in the design of model-based feedback control systems. These hindrances have paved the way for the use of reduced-order models (ROMs) in the design of controllers. Among the plethora of available techniques, Dynamic Mode Decomposition (DMD) [1], has gained a lot of attention as it accurately captures coherent structures that are dynamically relevant and contribute towards the long term dynamics of the system [2]. The primary assumption of the DMD technique is that the non-linear dynamics of a complex system can be represented in a linear form. But the dynamics of many chemical processes are influenced by the external inputs applied on them and therefore, Dynamic Mode Decomposition with Control (DMDc) technique [3] was developed to include the effect of the external inputs on the dynamics of the system. However, for a highly non-linear chemical process, a global linear representation may not accurately capture its dynamics considering that DMDc has very limited degrees of freedom. To overcome this limitation of DMDc, Local Dynamic Mode Decomposition with Control (Local DMDc) technique [4] was developed which utilizes unsupervised learning to cluster the sequential data, use DMDc to build local ROMs which individually capture the local dynamics accurately and together represent the dynamical system.

However, the domain of attraction (DOA) of Local DMDc is narrow; meaning, it will be less accurate when used for prediction purposes under conditions outside the training data. In this work, we overcome the problem of limited DOA of Local DMDc by using multiple training data sets obtained under distinct training conditions. Specifically, training data is divided into clusters such that each cluster represents certain unique, local dynamics of the system and a DMDc-based local ROM is built for each cluster. Also, a unique identity is built for each cluster by transforming the data through normalization, applying weights, and Principal Component Analysis (PCA). The identity for each cluster is the average of the PCA scores of all the data points it contains. During prediction, given a state and input, we utilize the k-nearest neighbors (kNN) technique to select the appropriate local ROM in order to calculate the future state of the system. We demonstrate the performance of our proposed algorithm by applying it to hydraulic fracturing, a highly non-linear and complex process represented by a system of nonlinear highly-coupled PDEs with time-dependent spatial domain [5]. We utilize its high-fidelity model to generate training data obtained by using distinct training inputs, build multiple local ROMs based on DMDc, and utilize kNN for the purposes of model selection. We demonstrate the enlarged DOA through increased applicability of the LDMDc models obtained using the proposed algorithm for a wide range of operating conditions

Literature cited:

[1] Schmid, P.J., 2010. Dynamic mode decomposition of numerical and experimental data. J. Fluid Mech. 656, 5–28.

[2] Ghommem, M., Carlo, V. M., Efendiev, Y., 2014. Mode decomposition methods for flows in high-contrast porous media. A Global approach. J. Comput. Phys. 257, 400-413.

[3] Proctor, J. L., Brunton, S. L., Kutz, J. N., 2016. Dynamic mode decomposition with control. SIAM J. Appl. Dyn. Syst. 15, 142-161.

[4] Narasingam, A., Kwon, J. S., 2017. Development of local dynamic mode decomposition with control: Application to model predictive control of hydraulic fracturing. Comput. Chem. Eng. 106, 501-511.

[5] Economides, M.J., Nolte, K.G., 2000. Reservoir stimulation. Chichester, Wiley.

564569 A Clustering Decomposition Algorithm for Energy Storage Design & Operation

Intermittent availability of solar and wind energy is a challenge for renewable power systems because they are often asynchronous with consumer demand. Increasing renewable penetration requires energy storage technologies to align varying supply with demand. Moreover, energy storage options give greater control and flexibility to grid operators by making solar and wind more dispatchable power sources.

Optimization-based design & scheduling models aim to minimize the power and storage capacities of renewable power systems to lower capital and operational costs [1]. Hourly time discretization in the schedule is often used to capture solar and wind dynamics, keep track of storage inventory levels, and model time-dependent operational decisions. As a result, large time horizons are characteristic of these problems and significantly increase the computational burden of solving them [2, 3]. In addition, multiple time series data coming from resource availability, demand loads, and prices grow the complexity as well.

In this work, a decomposition algorithm based on agglomerative hierarchical clustering (AHC) is proposed to alleviate the computational burden, where the optimization is performed over representative time periods. A key advantage for AHC compared to the popular K-means clustering approach is that the clusters maintain time chronology, which is important for analyzing inter-period energy storage [4]. An example case study on dense energy carriers for energy storage, performed in collaboration with Shell, is presented to demonstrate the algorithm's applicability.

[1] Zhang, Qi, Ignacio E. Grossmann, Clara F. Heuberger, Arul Sundaramoorthy, and Jose M. Pinto. "Air separation with cryogenic energy storage: optimal scheduling considering electric energy and reserve markets." *AIChE Journal* 61, no. 5 (2015): 1547-1558.

[2] Harjunkoski, Iiro, Christos T. Maravelias, Peter Bongers, Pedro M. Castro, Sebastian Engell, Ignacio E. Grossmann, John Hooker, Carlos Méndez, Guido Sand, and John Wassick. "Scope for industrial applications of production scheduling models and solution methods." *Computers & Chemical Engineering* 62 (2014): 161-193.

[3] Castro, Pedro M., Ignacio E. Grossmann, and Qi Zhang. "Expanding scope and computational challenges in process scheduling." *Computers & Chemical Engineering* 114 (2018): 14-42.

[4] Pineda, Salvador, and Juan M. Morales. "Chronological time-period clustering for optimal capacity expansion planning with storage." *IEEE Transactions on Power Systems* 33, no. 6 (2018): 7162-7170.

574862 Constrained Optimisation of Cell Culture Feeding Strategy and Temperature Shift Duration to Enhance Monoclonal Antibody Titre and Purity

Recombinant proteins have been extensively studied for their wide therapeutic and research applications. The main therapeutic product category is that of monoclonal antibodies, which have been approved to treat a variety of chronic and life-threatening diseases. Increasing production titre has been achieved mainly by cell culture medium improvement and genetic engineering. However, mathematical modeling of bioprocess dynamics is a valuable tool to improve industrial production at a much faster rate and lower cost in comparison to statistical design of experiment approaches. A holistic insight to comprehend the interplay between crucial processes variables can be gained through the application of process system engineering tools. In this study, a single stage population balance model has been built to capture mammalian cells behaviour in fed-batch bioreactors. The model represents two operating modes; the first at constant physiological temperature and the second with a shift to mild hypothermic conditions (32°C) from the fifth day onwards, which represents common industrial practice.

The model considers the dynamic profile of substrates and metabolites, product titre and one of the main cell-derived impurities, host cell proteins (HCPs), which can impact product immunogenicity and integrity. Certain HCPs are naturally secreted, while others are released as a result of cell death and lysis. Our model considers these three sources of HCPs to determine their overall concentration at harvest. Culture osmolality is also considered as a determining factor for cell growth rate and size distribution.

The model parameters have been estimated with high confidence using a recently published dataset for antibody-producing Chinese hamster ovary (CHO) fed-batch cell cultures for physiological temperature and with a shift to mild hypothermia in mid-exponential growth phase. The model was then used to optimise feeding volume and frequency as well as duration of temperature downshift subject to maintaining culture viability above 80%, as is current industrial practice. and this has been set as a constraint in all our model optimisation runs. Specifically, we looked at three scenarios: (a) increased feeding to sustain high viability for longer, which maintained titre but reduced cell death; (b) a 50% reduction in feed volume to avoid overfeeding, which led to a slightly higher titre compared to the control; (c) a shift in culture temperature on day 6 which gives the highest product/ HCP ratio.

In general, higher product titres and prolonged culture viability can be attained at the expense of higher feeding volume. However, when a constraint on HCP concentration is also applied model-based optimisation results in shorter culture duration and, in turn, overall lower antibody titre. This study shows the usefulness of mathematical modeling for exploring trade-offs in bioprocess performance. It is also a viable tool to systematically accelerate the duration of processes development and optimisation. Integrating this model with a downstream purification model to evaluate the cost of removing these fractions of impurities, can help determine what concentration of HCPs can be economically tolerated in the cell culture supernatant and aid whole bioprocess design.

564778 Control of Glycosylation and Titer in Fed-Batch Monoclonal Antibody Production

Therapeutic monoclonal antibodies (mAbs) are currently produced commercially in fed-batch cultures using media recipes and protocols based primarily on heuristics. The primary production objectives are maximizing mAb titer (productivity requirement) and achieving desired glycan distributions (quality attribute) that result from glycosylation—an enzymatic, post-translational process that attaches sugar molecules (glycans) to antibodies. The distribution of glycans is a critical quality attribute of mAb manufacturing, because it can either promote or inhibit desirable therapeutic effects of mAb drugs. However, controlling glycosylation is challenging primarily because it is a complex, non-template driven process that takes place within the cell. Consequently, in addition to the apparent chaotic nature of glycan attachment, the variables available for influencing glycosylation are far fewer than the controlled variables. Furthermore, as a consequence of the multifaceted nature of glycosylation, the process is difficult to model, and on-line sensors do not currently exist for characterizing glycan distribution frequently enough for effective on-line control. Nevertheless, recent advances in both glycosylation modeling and glycan assays have improved the possibility of implementing model-based control of glycosylation in place of the current heuristics-based approach.

Here we present an overall model-based strategy for meeting productivity and product quality requirements in the manufacture of mAbs. The proposed multivariate cascade control system consists of two nested control loops. The outer loop, which is responsible for controlling product attributes such as mAb titer, cell density, and the relative abundance of dominant glycans, uses a model predictive controller (MPC) to determine the setpoints for the inner loop. The inner loop in turn employs multiple PI controllers to sustain the cell culture by maintaining the levels of nutrients, pH, temperature, etc., at desired values. Because measurements of glycan distribution are not available frequently enough, the outer loop incorporates a state observer which uses available sensor measurements and model predictions to reconstruct the complete system state that the MPC uses, along with the process model, to determine the desired control policies.

As a first step toward implementing on-line control of industrial fed-batch mAb manufacturing processes, we tested the complete control system in a MATLAB simulation, investigating a variety of industrially relevant process conditions under several operating sensor configurations—frequencies, qualities, and delays. The simulation results offer insight into optimal sensing strategy as well as the usefulness and limitations of the proposed control system.

576555 Using High-Throughput Microbioreactors to Fit Raman Spectroscopy Model for Measuring Key Metabolites in Monoclonal Antibody Cell Culture Manufacturing Process

Raman spectroscopy is a process analytical technology for simultaneous in-line measurement of important process outputs in therapeutic protein production via mammalian cell culture, such as concentrations of key metabolites or even the protein being expressed. However, due to the complex and shifting nature of the various chemical species present in the cell culture system, Raman models will likely need to be at least partially trained on a process-specific basis. They must also be robust enough to maintain accuracy throughout the expected manufacturing process operational ranges. This presentation is a proof-of-concept study to demonstrate a reliable Raman model can be developed in parallel with standard process development experimentation for later deployment in a scaled-up manufacturing process.

Sartorius ambr® 250 high-throughput microbioreactors were used for cell culture process development studies for a CHO cell line expressing an IgG4 monoclonal antibody ("mAb1"). Process parameters including initial cell density, nutrient feed strategy, pH, and dissolved oxygen control for a fourteen-day bioprocess were examined, and standard off-line monitoring of all process outputs were carried out. Concurrently during this high-throughput study, Raman spectra were also periodically generated from bioreactor samples both before and after being spiked with known concentrations of key metabolites. These data were used to train a Raman model and verified against similar spectra obtained from larger bioreactors.

In conclusion, this study generated a scalable Raman spectroscopy model for various metabolites that can be used for monoclonal antibody process control and performance monitoring. A robust model can be trained over a wide range of processing conditions, in a manner that would not interfere with standard process development or require additional experiments. The comparison of the outputs from this Raman model to the standard off-line data will also be presented.

567854 Evaluation of Process Parameters Influencing Immobilized Metal Affinity Chromatography

The production of biopharmaceutical drugs requires high yielding and robust purification methods. Affinity chromatography is a preferred method to capture product from culture media due to its high selectivity. However, many protein therapeutics lack a suitable affinity capture modality and therefore require use of engineered tags for affinity purification. Immobilized metal ion affinity chromatography (IMAC) has been a popular choice for the purification of histidine-tagged molecules. One drawback of this technology is that process performance is often variable, making it difficult to predictably scale-up. This study contributes to the understanding of process parameters impacting performance and how to optimally control these parameters. Multivariate experiments were performed to investigate the binding of histidine-tagged proteins to commercial metal chelating resins. This study revealed that imidazole concentration, load capacity and metal binding ability of the resin influenced the protein binding capacity. This study also examined the impact of histidine tag structure on chromatographic performance. Adsorption isotherms were constructed to elucidate differences in equilibrium binding of multi-histidine tagged material. A mathematical model was then employed to describe the impact of changing process parameters on product quality. The findings of this study are useful to understanding factors impacting IMAC operation, including the relationship between process parameters, tag structure, process performance, and product quality.

566098 Investigation of Bio-Capacitance Driven Process Variation in a Site to Site Large Scale Technology Transfer Campaign

Bio-capacitance in cell culture processing allows for more automated process controls and on-line biomass monitoring. Here we presented a case study where the large-scale manufacture process (15,000L) is monitored and nutrient-fed based on the dual bio-capacitance automation driven protocol. In one of the batches, unexpected process variation was observed. Higher amount of feed, out of trend nutrient concentration and accumulative waste level were observed during the process which can be a signature of nutrient overfeed. Our investigation suggested that the primary and backup bio-capacitance reading have significant deviation especially at the earlier days. When bio-capacitance reading deviation happened, viable cell volume (VCV), calculated by Viable cell density (VCD) and cell diameter, was used to justify which probe signal to use. This is because capacitive signal is to measure the dielectric properties in the membrane of viable cell. Historical VCV/bio-capacitance ratio is referenced as process control strategy to translate the online bio-capacitance reading and offline Vi-cell measurement. In conclusion, the variation of Vi-cells to Vi-cells in addition to deviation of dual capacitance readings, created the signal to demand for higher feed volume. As a result, it might cause negative impact on the product quality such as glycosylation. Based on this large-scale process experience, we developed a more robust bio-capacitance control strategies to facilitate the site-to-site process transfer in the future.

566086 Characterization of Rotary Lobe Pump Damage in Large-Scale Cell Culture Perfusion Process

Perfusion bioprocessing is becoming more prevalent for increasing productivity and shortening seed train and bioreactor stages. When implementing the new perfusion platform in N-1 large-scale manufacturing process (3750L), unexpected viability drop was observed at day 1. This was not seen at the bench scale satellite runs using the same culture and medium. Higher turbidity and lactate dehydrogenase (LDH) were in the large-scale bioreactor also confirmed the cells were stressed more significantly. Towards the end of the process, the transmembrane pressure (TMP) increased significantly and fouled the hollow fiber membrane. Several investigations were conducted to understand the potential root cause including bioreactor sparging, hollow fiber filter shear, the piping of skid design and operation of the rotary lobe pump. A quick mitigation is to lower RPM in the rotary lobe pump which immediately reduced the cell damage. A follow up study to utilize a smaller scale of rotary lobe pump in the lab was conducted. RPM, flow rate, differential pressure across the pump were studied to identify the possible mechanism of cell damage. The results suggested that differential pressure across the pump was the primary cause of cell damage. A small-scale robustness model was developed to better predict manufacturing performance and improve the large-scale operation in the future.

577705 Data-Rich Development of a Novel Biocatalytic Aerobic Oxidation across Scales

Recent advances in the biocatalytic retrosynthetic toolbox have transformed the way Merck conceives of novel synthetic routes, particularly for drug candidates with significant structural similarity to naturally occurring biomolecules such as proteins, peptides and nucleosides. This is particularly evident in the ongoing development of the biocatalytic route to MK-8591, a nucleoside reverse transcriptase/translocation inhibitor (NRTTI) currently progressing through Phase II clinical trials, displaying great promise as a next-generation anti-HIV drug. Broadly speaking, this synthetic pathway represents a ground-breaking application of cascaded biocatalysis for the synthesis of a complex, clinically relevant small molecule compound.

Furthermore, the oxidation of alcohols to aldehydes represents key transformations in organic synthesis. Traditionally, many of these oxidation reactions have been carried out using stoichiometric amounts of transition metal-based oxidants while selective and green methods have remained largely underdeveloped. To surmount those obstacles, Merck has recently expanded its biocatalytic platform to include the copper-dependent galactose oxidases (GOases), which are highly efficient enzymes for the mild, regio- and stereoselective oxidation of a broad variety of primary and secondary alcohols, polyols and sugars. These enzymes are capable of catalytic oxidations in water, making direct use of the molecular oxygen from compressed air as the oxidant.

In this submission, we will showcase how collaborative efforts across Process Research and Development are contributing towards building deep, fundamental understanding of the oxidation of 2-ethynylglycerol, a key intermediate in the MK-8591 synthesis. We will highlight past and ongoing efforts centering on enzyme discovery and reaction engineering, underpinned by data-rich experimentation methods including; high-throughput experimentation and analysis, automation, and insitu reaction analysis using novel analytical tools. Furthermore, development of a robust process optimized for a commercially relevant synthetic chemistry reactor (vs. for example a more optimal bioreactor or fermentor) will be reviewed including demonstration across scales.

574224 Automated Platform for Experimental Execution and Model Regression of Volumetric Mass Transfer Coefficient and Vessel Characterization

Mass transfer is an important parameter when considering the scale up of aerobic oxidations or other reactions in which the delivery of gaseous reactants to a liquid phase impacts process performance. In the case of oxidations, the oxygen transfer rate from the gas phase to the liquid is dependent upon the volumetric mass transfer coefficient, kla, and the concentration driving force between the concentration of oxygen in the bulk liquid and the saturation concentration of oxygen at the gas-liquid bubble interface. For biocatalytic aerobic oxidations in particular, the reaction solvent is typically water, which is known to have a very low oxygen solubility relative to other solvent systems. As a result, the oxygen transfer rate can be dominated by the kinetics of mass transfer and is highly dependent on reactor geometry and process parameters. Proper characterization of kla under a given set of conditions becomes a critical activity in understanding and mitigating scale-up risks.

As an alternative to traditional pressurization/depressurization tests to determine kla, we have been refining experimental methods which leverage the dynamics of oxygen dissolution following a change in the equilibrium concentration of oxygen. These methods are often tedious to execute and require careful experimental planning and data capture and as a result, we sought to develop an automated platform for experimental execution, data stream integration and kla regression. Additionally, this automated platform was applied to characterize oxygen mass transfer in reactors of various geometries ranging in scale from 10 mL to 100 L to demonstrate the value of this platform for the scale up of aerobic oxidations.

561248 Continuous Manufacturing of Oral Disintegrating Films: Impact of Manufacturing Process Parameters on Product Quality

Continuous Manufacturing of Oral Disintegrating Films: Impact of Manufacturing Process Parameters on Product Quality

Sonal Mazumder*, Scott M. Krull, Nima YazdanPanah, Xiaoming Xu, Muhammad Ashraf, Celia N. Cruz, Thomas F. O'Connor, Naresh Pavurala

-

Office of Testing and Research, U.S. Food and Drug Administration, Silver Spring, MD

Continuous processing of pharmaceutical drug products has the potential to offer many advantages over established batch process methodologies. The objective of this work is to understand the impact of CM process variables on critical quality attributes (CQA's) of oral disintegrating films (ODFs). ODFs are popular dosage forms for geriatric and pediatric patients. This can aid the understanding of failure modes that provide the scientific foundation for systematic risk assessment.

A continuous tape caster with custom baffles was employed to manufacture Lamotrigine ODFs using the solvent casting method. The films were characterized for drug assay, content uniformity, disintegration, dissolution, thickness, % moisture content and mechanical properties. Central composite (CCD) design of experiments (DoE) with 16 runs were used to obtain a comprehensive understanding of the impact of critical process parameters (CPP) such as conduction temperature, residence time and air flowrate on quality and performance of ODFs. Dry thickness, tensile strength and young's modulus were impacted by the residence time of the film inside the dryer. The other CPP identified was air flowrate which impacted tensile strength. Dissolution and disintegration were not significantly impacted by the process.

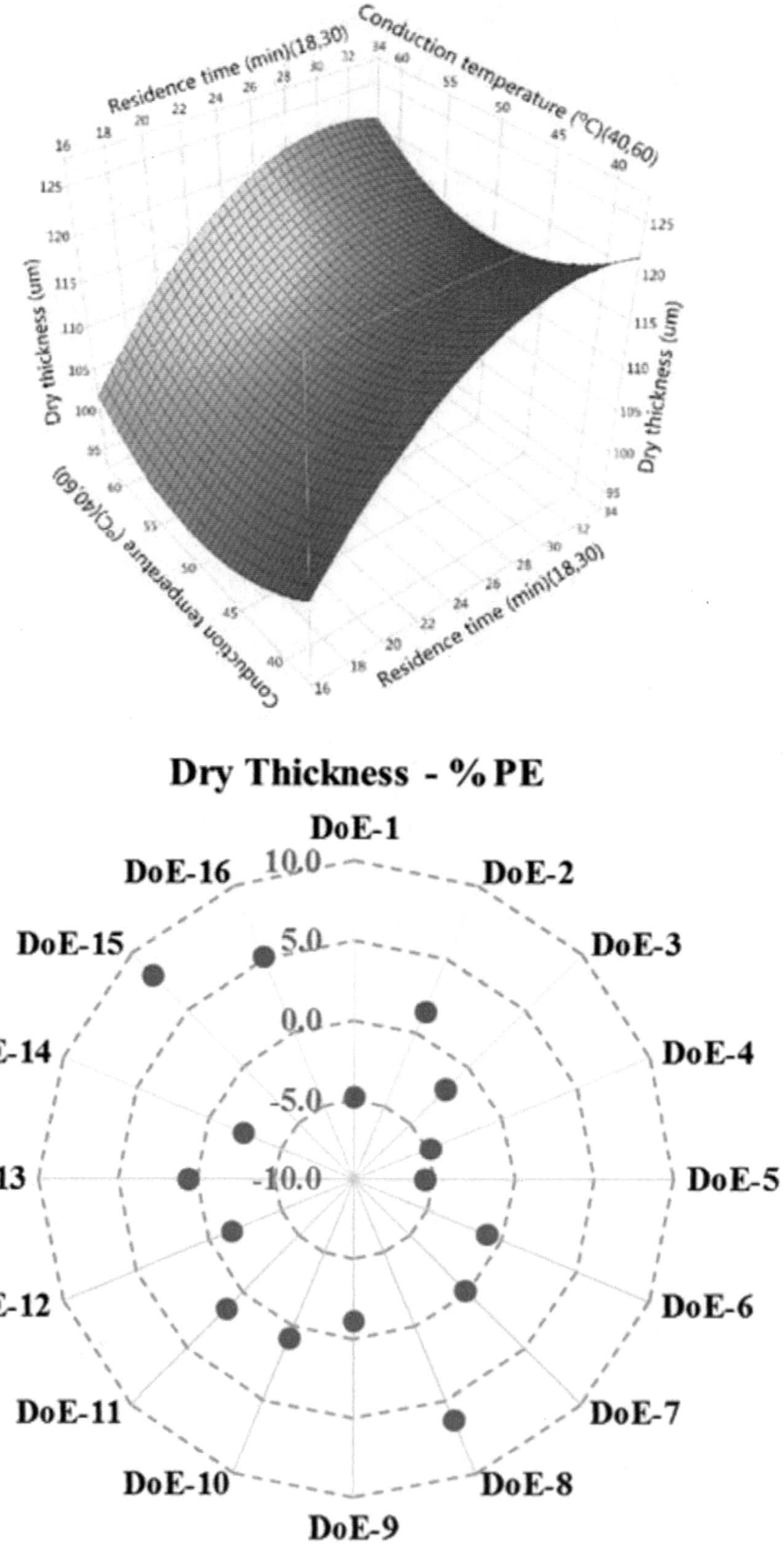

Figure 1. (A) 3D response surface plot showing the effect of residence time and conduction temperature on dry thickness when Linear Air Velocity is 0.6, (B) % Model prediction error for dry thickness

Response surface models were developed to predict the CQAs such as dry thickness (Figure 1A), tensile strength and dissolution time. The model equation developed was tested on the DoE data and the resulting model predictions for the CQA's were compared to the experimental data. The model prediction error was less than 10% for most of the DoEs (Figure 1B). A design space was established and verified. The design space experiments revealed that operating conditions conducive to producing ODFs with desirable CQAs are centered around 50 °C conduction temperature with 0.55 m/s convection air flowrate and 24 min residence time. This work helped to elucidate the relationships between process variables and CQAs for LMT ODFs which will provide insights for designing control strategies for ODFs produced via continuous manufacturing.

Disclaimer: This article reflects the views of the author and should not be construed to represent FDA's views or policies.

578872 The Development of a Predictive Tableting Platform in the Context of Continuous Direct Compression

The development of robust tableting processes in a timely manner is still challenging due to a lack of mechanistic process understanding, fundamental understanding of influence of raw material attributes, and limited usage of sophisticated process simulation tools. Experimentally determining the effects of involved process- and formulation parameters and thereafter optimization is labor-intensive, expensive and time-consuming. Therefore, raw material database management and process modeling using multivariate data analysis techniques and numerical simulations of the compaction process based on Finite Element Analysis (FEA) can provide a valuable and efficient alternative.

A set of 55 powders covering both excipients and active pharmaceutical ingredients (API) was characterized using over 20 techniques describing particle size and -shape, density, moisture content, powder flow, compressibility, aeration, surface area and triboelectric charging. Additionally, multiple mechanical properties, including plastic-, elastic, and brittle deformation, were determined based on in-die compaction tests. Principal Component Analysis (PCA) was then performed to elucidate correlations between the powders and their measured properties. Based on this analysis, formulation blends were selected containing multiple API's and fillers covering a maximal area of the material variability space determined via the PCA. Disintegrant (Sodium Crosscarmellose; 5%), glidant (Colloidal Silicon Dioxide; 0,5%) and lubricant (Magnesium Stearate; 0,75%) were kept fixed in the selected blends. Formulation blend bulk properties were characterized with a minimum number of relevant tests, as derived from the PCA model of the raw materials. These blends were subsequently compacted using the press type simulation tool on a Huxley Bertram (HB) 1088-C compaction simulator to evaluate tablet quality attributes under different process conditions using the Design of Experiments (DoE) approach.

T-PLS models were developed to link raw material properties, blending ratio's and process settings with the resulting tablet quality attributes. Based on these models it is possible to determine which formulation- and process parameters affect the direct compaction process and resulting tablet properties. By optimizing this model for new API's, it can be used to predict an optimal formulation and finetune the process settings based on a minimum of relevant raw material characterization techniques and compaction tests. So far, this approach has only been tested on a compaction simulator, for further implementation scale-up experiments are required.

These results contribute to a better understanding of the impact of powder properties and process settings on a direct compaction process and final properties of the produced tablets. Applying this knowledge can reduce consumption of expensive API's during product development phase. This predictive platform, which combines the use of PLS and FEA models, can be implemented for future development of formulations for direct compression and finding optimal process settings in a minimal amount of time. This platform can later be linked with additional unit operations such as feeding and granulation for implementation towards continuous manufacturing processes.

561578 Drug Form Considerations in Continuous Solution Processing of Oral Solid Dosage Products

The pharmaceutical industry is moving toward continuous manufacturing for small molecule API products and new technologies must be developed to process APIs and excipients into solid oral dosages such as tablets. Solution processing, where the drug and excipients are processed together as a liquid or suspension, has distinct advantages in improving mixing and reducing powder handling. Various techniques have been adopted from the plastics industry for solution processing of pharmaceuticals such as hot melt extrusion, electrospinning, film casting and 3D printing. In all of these, it is a challenge to control whether the drug is crystalline or amorphous and which polymorph results. We specifically focus on two manufacturing techniques, electrospinning and 3D printing, and show that we can prepare tablets containing crystalline or amorphous APIs as desired with acceptable dissolution properties in simulated gastric fluid. By starting with a high loading of API in a polymer solution with a non-solvent for the API, we can 3D print tablets containing crystalline API. Due to the high loading of large particles, the rheology of the suspensions is a crucial parameter for the printability and we have shown that this can be controlled through the particle loading and polymer molecular weight. We can also electrospin fibers containing crystalline API through a mixture of a non-solvent, polymer and API crystals, but by dissolving the API in a good solvent mixed with the polymer, we can prepare fibers containing amorphous API with similar processing conditions. These studies aim to improve control over drug form in solution processing, while expanding the utility of 3D printing and electrospinning for pharmaceuticals.

Formatting Issue – Best Available Paper/Graphic

578037 Half Tablet Methodology: Investigating the Relationship between Dissolution Behavior and Raman Chemical Imaging

There are known knowledge gaps in the current pharmaceutical manufacturing processes, particularly regarding the relationship between processing conditions, and resulting particle size and spatial distribution of drugs in solid dose products. These three aspects can be linked through a common feature of understanding the physical and chemical microstructure of a tablet. A better understanding of tablet microstructure has the potential to unveil the relationship between process parameters and performance attributes, mainly dissolution. Product structure understanding of this nature can aid in prediction of product performance, process development as well as deviation investigations. Understanding the microstructure of a tablet requires knowing the spatial distribution of the different ingredients and void space, which can be enabled by Chemical Imaging (CI). Raman imaging is a powerful tool for studying tablet structure, with superior spatial resolution and chemical specificity than NIR and is also less affected by physical characteristics of the sample.

This study focuses on using a commercial three-dimensional (3D) Raman Imaging system (H2Optx), as an off-line measurement system, to find a relation between chemical information obtained from Raman Imaging and API dissolution. Till now, extensive structural studies have not been conducted to study this relationship using Raman Imaging.

In this work, Ibuprofen (D50: 70 µm) is jet milled to two more particle size ranges (D_{50}: 28 and 11 µm) following by mixing with Fast Flo Lactose (filler) and Magnesium stearate (lubricant) a V-Blender with two levels (high and low) of shear. The six conditions were expected to have contrasting dissolution behaviors and Raman Chemical Images. Post blending, tablets were manufactured using an MCC Presster (a rotary tablet press simulator). An FT-NIR Transmittance based PLS regression model predicted the API content in each of the marked tablets. The 'deviated' and 'normal' tablets from each of the experimental conditions were cut in half with minimal powder loss. One-half (from each of the tablets) was used for dissolution test while the other half was used for 3D Raman Imaging. Statistical comparison of 3D Raman Images and dissolution profiles revealed the relationship of the chemical microstructure to the dissolution process. The overall methodology is described in the figure below:

560271 A Hybrid Modelling Approach Integrating First Principles Modelling with Subspace Identification

Good models are integral to process systems engineering, often being the core around which control and optimization formulations are designed. First principles or mechanistic modeling refers to the type of modeling where explicit knowledge of the process mechanism is present and utilized. Often times, the fewer the assumptions made, the higher the number of model parameters and the more complex the model structure. The task of building and maintaining a first principles model [1-2] continues to involve significant effort.

Increased availability of data, along with improved computational capabilities have made the data driven methods of modeling very attractive. In these approaches, a model structure is chosen a priori, and parameters are identified using available data. In this direction several statistical based approaches exist, distinguished by the kinds of model stricture being utilized, including latent variable based methods and subspace identification algorithms.

Statistical methods of process modeling are very well recognized nowadays. Projection to latent spaces (PLS) is one such statistical latent variable method where measurement data of high dimension is projected to lower dimensional space to create simpler and effective models. Their usefulness to batch processes [3] have been extensively demonstrated with algorithms utilized to handle batches of non-uniform lengths. Another statistical based modeling approach is subspace identification, which enable identifying models having state-space representation and thus are very effective for prediction and control purposes. Usefulness of subspace identification scheme to batch processes [4] have been explored in great detail recently.

However, integration of these statistical modeling techniques in synergy with first principles models have not been demonstrated yet. There have been some examples in the past that utilize a priori process knowledge to create hybrid or grey-box models [5-6], but these have artificial neural networks (ANN) as the data driven model.

This presentation addresses the problem of synergizing first principles models with subspace identified models. This is achieved by building a hybrid model where the subspace model identification algorithm is used to create a model for the residuals (mismatch in the outputs generated by the first principles model and the plant output) rather than being used to create a dynamic model for the process outputs. A continuous stirred tank (CSTR) setup is used to illustrate the proposed approach on a continuous system. The further evaluate its efficacy, the proposed methodology is applied on a batch polymethyl methacrylate (PMMA) polymerization reactor and the predictions are compared with that of first principles modelling and data driven approach alone. The presentation demonstrates the improved modeling capability of the hybrid model over either of its components.

References:

[1] Pinto, J.; Ray, W. The dynamic behavior of continuous solution polymerization reactorsa .VII. Experimental study of a copolymerization reactor. Chemical Engineering Science 1995, 50, 715 _ 736.

[2] Poyton, A.; Varziri, M.; McAuley, K.; McLellan, P.; Ramsay, J. Parameter estimation in continuous-time dynamic models using principal differential analysis. Computers & Chemical Engineering 2006, 30, 698-708.

[3] Flores-Cerrillo, J.; MacGregor, J. F. Iterative Learning Control for Final Batch Product Quality Using Partial Least Squares Models. Industrial & Engineering Chemistry Research 2005, 44, 9146_9155.

[4] Corbett, B.; Mhaskar, P. Subspace identification for data-driven modeling and quality control of batch processes. AIChE Journal 62, 1581_1601.

[5] Su, H.-T.; Bhat, N.; Minderman, P.; McAvoy, T. Integrating Neural Networks with First Principles Models for Dynamic Modeling. IFAC Proceedings Volumes 1992, 25, 327 _332, 3rd IFAC Symposium on Dynamics and Control of Chemical Reactors, Distillation Columns and Batch Processes (DYCORD+ '92), Maryland, USA, 26-29 April.

[6] Wilson, J.; Zorzetto, L. A generalised approach to process state estimation using hybrid artificial neural network mechanistic models. Computers & Chemical Engineering 1997, 21, 951-963

575016 Revisiting Hybrid Modeling: Integrating Machine-Learning Models with Physical Constraints

Data science has demonstrated resounding success in solving complex problems within ChemE disciplines from material discovery to reaction engineering. Process systems engineering (PSE) has likewise seen many uses of machine learning (ML) tools, from modeling unit operations to entire processes. However, practitioners remain hesitant to incorporate machine-learning models for PSE as their inherent black-box nature makes their interpretation difficult and their performance unreliable for predicting system behavior outside the range of data collected for model fitting. Such situations arise frequently in PSE with common examples including process scale-up, changes in feedstock, process deviation, equipment degradation, and design of experiments leading to process optimization.

Hybrid "semi-parametric" modeling offers a promising alternative to purely data-driven process models by merging the flexibility of data-driven models and the reliability of physical knowledge. [1, 2] By constraining the data-driven model to mechanistic constraints, several authors from disparate fields have shown the superior performance of hybrid models over purely data-driven models in terms of interpretability, extrapolation, and data-dependency. [3] Nevertheless, due to its interdisciplinary nature, hybrid modeling as a tool still remains a black-box to most process engineers with examples of incorporating hybrid modeling techniques scarce in industrial practice.

Recent and ongoing advances of ML methods in terms of availability (via open-source, user-friendly software) and power (computational efficiency) merits revisiting the hybrid modeling paradigm. Herein we examine classical approaches to hybrid modeling and employ open-source software environments which enable scalable construction and optimization with hybrid models. A workflow will be presented through which classical hybrid modeling frameworks can be constructed, solved and validated, with an emphasis on serial and parallel approaches as defined by von Stosch et al. [3]. The presentation will further emphasize the difference between approaches that simultaneously solve the data-driven and mechanistic model ("coupled solution") vs. those approaches wherein the data-driven and mechanistic model are solved separately ("uncoupled solution"), a distinction not readily apparent in hybrid modeling literature.

Finally, we employ case studies in PSE to evaluate the performance of hybrid modeling frameworks with their data-driven and mechanistic counterparts. Specifically, studies will characterize metrics for model performance where data is scarce and noisy and mechanistic knowledge is available at varying levels of sophistication and certainty. While much work has been done on comparing the performance of various data-driven models within hybrid models, results are inconclusive with no candidate demonstrating consistently superior performance. Therefore, recent work in identifying an 'optimal' data-driven model for PSE will be presented, considering data-driven models typical in hybrid modeling as well as modern ML models that have thus far received little attention in hybrid modeling literature. We present the potential advantages of incorporating various data-driven models to represent the black-box component of the hybrid model.

References

1. Thompson, M.L. and M.A. Kramer, *Modeling chemical processes using prior knowledge and neural networks.* AIChE Journal, 1994. **40**(8): p. 1328-1340.
2. Psichogios, D.C. and L.H. Ungar, *A hybrid neural network-first principles approach to process modeling.* AIChE Journal, 1992. **38**(10): p. 1499-1511.
3. von Stosch, M., et al., *Hybrid semi-parametric modeling in process systems engineering: Past, present and future.* Computers & Chemical Engineering, 2014. **60**: p. 86-101.

570429 A Hybrid Modeling Architecture

There is a growing interest in exploring new architectures and approaches in artificial intelligence that incorporate prior knowledge, such as known physical and chemical laws, to augment sparse data and to ensure robust predictions. There are many robust known physical and chemical laws at different time and length scales that can be used to develop hybrid models that combine artificial intelligence with first-principles (white box) models.

The general problem that a hybrid model needs to solve, is the prediction of multiple output variables such that the predictions agree well with measurement data while not violating any first principles. There are currently four main architectures for the creation of hybrid models. The first one is a parallel architecture in which a white-box (WB) model and a black-box (data-driven) model are used in parallel to make predictions and then some sort of prediction averaging is carried out to predict the desired output variables. However, this approach does not guarantee no violation of first principles. The parallel architecture can also be used to predict some output variables using a WB model and to predict the remaining output variables using a black-box (BB) model. The second one is a WB/BB serial architecture in which a WB model and a BB model are used in series to make predictions. The third one is a BB/WB serial architecture in which a BB model and a WB model are used in series to make predictions. Finally, the fourth one is a WB-embedded-in-BB architecture, in which known physical and chemical laws are embedded in a BB model as equality and inequality constraints.

In this paper, we propose and implement a hybrid model architecture in which BB models are used to predict unknown/uncertain parameters and inputs of a WB model. The WB model then predicts desired output variables based on the BB-model predictions and values of known/certain inputs. The application and performance of this hybrid model architecture are shown using numerically-simulated chemical process examples.

575017 Data Driven Parameter Reduction

Traditional model reduction techniques build parsimonious representations of a system by (a) neglecting states expected to have minor effects on the system behavior or (b) by discovering linear or nonlinear combinations of the states that influence behavior. Traditionally, the reduction can be achieved by analyzing the mathematical equations modeling the system. Time-scale analysis underpins one of the well-known reduction methods, exploiting a disparity of scales in chemical kinetics (e.g. Quasi-Steady State (QSS) approximation).

In addition to its states, however, a system is also influenced by several parameters. In many multiparameter models, there exist a set of so-called effective parameters, which are those combinations of the model parameters that dictate the system behavior (such as the Reynolds number in fluid mechanics). The number of effective parameters may be much lower than the number of model parameters in the system original description. If the mathematical model of the system is available, one can traditionally perform (a) sensitivity analysis to discriminate the importance of every single parameter and (b) dimensional analysis to extract the effective parameters.

Here, we present a data-driven approach to find the effective parameters of a system, and thus reduce the dimensionality of its parameter space. This data-driven approach does not require access to the model's mathematical description, and can even work if only partial (but rich enough) observations of the system are available. Specifically, we modify a manifold learning technique, namely Diffusion Maps, through the use of an output-informed kernel; this allows us to find nonlinear directions in parameter space corresponding to a reduced, effective parameter set. In the same step, we discover parameter directions in which the system behavior does not change. This framework will be illustrated through prototype examples, such as singular and regular perturbations, and then through effective parameter identification in coupled oscillator networks.

574123 Deep Learning Based Hybrid Model of Hydraulic Fracturing Process

Many chemical processes undergo complex dynamics which can be explained using a model of the process. Models can either be based on first principle techniques (white box models) or can be data driven (black box models) [1]. One of the key difference between the methods is in the nature of the parameters. In first principles based models, the number of parameters are fixed and they carry physical meaning related to the process. On the contrary, in data-based models the number of parameters vary and their nature is flexible [2]. Hybrid models are a combination of the first principles and data based sub-models, and were first built using neural networks [2, 3]. The key idea was that the first principles knowledge about the process being modeled can be integrated with a neural network which can approximately estimate the unmeasured process parameters [3]. Since then, the field of neural networks has transformed from a single hidden layer to multiple hidden layers in a neural network which has shown tremendous ability to capture and represent complex data [4]. In this work, we combine the power of deep learning and first principles based knowledge of a hydraulic fracturing process to build a hybrid model which balances the advantages and disadvantages of a strictly first principles or neural networks based model.

Hydraulic fracturing is a highly complex and non-linear process of extracting oil/gas from rock formations which have low porosity and permeability. Its first principles based models consist of a system of nonlinear highly-coupled PDEs with time-dependent spatial domain and usually assume homogeneity of reservoir rock properties [5, 6]. But in reality the rock properties such as Young's modulus, rock permeability and porosity etc. vary even within the same formation. Other process parameters such as fluid leak-off coefficient, proppant particle size are assumed to be constant throughout the process which is not true. The effect of these uncertainties cannot be undermined and they need to be integrated with the first principles based model in order to improve its prediction accuracy during the propagation of hydraulic fracture. We aim to achieve this by developing a hybrid model which includes a deep neural network that will approximate the uncertainties within the rock formation and a first principles based model that will explain the dynamics of the process. The overall structure of the hybrid model is similar to the first principles model which helps in interpreting the quantified uncertainties. We will demonstrate the superiority of this hybrid model in comparison to the strictly first principles based model in terms of prediction accuracy and also in comparison to strictly deep neural network based model in terms of reliability and extrapolation.

Literature cited:

[1] von Stosch, M., Oliveira, R., Peres, J., Feyo de Azevedo, S., 2014. Hybrid semi-parametric modeling in process systems engineering: Past, present and future. Comput. Chem. Eng. 60, 86-101.

[2] Thompson, M.L., Kramer, M.A., 1994. Modeling chemical processes using prior knowledge and neural networks. AIChE J. 40, 1328–1340.

[3] Psichogios, D.C., Ungar, L.H., 1992. A hybrid neural network-first principles approach to process modeling. AIChE J. 38, 1499-1511.

[4] Venkatasubramaniam, V., 2018. The promise of artificial intelligence in chemical engineering: Is it here, finally? AIChE J. 65, 466-478.

[5] Yang, S., Siddhamshetty, P., Kwon, J.S., 2017. Optimal pumping schedule design to achieve a uniform proppant concentration level in hydraulic fracturing. Comput. Chem. Eng. 101, 138-147.

[6] Economides, M.J., Nolte, K.G., 2000. Reservoir stimulation. Chichester, Wiley.

568692 Randomized Rounding for Best Subset Selection Regression

Learning algebraic models from data is becoming an increasingly popular method to model complex chemical engineering systems [1]. This trend is driven by the difficulty of constructing and optimization physics-based models, increases in available system and experimental data, and improved strategies for regression. In this work, we focus on predicting the value of a single output variable using a linear combination of features. These features include the original regression variables as well as nonlinear transformations of the regressors.

Well-known strategies for linear regression include Forward Stepwise Selection, the LASSO, and Best Subset Selection (BSS) [2]. BSS seeks to features to minimize the sum of squared error between the actual and model-predicted values of the output while also penalizing the inclusion of features in the objective. This task can be posed as a mixed-integer quadratic program (MIQP) and leads to models with low complexity and good accuracy [3, 4]. In order to capture the nonlinearities in the data-generating function, it is necessary to consider a wide-variety of transformations of the original variables. However, as the number of considered variables grows, the resulting combinatorial explosion makes the MIQP problem intractable to solve in a reasonable amount of time [5].

In this work, we propose a new strategy to construct a solution to the Best Subset Selection MIQP. Our algorithm cleverly applies a randomized approach to rounding the continuous relaxation of the integer variables. The coefficients of the variables in the active set can then be quickly calculated using basic linear algebra techniques. The resulting algorithm is orders of magnitude faster than solving large problems to global optimality via the MIQP and results in solutions that are near-optimal for a wide-variety of synthetic and real data.

Additionally, we propose a procedure that incorporates our heuristic strategy into a larger framework that uses insights gained from the approximate solution—along with work done in the statistics community to solve convex quadratic programs—to develop a faster branch-and-bound strategy to solving the Best Subset Selection problem to optimality.

1. Cozad , N. V. Sahinidis , and D. C. Miller. Learning surrogate models for simulation-based optimization. AIChE Journal , 60(6):2211–2227, 2014.
2. Tibshirani. Regression Selection and Shrinkage via the Lasso. Journal of the Royal Statistical Society, 58(1):267–288, 1996.
3. Cozad, N. V. Sahinidis, and D. C. Miller. A combined first-principles and data-driven approach to model building. Computers & Chemical Engineering, 73:116–127, 2015.
4. T. Wilson and N. V. Sahinidis. The ALAMO approach to machine learning. Computers & Chemical Engineering, 106:785–795, 2017.
5. Bertsimas, A. King, and R. Mazumder. Best Subset Selection via a Modern Optimization Lens. The Annals of Statistics, 44(2):813–852, 2015.

569164 Data-Driven Process Synthesis, Design and Intensification

The complex physicochemical phenomena underpinning a majority of chemical processes can be accurately represented by equivalent physics-based first-principles models. The resulting ordinary differential equation (ODE)- or partial differential equation (PDE)-based models are adept at computing spatial and/or temporal variations in chemical systems. However, performing optimization studies while satisfying the first-principles models is computationally challenging as discretization of differential equations in space and/or time often result in large-scale nonlinear problems. Through a variety of chemical engineering case studies, we demonstrate how effective data-driven strategies can be conceptualized and implemented to obtain computationally-tractable optimization problems which can then be solved to obtain optimal solutions.

Firstly, we consider methanol and ammonia production flowsheets where the reaction unit operations are modeled using pseudo-homogeneous, one-dimensional, non-isobaric and non-isothermal ODE-based models [1,2]. The major objective is to obtain optimal reactor design and operation for which accurate surrogate models are developed using ALAMO [3]. The inputs and outputs of the reactor units are judiciously selected such that the accuracy of approximated models could be preserved while incorporating the crucial design and operation decision variables. The developed surrogate models are cross-validated and subsequently incorporated in the overall optimization formulation, which is solved to obtain globally-optimal solutions.

Secondly, we study dynamic process intensification (DPI) case studies where reaction and adsorption phenomena are combined in a single intensified column to yield cost-effective, efficient and sustainable processes compared to their non-intensified counterparts. The hybrid adsorption-reaction systems are represented through PDE-based one-dimensional, pseudo-homogeneous, non-isothermal, non-adiabatic, and non-isobaric models [4]. Due to the complexity of underlying physics and the resulting PDE-based model, it is difficult to develop accurate approximated surrogate models. Furthermore, as performing the simulations are computationally expensive, the optimization of DPI systems must be performed using as few simulations as possible. To achieve it, we develop an in-house simulation-based data-driven optimization algorithm which comprises of feasibility and optimization phases [5]. Both phases develop surrogate models on the fly for approximating black-box objective and constraints. A novel sampling strategy is also incorporated which can handle hard constraints and reduce overall number of simulations performed. The developed algorithm is applied to optimize several DPI case studies including sorption-enhanced steam methane reforming, sorption-enhanced methanol synthesis and carbon dioxide capture and conversion [6,7].

References

[1] A. Arora, J. Li, M.S. Zantye, M.M.F. Hasan, Process Design Frameworks for Economic Utilization of Small-scale and Unconventional Feedstocks, Found. Comput. Process Des. Accepted (2019).

[2] A. Arora, J. Li, M.S. Zantye, M.M.F. Hasan, Functionality-based Design Framework for Reducing Capital Intensity of Small-scale, Modular Processes, Submitted. (2019).

[3] Z.T. Wilson, N. V Sahinidis, The {ALAMO} approach to machine learning, Comput. Chem. Eng. 106 (2017) 785–795.

[4] A. Arora, S.S. Iyer, M.M.F. Hasan, GRAMS: A General Framework Describing Adsorption, Reaction and Sorption-Enhanced Reaction Processes, Chem. Eng. Sci. 192 (2018) 335–358.

[5] I. Bajaj, S.S. Iyer, M.M.F. Hasan, A trust region-based two phase algorithm for constrained black-box and grey-box optimization with infeasible initial point, Comput. Chem. Eng. 116 (2018) 306–321.

[6] A. Arora, I. Bajaj, S.S. Iyer, M.M.F. Hasan, Optimal Synthesis of Periodic Sorption Enhanced Reaction Processes with Application to Hydrogen Production, Comput. Chem. Eng. 115 (2018) 89–111.

[7] A. Arora, S.S. Iyer, I. Bajaj, M.M.F. Hasan, Optimal Methanol Production via Sorption Enhanced Reaction Process, Ind. Eng. Chem. Res. 57 (2018) 14143–14161.

565884 Parameterizations of Data-Driven Process Models for Integrated Stochastic Scheduling and Process Control

Modern chemical manufacturing plants increasingly seek to integrate decision-making processes in real time ("enterprise-wide optimization", "Industry 4.0", "smart manufacturing"), in response to competitive global market conditions and aided by advances in hardware and algorithms. Process scheduling, which is concerned with setting production amounts/sequences and allocating plant resources, is a natural target for (real-time) optimization. For example, under demand response programs, processes which consume electricity as a primary feedstock can modulate their production schedule in response to electricity prices; these decisions are generally made hourly or daily, and looking several days into the future. The optimal production schedules are then implemented in the plant by a process control system, which operates at a higher frequency and over a shorter prediction horizon than the scheduling layer.

As production schedules are adjusted more frequently under paradigms like demand response, it is no longer possible to assume that the process operates primarily at steady-state when solving the scheduling optimization problem; the dynamic response of the closed-loop process should be explicitly considered. However, integrating scheduling and control results in a computationally challenging optimization problem, especially for large, multivariable industrial processes. Computational complexity increases further if (inevitable) uncertainty in problem parameters is to be considered. As a result, parsimonious models that capture scheduling-relevant process dynamics (which we refer to as *"scale-bridging models")* are needed [1].

In this work, we build on prior developments to introduce a framework for computationally efficient integrated scheduling and control under uncertainty. As in [1], we build scale-bridging models, which represent the closed-loop dynamic response to of a process changes in *setpoints/targets* imposed by the scheduling layer, noting that closed-loop data of the type required for this effort are readily available at most chemical plants. The models are in Hammerstein-Wiener (HW) [2] form, consisting of static nonlinear input and output blocks flanking a linear dynamic block. The novel aspect of the current contribution is that we exploit features of the HW model structure to parameterize the scale-bridging models, and therefore to significantly reduce the dimension of the integrated scheduling and control problem. We show that dimensionality reduction is possible due to the linear nature of the HW dynamic block, and depends on the form of the static input/output nonlinear blocks, which are assumed to be piecewise linear (PWL) in this work. Our framework considers parameterizations of path constraints, accumulative (integrated) terms, and constraints relating to transition times. We demonstrate its efficacy on a case study considering demand-response scheduling of a chlor-alkali process; the reduced problem size enables order-of-magnitude reduction in solution time compared to a reference problem [3], and enables the formulation and solution of (previously intractable) stochastic formulations that account for uncertainty.

[1] Du, J., Park, J., Harjunkoski, I., & Baldea, M. (2015). A time scale-bridging approach for integrating production scheduling and process control. *Computers & Chemical Engineering*, *79*, 59-69.

[2] Zhu, Y. (2002). Estimation of an N–L–N Hammerstein–Wiener model. *Automatica*, *38*(9), 1607-1614.

[3] Otashu, J. I., & Baldea, M. (2019). Demand response-oriented dynamic modeling and operational optimization of membrane-based chlor-alkali plants. *Computers & Chemical Engineering*, *121*, 396-408.

AUTHOR INDEX

AUTHOR INDEX

AUTHOR INDEX

AUTHOR INDEX

AUTHOR INDEX

AUTHOR INDEX

AUTHOR INDEX

AUTHOR INDEX

AUTHOR INDEX

AIChE
120 Wall Street, FL 23
New York, NY 10005-4020

ISBN 978-1-7138-0547-2

CURRAN ASSOCIATES INC.
proceedings
.com